OPTIMIZATION IN PRACTICE WITH MATLAB®
FOR ENGINEERING STUDENTS AND PROFESSIONALS

Optimization in Practice with MATLAB® provides a unique approach to optimization education. It is accessible to ***junior and senior undergraduate, and graduate students***, as well as industry practitioners. It provides a strongly practical perspective that allows the student to be ***ready to use optimization*** in the workplace. It covers traditional materials, as well as important topics previously unavailable in optimization books (*e.g.*, Numerical Essentials – for successful optimization).

Outstanding features include:

- Provides practical applications of real-world problems using **MATLAB**.
- Each chapter includes a ***suite of practical examples and exercises*** that help students ***link the theoretical, the analytical and the computational.*** These include a robust set of real-world exercises.
- Provides ***supporting*** MATLAB ***codes*** that offer the opportunity to apply optimization at all levels, from students' term projects to industry applications.
- Offers instructors a comprehensive ***solution manual*** with solution codes along with **lectures in PowerPoint** *with animations* for each chapter. The MATLAB m-files are available for download from the book's website.
- ***Instructors have the unique flexibility*** to structure one- or two-semester courses that may range from gentle introductions to highly challenging, for undergraduate or graduate students.

Dr. Achille Messac received his BS, MS and PhD from MIT in Aerospace Engineering. Dr. Messac is a Fellow of the American Institute of Aeronautics and Astronautics (AIAA) and the American Society of Mechanical Engineers. He has authored or co-authored more than 70 journal and 130 conference articles, chaired several international conferences, delivered several keynote addresses, and received the prestigious AIAA Multidisciplinary Design Optimization Award. He has taught or advised undergraduate and graduate students in the areas of design and optimization for more than three decades at Rensselaer Polytechnic Institute, MIT, Syracuse University, Mississippi State and Northeastern University.

Optimization in Practice with MATLAB® for Engineering Students and Professionals

Achille Messac, PhD

CAMBRIDGE
UNIVERSITY PRESS

CAMBRIDGE
UNIVERSITY PRESS

University Printing House, Cambridge CB2 8BS, United Kingdom

One Liberty Plaza, 20th Floor, New York, NY 10006, USA

477 Williamstown Road, Port Melbourne, VIC 3207, Australia

314-321, 3rd Floor, Plot 3, Splendor Forum, Jasola District Centre, New Delhi - 110025, India

79 Anson Road, #06-04/06, Singapore 079906

Cambridge University Press is part of the University of Cambridge.

It furthers the University's mission by disseminating knowledge in the pursuit of education, learning and research at the highest international levels of excellence.

www.cambridge.org
Information on this title: www.cambridge.org/9781107109186

First published 2015
Reprinted 2016

A catalogue record for this publication is available from the British Library

ISBN 978-1-107-10918-6 Hardback

Additional resources for this publication at www.cambridge.org/Messac

Contents

List of Figures

List of Tables

Preface

Contacting the Author Regarding this Book

I am delighted that you are using this book in your study of, or involvement with, optimization. I would very much welcome your comments, particularly regarding this first edition, and any suggestions that you might have for the next edition. For your comments regarding this book, I will be happy to receive your direct email at OptimizationInPracticeMessac@google.com.

Book Website

The website www.cambridge.org/Messac will be maintained for this book. Information for instructors and students will be separately provided. Software for various problems will be provided in this website. I expect it to be a dynamic website, where the information available will evolve over time to be responsive to readers' requests and feedback.

Book Organization

This book is intended to be used by undergraduate and graduate students in the classroom, or by industry practitioners learning independently. Its organization suits these objectives. Following are messages specifically tailored for students, for industry practitioners and for instructors. The book has five parts:

Part I   *Helpful Preliminaries*
Part II   *Using Optimization—The Road Map*
Part III *Using Optimization—Practical Essentials*
Part IV *Going Deeper: Inside the Codes and Theoretical Aspects*
Part V   *More Advanced Topics in Optimization*

Part I has two chapters that present prerequisite material explaining how to use MATLAB and some useful mathematical information. Most of this book assumes knowledge of undergraduate calculus and elementary linear algebra. The second chapter provides a brief review of the math needed. In **Part II**, three chapters introduce the world of optimization in the form of a road map. This part provides the basics of what should be known about optimization before attempting to use it. In **Part III**, there are five chapters that teach the basic use of optimization. In fact, learning the material up to Part III provides the practitioner with sufficient information for solving practical problems. In doing so, student users will not be experts on how optimization *works under the hood*, so to speak, but will have the ability to use optimization in general practical contexts. Stated differently, while not being an expert about what is *under the hood*, the student will be a pretty good *driver* and should be able to *drive* to the desired destination (*i.e.*, the optimal design). In **Part IV**, three chapters provide a meaningful understanding of the computational and theoretical optimization process for linear programming and nonlinear programming with and without constraints. This is equivalent to learning the basics of what is **under the hood** of a car. **Part V** builds on the first parts of the book to provide a foundation for more advanced studies in optimization. In Part V, we learn optimization at a deeper level, where advanced topics are introduced over six chapters.

A Message to Students

As this book reaches your hands, you may have already signed up to take an optimization course, you may be in your junior or senior year or a graduate student, or you may already be in industry and feel that optimization may be useful to you. Regardless of what the case may be, you probably have two important questions in your mind: (i) Why should I learn optimization? and (ii) Is this a good book for me to use to learn optimization? Let me provide you with some objective comments, together with some subjective thoughts.

Why should you learn optimization? The truth is this. If you are an engineer or about to become one, if you are a financial analyst, or if you deal with numbers to determine how desirable or undesirable the performance of a system or a design is, then optimization will almost certainly be able to help you do a better job—without fail. Optimization applies to most engineering activities, management operations activities and numerous other fields where performance (or *goodness*) can be numerically quantified.

More importantly, if you are trying to design a system that must perform a certain way, you have unsuccessfully tried all that you know, and you have already asked the experts for help without success, then there is a good chance that optimization will help you succeed. Interestingly, it would not be at all surprising that the expert may find it too difficult to obtain an adequate answer, when optimization can be successfully used to find one. This is, in part, because optimization can intelligently examine thousands of design alternatives in less time than it takes the expert to examine a single design alternative. Optimization can also perform an intelligent search in a complex environment that may not be clear to the human mind. Optimization essentially makes it obsolete to engage in the typical trial-and-error process, as we search for a good design.

The following are some important comments regarding the rapidly growing popularity of optimization in recent years, and about how this book offers a unique approach to bringing optimization to a broader audience.

Until recently, *the cost of computing* was a critical issue that hindered the broad application of optimization. Fortunately, with the revolutionary decrease in the cost of computing in recent years, a desktop computer is often all that is needed to solve many practical optimization problems—making the application of optimization dramatically more practical.

The application of optimization in practical settings is increasingly becoming commonplace. A growing number of software developers have begun to include optimization capabilities as they respond to growing demand. They realize that it is not sufficient to determine how a structure deforms under given loads. It is also important, and of interest, to *discover* how to change the design in order to reduce that deformation. Optimization provides a reliable and systematic way to obtain this reduction. Importantly, this powerful benefit of optimization applies to a plethora of analysis software in engineering, management, finance and numerous other fields.

Let us now turn our attention to the second question: Is this a good book for you to use to learn optimization? This book will provide you with a unique combination of desirable attributes. It will provide you with the knowledge to start using optimization software in general. In particular, you will have the ability to use MATLAB for this purpose with effectiveness and efficiency. *Note that any other optimization software could have been used, but we use MATLAB for its broad popularity, effectiveness and convenience.*

This book focuses primarily on the material that is required for the practical application of optimization: more time spent on the computational analysis and optimization of real-world problems, and less time on the inner theory and mathematics of optimization. This approach is beneficial to students who wish to initiate research studies in optimization, as well as to those students who primarily wish to use optimization for practical purposes. This book is explicitly intended to teach you *how to use optimization successfully*, and to do so while avoiding unnecessary mathematics. The first parts of this book can be used for a one-semester course that is fully accessible to junior-level undergraduate students. Those interested in advanced topics are referred to Part V of this book.

To make this practical learning possible, after the preparatory material of Part I, realistic solution approaches that involve the use of a computer are provided in Part II. With our ability to use an optimization and modeling code, such as MATLAB, we will be able to readily apply what we learn to real-life problems. The approaches do not fundamentally change with different optimization codes. The changes that occur from problem to problem are readily handled with any optimization code, such as the size and other generic features of the problem. MATLAB is an easy-to-use and very popular software that is useful in all areas of engineering. It is also used in a growing number of non-engineering fields. If you don't yet know MATLAB, that is fine. You will learn it as we go along. The first chapter of this book provides an introduction to MATLAB that focuses on the material that we will need.

Your prospective learning of optimization in a way that focuses on practical applications is timely in view of the increasingly computational world. It will prove to be an important component of your education. Ultimately, learning optimization will provide you with a truly powerful tool to do things more successfully than could be done without optimization.

A Message to Industry Practitioners

To obtain an overall idea about the objectives and intent of this book, I suggest that you also read the messages to students and instructors. Regarding the particular needs of industry practitioners, this book is deliberately designed to quickly get to the point. Many software products are coming to market with optimization capabilities, and it is wise to acquire the appropriate background to start using these capabilities. This book directly provides the required knowledge in a way that is unencumbered by unnecessary math, and focuses on such practical aspects as: (i) How do I make my design lighter, stronger and cheaper? (ii) Once I get an answer from my analysis code, what is the next step to improving that answer?

In addition, even if you are not using an analysis software (but can compute the performance of your system), you may ask how to modify your system to improve its performance, and how to do so systematically without the usual manual trial and error. Optimization is a powerful way to accomplish these objectives, and this book provides the required knowledge in a practical and accessible way. As mentioned earlier, those interested in more advanced topics are referred to Part V of this book.

A Message to Instructors

This book takes a novel pedagogical approach to the teaching of optimization. It takes a different perspective regarding what material should be included in a first optimization course; and it provides the means to teach optimization to juniors, seniors, and graduate students who are taking their first course in optimization. The material in this book can be divided into two self-contained one-semester courses. The first course can be offered at the undergraduate or graduate level. At the undergraduate level, Parts I, II, and III could be covered. At the graduate level, (i) Parts II, III, and IV could be nominally covered, (ii) Part II can be covered quickly, (iii) Part V can be covered as part of a challenging first course or as part of a second course, and (iv) a term project could be assigned potentially based on some advanced problems in the book. The problems at the end of each chapter are divided into "Warm-up," "Intermediate," and "Advanced," reflecting their respective levels of difficulty. Some "Graduate Level" problems are provided as well. This flexibility can be exploited to address the diverse skills of students cohorts and of undergraduate/graduate students. The book is structured to provide full flexibility to accommodate the objectives of the instructor, as well as the skills and interests of the students.

Importantly, this book is also structured with the instructor in mind to facilitate the pedagogical process. In particular, the book's website (`www.cambridge.org/Messac`) provides a comprehensive set of materials that support the instructor's needs. A 270-page solution manual is provided. A comprehensive set of lecture materials is provided *in editable form,* which comprises approximately 850 PowerPoint slides, thereby providing the flexibility to suit the instructor's goals and pedagogical style. The book's website is also provided to enrich the coverage of certain topics and to allow us to pose larger practical problems that might not otherwise be considered. The book MATLAB codes are also available in the book's website. The instructor is invited to visit the website and to contact me at `OptimizationInPracticeMessac@google.com` for any suggestions for the website or any other aspects of the book.

This book provides an introduction to the practical application of optimization as a potentially last optimization course. Alternatively, this book can be used as a more accessible introduction to the subject of optimization, to be followed by courses that cover more advanced topics.

Regarding the philosophy of the material covered in this book, we critically assess the need to include candidate topics by asking the question: Is this material needed for someone who primarily wishes to *use* optimization, or is it primarily required for someone who intends to *develop* optimization algorithms? Priority is given to topics that directly contribute to the successful *use* of optimization. For example, is it necessary for an introductory user to learn sequential quadratic programming? Is this knowledge necessary for the successful application of optimization, when robust implementations are broadly available? The book is structured to easily allow the instructor to potentially leave that topic to a second optimization course (using this book), where more theoretically advanced subjects can be presented. This singular philosophy helped prioritize the candidate topics for inclusion (*and where*) in the book. The net result is a text that appreciably departs from tradition, but that we believe makes a novel contribution to the teaching of optimization. It provides the material to teach optimization in the traditional way, but with the flexibility/option to employ a more pragmatic approach that may be more inviting and effective in an introductory undergraduate or graduate first or second course.

A top-level view of the five parts of the book is provided as follows:

Part I *Helpful Preliminaries*

1. MATLAB as a Computation Tool
2. Mathematical Preliminaries

Part II *Using Optimization - The Road Map*

3. Welcome to the Fascinating World of Optimization
4. Analysis, Design, Optimization, and Modeling
5. Introducing Linear and Nonlinear Programming

Acknowledgements

I dedicate this book to my parents, Marie and Achille; my sons, Owen, Luke and Patrick; and my friends and colleagues Mel Payne and Thomas Mensah.

I would like to express my deepest thanks and appreciation to my former doctoral students Christopher Mattson, Ritesh Khire, Sirisha Rangavajhala, Anoop Mullur, Junqiang Zhang, Souma Chowdhury, Weiyang Tong and Ali Mehmani who have provided invaluable assistance and insight in the preparation of this book. I also express my appreciation to the many students, undergraduates and graduates, who have used earlier evolving versions of this book for approximately a decade. They have helped fashion the present form of this book.

To the many colleagues who have been part of my exciting involvement with the field of optimization and design for more than three decades, and who have knowingly and unknowingly influenced the contents and philosophy of this book, I would like to express my deepest gratitude!

PART 1

HELPFUL PRELIMINARIES

Part I provides prerequisite information for using optimization in practice. Specifically, we need to use software and we need to have some basic knowledge of mathematics. This book uses MATLAB as a computational tool. If you are new to MATLAB or do not know it yet, Chapter 1 will provide you with the preparatory information to get started. If your mathematics knowledge is a bit rusty, Chapter 2 will provide you with the preparatory knowledge for optimization application.

Specifically, the topics presented, with the chapter numbers, are given below:

1. MATLAB® as a Computation Tool

2. Mathematical Preliminaries

1

MATLAB® as a Computational Tool

1.1 Overview

Optimization can be viewed as a process that searches methodically for better answers, better solutions, or better designs that a human being may not be able to find through experience, intuition, or courageous trial-and-error. Optimization can be defined as the art of making things better. Interestingly, optimization very often does not simply allow us to do something better, but it may also make it possible to do something that we did not otherwise know how to do. To take full advantage of the power of optimization in practice, there is no choice but to use a computer.

The study of optimization typically takes a theoretical and/or computational approach. The theoretical approach is highly useful when the objective is to develop new optimization methods or to assess how the current methods work [1]. Additionally, the study of optimization often focuses on the understanding of various search algorithms for optimization. A book by Reklaitis and co-authors [2] is an example of methods-based optimization presentation. While these books play an important role in the general study of optimization, we pursue a different approach. The objective of this book takes on a practical perspective. The current interest is in the immediate ability to apply optimization in practice. In order to reach this objective, and to have the ability to apply optimization to real world problems, we use the approach that is almost always required in the application of optimization. We use the power of the computer, in conjunction with the study of different important practical aspects of optimization.

In order readily apply optimization in practice, the focus will be on the computational application of what we will learn, as we learn it. To do so, we use a computational modeling and coding tool that is powerful, easy to use, and one that is widely applied in engineering and other fields. MATLAB is an excellent tool choice and will be used in this book. MATLAB has arguably become the most popular tool for computational modeling worldwide. Using MATLAB will enable you to optimize both simple and complex systems or designs with effectiveness and efficiency.

MATLAB is a registered trademark of The MathWorks, Inc.

This chapter provides a brief introduction to MATLAB, which is primarily for those who have little or no experience with MATLAB. Previous users of MATLAB may find this introduction to be a useful review and a way to learn about the MATLAB optimization capabilities. This chapter has six sections. Section 1.2 defines MATLAB. Section 1.3 provides a basic introduction of MATLAB, while Sec. 1.4 goes beyond the basics. Section 1.5 focuses on the MATLAB plotting capabilities. In Sec. 1.6, the MATLAB nonlinear and linear optimization capabilities are introduced. For historical reasons, the terminology "linear programming" and "nonlinear programming" is often used synonymously with "linear optimization" and "nonlinear optimization," respectively. In Sec. 1.7, a list of useful functions is presented.

1.2 MATLAB Preliminaries—Before Starting

This section provides useful information about MATLAB as a software tool. It includes the following components:

1. What is MATLAB?
2. Why MATLAB?
3. MATLAB Toolboxes
4. How to Use MATLAB in This Book?
5. Acquiring MATLAB
6. MATLAB Documentation
7. Other Software for Optimization

1.2.1 What Is MATLAB?

MATLAB is a very popular high level language for computation. It is used extensively both in industry and in universities worldwide. It is much easier to use than other popular programming languages such as Fortran or C. It takes a very short time to start becoming productive with MATLAB. Mathematical expressions are evaluated much the same way as they would be written in text form. MATLAB is used for a wide variety of activities, including computation, algorithm development, modeling, simulation, prototyping, data analysis, visualization, engineering graphics, and graphical user interface building (Ref. [3]).

In this book, the use of MATLAB is limited to the context of the application of optimization. In doing so, we will be able to optimize practically any system. This is because the MATLAB environment is very powerful. We will be able to optimize any system that is modeled in the MATLAB computational environment using this newly acquired optimization knowledge.

Over the years, various parties have developed MATLAB tools that are applicable to specific fields. These tools essentially constitute a set of functions that work together to perform powerful tasks in specific technical areas, such as control, dynamics, financial analysis, and signal processing. Some are available through private parties, while others are available through the MATLAB developer (The MathWorks, Inc.) [4]. Information concerning MATLAB can be obtained from its developer or at the website `www.mathworks.com`.

MATLAB is organized as a collection of several independent components that work together harmoniously. The central component is the basic MATLAB software, which can be used for most general computation and algorithm development. When the needs become more advanced and specific, we acquire MATLAB Toolboxes. These toolboxes are a collection of MATLAB function codes that perform tasks in given technical areas. To perform advanced optimization, the basic MATLAB software is needed, together with the "Optimization" Toolbox.

MATLAB can be acquired in the form of professional and educational versions [4]. The latter has some reduced capabilities, but should be able to perform most required tasks for moderately sized problems.

1.2.2 Why MATLAB?

It is important to keep in mind that MATLAB is in no way required to perform computational optimization. Several other codes could be used to perform all of the tasks for which MATLAB can be used. However, for the purpose of this book, MATLAB is an excellent choice. It is also a recommended software for future activities after completing the study of optimization using this book. Related information is also discussed in the Preface.

1.2.3 MATLAB Toolboxes

MATLAB Toolboxes provide useful functions for research and development in technical fields. Specific tasks can be performed using these toolboxes to satisfy users' requirements through user-friendly commands or visual interfaces. These toolboxes are convenient to use, as well as powerful. They provide functions that can be called by the MATLAB code written by users. The toolboxes support various functionalities for a broad range of applications. They are available for applications in (1) parallel computing, (2) mathematics, statistics, and optimization, (3) control system design and analysis, (4) signal processing and communications, (5) image processing and computer vision, (6) test and measurement, (7) computational finance, (8) computational biology, and (9) database connectivity and reporting. A complete list of MATLAB Toolboxes is available at the MathWorks website. The list includes the following toolboxes.

1. Parallel Computing
   - Parallel Computing Toolbox
2. Math, Statistics, and Optimization
   - Symbolic Math Toolbox
   - Partial Differential Equation Toolbox
   - Statistics Toolbox
   - Curve Fitting Toolbox
   - Optimization Toolbox
   - Global Optimization Toolbox

- Neural Network Toolbox
- Model-Based Calibration Toolbox

3. Control System Design and Analysis

- Control System Toolbox
- System Identification Toolbox
- Fuzzy Logic Toolbox
- Robust Control Toolbox
- Model Predictive Control Toolbox
- Aerospace Toolbox

4. Signal Processing and Communications

- Signal Processing Toolbox
- DSP System Toolbox
- Communications System Toolbox
- Wavelet Toolbox
- Fixed-Point Toolbox
- RF Toolbox
- Phased Array System Toolbox

5. Image Processing and Computer Vision

- Image Processing Toolbox
- Computer Vision System Toolbox
- Image Acquisition Toolbox
- Mapping Toolbox

6. Test and Measurement

- Data Acquisition Toolbox
- Instrument Control Toolbox
- Image Acquisition Toolbox
- OPC Toolbox
- Vehicle Network Toolbox

7. Computational Finance

- Financial Toolbox
- Econometrics Toolbox
- Datafeed Toolbox
- Database Toolbox
- Financial Instruments Toolbox

8. Computational Biology

- Bioinformatics Toolbox

9. Database Connectivity and Reporting

- Database Toolbox

MATLAB users can develop unique toolboxes for specific purposes. Many of these user-developed toolboxes are available online for download and use.

This book focuses on optimization using MATLAB Toolboxes. The optimization toolbox and the Global Optimization Toolbox are used for the study of optimization.

To access the MATLAB Toolboxes, you can click the APPS tab (Fig. 1.1) at the top menu of the MATLAB Desktop. The toolboxes have graphical user interfaces. A more advanced way to use the MATLAB Toolboxes is to call their functions using MATLAB codes. In this book, different MATLAB optimization functions are used to solve different types of problems.

1.2.4 How to Use MATLAB in this Book

This book provides a brief introduction to MATLAB to help with the initial use of MATLAB. Also provided is information regarding how to perform optimization using MATLAB. When it is deemed helpful and practical, the actual MATLAB code will be provided. By examining these preliminary coding examples, it will become clear how to write other more complicated code. Some code will be provided in the text, while others will be provided in the media device that accompanies this book. It is assumed that the PC Windows version of MATLAB 2013 is being used. The distinctions between the different platform versions of MATLAB are minor. The basic code is almost always identical across platforms, except for some tasks, such as file manipulations.

1.2.5 Acquiring MATLAB

MATLAB is widely available in most engineering firms and universities, as well as many financial institutions. It can be purchased from the MATLAB developer [4]. This book makes use of the basic MATLAB software and of the optimization toolbox software. Both are required to perform the computational portion of this book. It should also be noted that several other software options can be used to do this computational optimization work.

1.2.6 MATLAB Documentation

Several forms of documentation are available for the MATLAB user. An abundance of information is available on the web and on the developer's website. Several books that document the different components of MATLAB are sold by The MathWorks. Many of these books can also be downloaded from The MathWorks' website [4].

MATLAB documentation books can be downloaded from the website [4]. The powerful capabilities of the MATLAB graphics are presented in the book entitled *MATLAB Graphics* [5]. A book entitled *MATLAB Primer* provides a handy documentation of MATLAB [6] and, for the novice, a good basic introduction of MATLAB is given in the book entitled *Basics of MATLAB and Beyond* (Ref. [7]) or *Getting Starter with MATLAB: a Quick Introduction for Scientists and Engineers* (Ref. [8]). We note that the information provided here is fluid and dynamic. A visit to the developer's website [4] and a web search will provide the latest information.

1.2.7 Other Software for Optimization

MATLAB is not unique in its ability to perform computational optimization. It is chosen here because it is a highly effective software that is user-friendly and widely used. The skills that will be developed are useful in numerous other fields.

By using the MATLAB Toolboxes, it is possible to perform optimization in a number of different technical areas. This is because these toolboxes provide the means to model the performance behavior of the pertinent systems.

In certain situations, it is more appropriate to use other software, either independently or in conjunction with MATLAB. When used in conjunction with MATLAB, the external interface capability of MATLAB is used. Pertinent documentation is available from the MATLAB help menu or from The MathWorks' website.

There is a plethora of optimization software available to perform optimization independently. A book entitled *Optimization Software Guide* [9] catalogs a large number of these software tools. Notable optimization software products in the area of structural optimization are GENESIS [10], MSC/Nastran [11], and Altair [12]. Genesis was explicitly developed for the purpose of structural optimization and is considered a powerful tool.

1.3 Basics of MATLAB—Getting Started

This section provides a brief description of how to get around MATLAB and of how to use MATLAB at an introductory level. Specifically, the following topics are presented:

1. **Starting and Quitting Matlab**
2. **Matlab Desktop:** Basic MATLAB Graphical User Interface
3. **Matrices and Variables:** Matrices and Variables Operations
4. **Expressions:** Evaluations of Mathematical Expressions
5. **Control Flow Statements:** Using the `for` loop, the `while` loop, the `if` statement, and others
6. **Input and Output, Directories, and Files:** Input and Output of Data and Editing Commands, Directories, and Files
7. **Script File:** File that has MATLAB Commands
8. **Function File:** MATLAB File that Performs Independent Tasks
9. **Plotting:** Elementary Plotting Capabilities of MATLAB

1.3.1 Starting and Quitting MATLAB

To start MATLAB in Microsoft Windows, double-click on the MATLAB icon on the Windows desktop. MATLAB can also be started by selecting MATLAB from the `Start` menu. On a UNIX platform, type `matlab` at the operating system prompt. Once MATLAB is launched, a MATLAB Desktop appears on the screen.

To quit MATLAB, click on the top right close window button (Fig. 1.1). Alternatively, select `Exit` from the `File` menu in the desktop, or type `exit` or `quit` in the Command Window.

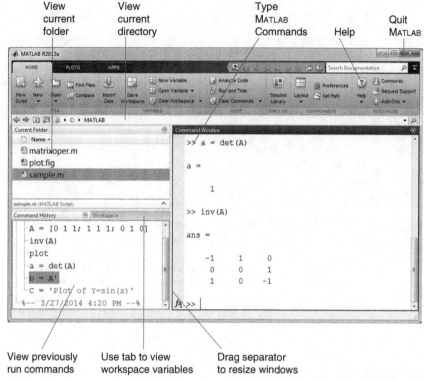

Figure 1.1. MATLAB Desktop

1.3.2 MATLAB Desktop: Its Graphical User Interface

The MATLAB Desktop (Fig. 1.1) appears when MATLAB is started. It provides a Graphical User Interface (GUI) that facilitates various MATLAB functions, such as managing files, variables, and applications.

The first time MATLAB starts, the desktop appears as shown in Fig. 1.1, although the desktop may have been customized to contain fewer components. Customize the desktop by opening, closing, moving, docking, and resizing the tools in it. Use `Preferences` from the `File` menu to specify features of the desktop. The MATLAB desktop environment provides useful tools that can be used for various purposes.

Command Window

The Command Window (Fig. 1.2) is used to enter variables, evaluate MATLAB commands, and run M-files or functions. M-files are the programs written to run MATLAB functions.

Command History

The Command History window (Fig. 1.3) is used to view previously used functions, copy, and execute selected lines from those functions. The lines entered in the Command Window at the command prompt are logged into Command History.

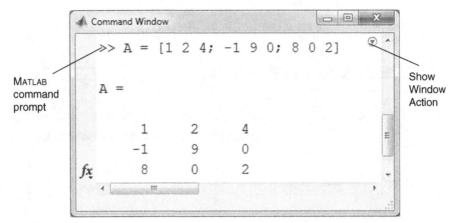

Figure 1.2. MATLAB Command Window

Select one or more lines and right click to copy,
evaluate or create an M-file from the selection

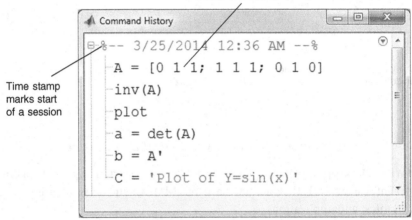

Figure 1.3. MATLAB Command History

Current Directory Browser

The Current Directory browser (Fig. 1.4) can be used to view, open, and make changes to MATLAB related directories and files. You can also use the commands `dir`, `cd`, and `delete` at the command prompt to view, change, and delete directories, respectively. MATLAB uses the current directory and the search path as a reference point to run and save files.

Any file you wish to run must be in the current directory or on the search path. A quick way to change the current directory is to use the `Current Directory` menu in the Desktop shown in Fig. 1.1.

Search Path: MATLAB uses a search path to find and execute the M files/functions you call. It also uses the search path to find other necessary MATLAB files, which are organized in the directories in the file system. By default, the files supplied with

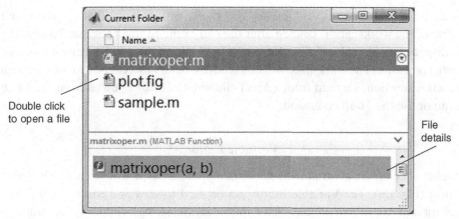

Figure 1.4. MATLAB Current Directory

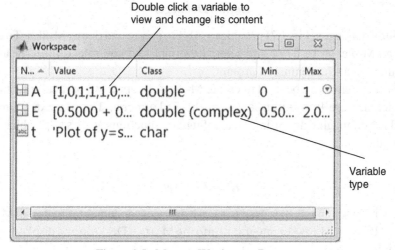

Figure 1.5. MATLAB Workspace Browser

MATLAB and Mathworks Toolboxes are included in the search path. To see or modify the current search path, select Set Path from the File menu in the Desktop and use the Set Path dialog box. Use the path command at the command prompt to view the current search path, addpath to add directories to the path, and rmpath to remove directories from the path.

Workspace Browser

The Workspace Browser (Fig. 1.5) is used to view the workspace and information about each variable. The MATLAB workspace consists of stored variables that are built up during a MATLAB session by running functions, M-files, and loading saved workspaces. At the command prompt, use the commands who and whos to view the workspace. To delete variables from the workspace, select the variable and select Delete from the Edit menu. Use the clear command at the command prompt. The workspace is deleted after the end of the MATLAB session. To save the workspace to a file, select Save Workspace As from the File menu, or use the save command. The workspace will be saved to a binary file, called a MAT-file, with a .mat extension. To read from a MAT-file, select Import Data from the File menu or use the load command.

Variable Editor

Double-clicking on a variable in the Workspace Browser will open the Variable Editor (Fig. 1.6). The Variable Editor can be used to view and edit a visual representation of one or two dimensional numeric arrays, strings, and arrays of strings in the workspace.

Editor/Debugger

The Editor/debugger (Fig. 1.7) provides a GUI to create and debug M-files. To create or edit an M-file, go to File and select New, or File and select Open, or use the edit function at the command prompt.

Any text editor can be used to create M- files. To specify a particular text editor as the default, use Preferences from the File menu. The MATLAB Editor can be used for debugging and using such debugging functions as dbstop, which sets a breakpoint.

Help Browser

The Help browser (Fig. 1.8) is used to search and view documentation for all MATLAB products. It is a web browser integrated into the MATLAB Desktop and displays HTML documents.

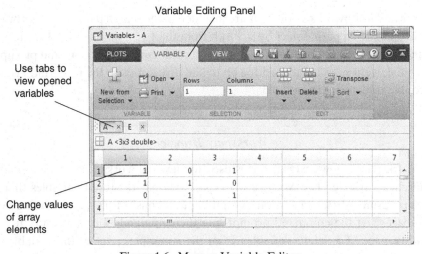

Figure 1.6. MATLAB Variable Editor

Use document bar to review
other open files in editor

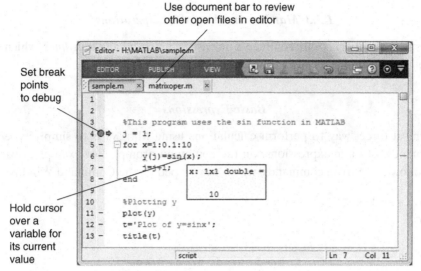

Set break
points
to debug

Hold cursor
over a
variable for
its current
value

Figure 1.7. MATLAB Editor or Debugger

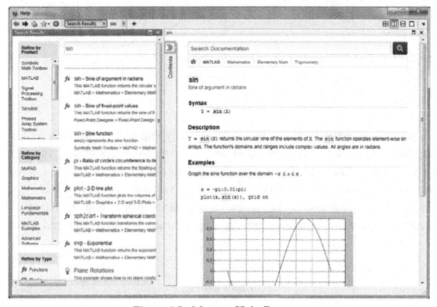

Figure 1.8. MATLAB Help Browser

There are several ways to open the Help browser: (i) Click the Help button in the MATLAB Desktop (shown in Fig. 1.1), (ii) Type helpbrowser at the command prompt, or (iii) Launch help from the Start button. The Help Browser consists of two panes: the Help Navigator pane, which is used to find information and type search keywords, and the Display pane, where the information can be viewed.

More detailed information on the MATLAB Desktop and the desktop tools can be found in Refs. [4, 6], or by simply exploring it in the Help section of MATLAB.

1.3.3 Matrices and Variables Operations

This subsection begins with comments on *basic expressions* and *long lines*, which will be followed by matrices and variables operations.

Basic Expressions

The most direct way to perform calculations using MATLAB is to simply type the commands, or basic expressions, on the command prompt. For example, observe the following MATLAB commands, which were typed on the Command Window:

```
>> a = 1 + 2
a =
        3
>> b = 4
b =
        4
>> c = sqrt(a^2 + b^2)
c =
        5
>> x = a + b;
>> y = 2*x
y =
        14
```

A few basic observations can be made. After typing a=1+2, MATLAB writes the result. After typing b=4, MATLAB again writes the result. After writing the expression for c, the result is again written on the screen. In the case of the variable x, the result is not written on the screen. This is because a semicolon is typed at the end of the expression. The value of the variable x is now available in the MATLAB **workspace** for further computation. The value of x is available and is used to evaluate y.

Long Lines

Very long commands or expressions can be typed on several lines. To do this, begin by writing the expression on one line, then type three periods followed by **Enter**, and continue to type the expression on the following line. Below is an example.

```
>> a_long_variable = 2
a_long_variable =
     2
>> another_long_variable = 3
another_long_variable =
     3
```

```
>> x = a_long_variable + a_long_variable^2 ...
+ another_long_variable
x =
    9
```

Typing Matrices

Recall that a matrix is a multidimensional variable. A 3 by 4 matrix has 3 rows and 4 columns. Define a 3 by 4 matrix M by typing in the Command Window:

```
>> M = [ 1 2 3 4; 5 6 7 8; 9 10 11 12]
M =
     1     2     3     4
     5     6     7     8
     9    10    11    12
```

Note that entries within a row can be separated by a space (or a comma), and that a semicolon ends a row. Expressions that include matrices can be formed, such as:

```
>> v=[11 ; 22 ; 33]
v =
    11
    22
    33
>> mv = [M v]
mv =
     1     2     3     4    11
     5     6     7     8    22
     9    10    11    12    33
```

where we concatenate two matrices in a row (with compatible dimensions – same number of rows). Similarly, two matrices with an equal number of columns can be vertically concatenated as

```
>> h = [11 22 33 44]
h =
    11    22    33    44

>> mh = [M ; h]
mh =
     1     2     3     4
     5     6     7     8
     9    10    11    12
    11    22    33    44
```

Matrices Generators

MATLAB makes it convenient to generate certain commonly used matrices. Note the following statements:

```
>> ones(3)
ans =
       1       1       1
       1       1       1
       1       1       1
>> ones(1,3)
ans =
       1       1       1
>> zeros(2)
ans =
       0       0
       0       0
>> zeros(1,3)
ans =
       0       0       0
>> A = eye(3)
A =
       1       0       0
       0       1       0
       0       0       1
```

Observe that the command ones(3) produces a 3 by 3 matrix of ones. Note that the command ones(3,3) will produce the same matrix. Similar comments apply to the command zeros(2). The command eye(n) will generate an identity matrix of dimension n by n.

Recall that if more information is needed, typing the commands help zero at the MATLAB prompt will provide more information about the command.

Subscripting

For the following matrix

```
>> A = [1 3 5; 4 6 8]
A =
       1       3       5
       4       6       8
>>
```

the following commands produce insightful results.

```
>> A(1,2)
ans =
 3
>> A(2,end)
ans =
        8
```

Note the use of the word end above as a subscript.

Matrix Arithmetic

Vector and matrix operations can be written in the usual way. For example, define the vector

```
>> v=[1;1;1]
v =
        1
        1
        1
```

Using the matrix A, just defined, we can evaluate

```
>> A*v
ans =
        9
       18
```

Recall that the dimensions of the matrices must be compatible for the matrix multiplication to take place (see Chapter 2).

Colon Operator

Powerful matrix manipulations can be performed by using the colon (:) operator. The most direct way to use the colon operator is to type

```
>> a = 2:8
a =
        2     3     4     5     6     7     8
```

where a list of numbers is generated from the lowest to the highest given numbers. Also write

```
>> 1:2:9
ans =
        1     3     5     7     9
```

where the middle number is used as an increment, 2. The increment can also be negative as in

```
>> 50:-5:30
ans =
    50    45    40    35    30
```

The other common use of the colon operator is to refer to a portion of a matrix. If we define the matrix

```
>> P=[1 2 3 4 ; 5 6 7 8]
P =
     1     2     3     4
     5     6     7     8
```

then we can retrieve the second through fourth columns of the second row using the command

```
>> pp = P(2, 2:4)
pp =
     6     7     8
```

Transpose

The transpose of a matrix can be obtained as follows

```
>> a=[1 2 3;4 5 6]
a =
     1     2     3
     4     5     6
>> b = a'
b =
     1     4
     2     5
     3     6
```

where b is the transpose of a, which is evaluated by a'.

The Command linspace

The linspace command is a quick way to generate a row vector of 100 linearly spaced points between two given numbers. The syntax is

```
linspace(100,200)
```

The above command generates 100 linearly spaced points between 100 and 200.

1.3.4 More MATLAB *Expressions*

Below are some more examples of MATLAB expressions. More information regarding these functions can be obtained by typing help followed by the function name.

```
>> x = 2
x =
    2
>> y = exp(x) + log(x)
y =
    8.0822
>> z = (-sin(y)) + abs(-x)
z =
    1.0259
>> complex_number = 4 + 3i
complex_number =
4.0000 + 3.0000i
>> magnitude = abs(complex_number)
magnitude =
    5
```

The log function evaluates the natural logarithm of a number, while the abs function evaluates the magnitude of a real or complex number.

1.4 Beyond the Basics of MATLAB

This section provides important information about MATLAB that you will need to know that is beyond the basics that we are covered so far. It will all become much easier as you get more practice. The following topics are presented:

1. Input and Output, Directories, and Files
2. Flow Control, Relational, and Logical Operators
3. M-files
4. Global and Local Variables
5. MATLAB Help

1.4.1 Input and Output, Directories and Files

The MATLAB environment can be easily manipulated to make file handling convenient. Some useful features are discussed below.

Current Directory

The current directory is generally displayed in a text box toward the top of the MATLAB screen. Alternatively, type pwd at the command prompt to display the current directory.

Setting the Path

The path to directories that are often used can be set. The files in these directories can then be directly accessed from any other directory. The path can be set from the `File » Set Path` menu.

Saving and Loading Variables

Any variable in the workspace can be saved to the hard disk using the `save` command.

```
save FILENAME x y
```

The above command saves the workspace variables `x` and `y` to a file named `FILENAME.mat` in the current directory. Note that there is no comma in the above syntax. To retrieve these variables, type the command

```
load FILENAME x y
```

1.4.2 Flow Control, Relational and Logical Operators

The MATLAB flow control statements operate similarly to those in most programming languages.

The for Loop

The `for` loop executes a set of statements for a specified number of times.

```
>> for i = 1:5
      x(i)=1;
   end
```

The above `for` loop generates an array `x` of five elements, each equal to 1.

```
>> x
x =
      1      1      1      1      1
```

MATLAB also provides nesting of `for` loops.

The while Loop

The general form of the `while` loop is

```
while EXPRESSION
STATEMENTS
end
```

The STATEMENTS will be executed as long as the EXPRESSION is true. For example, if the expression is A<B, then the statements will be executed while this condition holds true, and will stop executing as soon as A >=B.

The for and while loops can be terminated using the break command. The continue command passes control to the next iteration of the loop.

The if *Statement*

The if statement allows execution of statements provided certain conditions hold. The general form of the if statement is:

```
if EXPRESSION
    STATEMENTS
else
    STATEMENTS
end
```

The statements in the if part will be executed only if the EXPRESSION is true. Otherwise, the statements in the else part will be executed. The else portion is optional.

The switch-case *Statements*

Switch provides a way to switch between several cases based on an expression. For example, if a variable x in the workspace has a value of 2, then the following statements

```
>> switch x
     case 1
         disp('x is equal to 1')
     case 2
         disp('x is equal to 2')
   end
```

will yield

```
x is equal to 2.
```

Logical and Relational Operators

MATLAB provides a number of different logical and relational operators. These operators can be used in conjunction with variables to create expressions for use in if statements, while loops, and other flow control statements. A complete list of operators can be obtained by typing help ops at the command prompt. Some commonly used relational operators are EQUAL (==), GREATER THAN (>), LESS THAN (<), and NOT EQUAL (~=). Some commonly used logical operators are: AND (&), OR (|), and NOT (~).

1.4.3 M-files

MATLAB provides the powerful M-file feature. Using this feature, a sequence of MATLAB commands can be saved as an .m file, and can be executed as a batch process by simply typing the name of the M-file.

Script M-files

Script files are a series of MATLAB commands stored in a .m file. They do not necessarily take any input or yield any output. The variables generated during the execution are stored in the MATLAB workspace. A sample file myMfile.m is shown below.

```
% This is my first .m file
var = 5; new_var = var^2;
if new_var > 5
    disp('My first output');
end
```

This file can be generated using any external text editor, such as Notepad, or can be created using the MATLAB editor and debugger. The file is saved as myMfile.m in the current directory. Typing the name of the file at the command prompt will execute the commands in the file sequentially.

```
>> myMFile
My first output
```

Note that the file name is not allowed to have spaces, $-$, $+$, $/$, $*$, $\hat{}$, or any other mathematical symbols. Alternatively, right-clicking on its filename in the Directory Browser and clicking on Run will also execute the file.

Function M-files

Function M-files are similar to the *script* M-files in that they consist of a sequence of MATLAB commands. The difference between the two is that *function* M-files can receive one or more variables as inputs, and can return one or more variables as outputs. The input and output variables are available for use outside of the M-file function, but the variables or parameters used inside the function are not available outside the function. All variables or parameters that are used inside a script M-file are available in the environment that calls that script M-file. The input and output variables are called arguments. Creating or modifying variables within a function M-file does not affect the workspace, unless these variables are also output variables. To use an existing function, first define the numerical values of the input arguments that will be used, then define the list of output arguments in the function call.

Here is a representative function file called cuberoot.m, which has a single variable as an input argument, and returns its cubic root.

```
function output_var = cuberoot(input_var)
output_var = input_var^(1/3);
```

The first line is the syntax definition. The name of the output variable is output_var, input_var is the name of the input variable, and cuberoot is the name of the M-file. This file should be saved as cuberoot.m in the current directory.

The function cuberoot can be directly called from the command prompt. For example, to calculate the cube root of the number 5, type cuberoot(5) at the command prompt, which yields

```
ans =
    1.7100
```

One can also assign the M-file function output to a workspace variable, say x, by typing x = cuberoot(5). The function M-file can be called from within a script M-file.

Subfunctions

Subfunctions can be declared following the definition of the main function in the same M-file. Subfunctions are visible to the main function and other subfunctions in the same M-file, but are not visible outside the M-file in which they are declared.

1.4.4 Global and Local Variables

All the variables created within a function M-file are local to that function, and cannot be accessed from outside the function. Similarly, workspace variables are local to the workspace, and are not available to any function. To make a workspace variable, x, globally available, use the global command as follows.

```
global x
```

The above command should be used at the beginning of every script M-file and function M-file where the global variable needs to be accessed.

1.4.5 MATLAB Help

MATLAB help can be obtained from various sources. A good way to get started is to read this introductory chapter. This introduction can be followed by reading the book MATLAB *Primer* [6]. Comprehensive information about MATLAB can be obtained from the many documentation books (*i.e.*, [5]). The Mathworks' website [4] has a large amount of information that will address almost any issue. In situations where there is no access to the web, it is possible that all the needed information is already available on the computer where MATLAB is installed. Clicking on the Help menu will indicate the extent of the help information already installed.

A direct way to get help is to type at the MATLAB prompt

- `helpdesk`, which opens a MATLAB help GUI,
- `helpwin`, which opens a hypertext help browser,
- `demo`, which starts the MATLAB demonstration,
- `help`, which prints on the screen the various help topics available,
- `help` followed by a help topic or any function name, which provides help on the requested topic or function, or
- `lookfor` followed by a topic keyword, which gives the names of all the MATLAB functions that have that keyword on the first help line (that keyword does not have to be the name of a function, unlike for the help command).

As always, perhaps the most interesting way to get help is to ask, or work with a friend who might have more experience with MATLAB.

1.5 Plotting Using MATLAB

A picture is worth of thousand words. This is very true in optimization as well. Optimization methods are understood more clearly when presented in graphical form. At every stage of working life, presenting work using graphs and charts is a critical activity. In this endeavor, MATLAB can be used with its extensive set of functions to help develop the required graphs and charts (Ref. [13]). Following are pertinent important information:

1. Basic Plots
2. Special Plots: Contour, Scatter, fplot
3. 3-D Mesh and Surface Plots
4. Using the Plot Editing Mode

1.5.1 Basic Plots

First, consider the simple commands that generate two dimensional (2D) plots.

Use of Plot Command

`Plot` is one of the simplest graphics commands available in MATLAB. The following sample code will generate a sine curve. Figure 1.9(a) shows the plot generated by MATLAB.

```
x = 0:pi/100:2*pi;
y = sin(x);
plot(x,y);
```

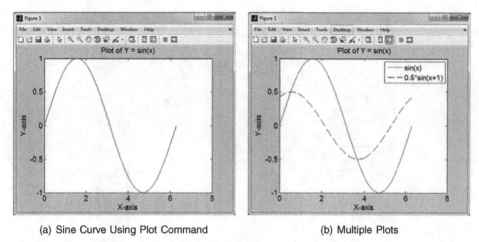

(a) Sine Curve Using Plot Command   (b) Multiple Plots

Figure 1.9. Single and Multiple Plots

Axes and Labels

Notice that the plot generated by the above set of commands does not generate any axes labels or a title. These can be added by using following set of commands.

```
xlabel('X-axis')
ylabel('Y-axis')
title('Plot of Y = sin(x)')
```

Multiple Plots on the Same Figure

To show how to put several plots on the same figure, a second line, given by $y = 0.5sin(x+1)$, will be plotted on the same figure. The following code can be used for this operation.

```
clc
clear
x = 0:pi/100:2*pi;
y = sin(x);
y2 = 0.5*sin(x+1);
plot(x,y);
plot(x,y,x,y2,'--');
xlabel('X-axis')
ylabel('Y-axis')
title('Plot of Y = sin(x)')
```

The clc and clear commands are introduced above. The clc command removes the content of the workspace, while the clear command clears the memory of MATLAB. Figure 1.9(b) shows the plot generated by MATLAB.

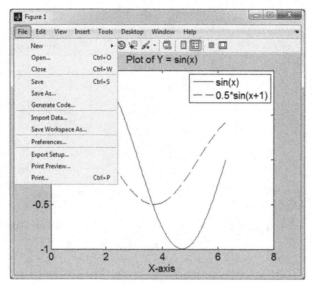

Figure 1.10. Printing and Saving

Generating Legends for the Plot

In Fig. 1.9(b), the legend has been added using the following command.

```
legend('sin(x)','0.5*sin(x+1)')
```

Printing and Saving Plots

Once the plot is ready, it can either be saved in a computer file or be printed. Here is an example of how to perform these two operations. To print the plot, click on the `File` menu, as shown in Fig. 1.10. A column menu will appear. By clicking on the `Print` item on this column menu, the plot will be printed. On the same column menu is the `Export Setup` item. This dialog window provides options to specify attributes of the output file (*e.g.*, the figure size, fonts, line width, and format). One click on this item will yield a standard save window. Two things need to be done in that window. First, name the file and, second, choose the file format of the picture, such as jpg, bmp or eps.

1.5.2 Special Plots: Contour, Scatter, fplot

In the previous subsection, some MATLAB basic plotting techniques were used. In this subsection, some specialized plotting techniques available in MATLAB will be discussed.

Contour Plot

The contours of an equation can be plotted using the `contour` command. The contours of an ellipse given by the equation $3X^2 + 4Y^2 = C$ will be plotted, where C can take on different values. It is also possible to generate and plot contours for

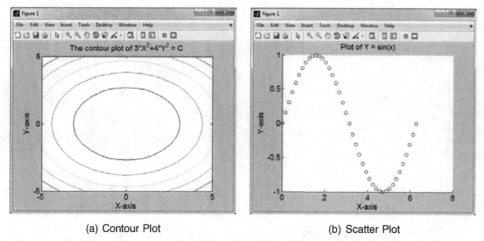

(a) Contour Plot   (b) Scatter Plot

Figure 1.11. Contour and Scatter Plots

specific given values of C. The following sample code will generate the required contour plot, and the actual plot is shown in Fig. 1.11(a).

```
[X,Y] = meshgrid(-5:.5:5,-5:.5:5);
Z = (3*X.^2+4*Y.^2);
[C,h] = contour(X,Y,Z,5);
xlabel('X-axis')
ylabel('Y-axis')
title('The contour plot of 3*X^2+4*Y^2 = C')
```

Note that the dimensions of the quantities X, Y, and Z are automatically determined by MATLAB.

Scatter Plot

The plot command generates a smooth curve passing though all the points that are represented by vectors x and y. The scatter command will generate markers at the locations specified by the vectors x and y, instead of a curve. The code below will generate a scatter plot for the sine curve.

```
x = 0:pi/20:2*pi;
y = sin(x);
scatter(x,y);
xlabel('X-axis')
ylabel('Y-axis')
title('Plot of Y = sin(x)')
```

Figure 1.11(b) shows the scatter plot generated using this code.

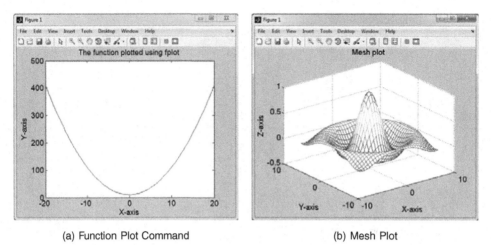

(a) Function Plot Command   (b) Mesh Plot

Figure 1.12. Function and Mesh Plots

fplot: Function Plots

The special command `fplot` plots a function between specified limits. The function must be specified as $y = f(x)$. It is required to specify the two end points for x. The sample code for generating a parabola given by $y = x^2 + 10$ is given here. The plot generated by the `fplot` command is given in Fig. 1.12(a).

```
fplot('x^2+10',[-20 20])
xlabel('X-axis')
ylabel('Y-axis')
title('The function plotted using fplot')
```

1.5.3 3-D Mesh and Surface Plots

Thus far, some of the two dimensional plotting techniques available in MATLAB have been discussed. Next, our attention is focused on some of the three dimensional plotting techniques.

Mesh Plot

Up to this point, plots have been demonstrated that are made of curves. The mesh plot generates a surface specified by the matrices X, Y, and Z. Plot the *sine* function using the `mesh` command. The mesh plot of the *sine* function can also be seen in the MATLAB help manual. The sample code for this plot is as follows.

```
[X,Y] = meshgrid(-8:.5:8);
R = sqrt(X.^2 + Y.^2) + eps;
Z = sin(R)./R;
mesh(X,Y,Z)
```

```
xlabel('X-axis')
ylabel('Y-axis')
zlabel('Z-axis')
title('Mesh plot')
```

The mesh plot is illustrated in Fig. 1.12(b). MATLAB also has a command called surf to generate surface plots. Replace the mesh command in the above sample code with the surf command. The reader is encouraged to practice the surf command.

1.5.4 Using the Plot Editing Mode

Some of the frequently used commands for generating plots in MATLAB have been discussed. Next, practice editing a plot using the options available on the plot window. These editing options are shown in Fig. 1.13. The plot window is shown at the top of Fig. 1.13. The bottom part of this figure explains each of these editing options, and will enable the user to edit the plots for reports or live presentations.

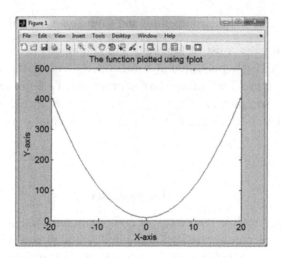

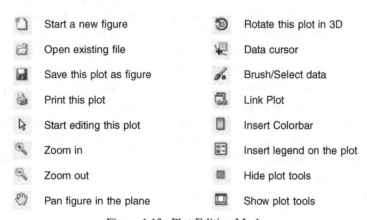

Figure 1.13. Plot Editing Mode

1.6 Optimizing with MATLAB

This section provides a basic guide for solving numerical optimization problems using MATLAB. The MATLAB Optimization Toolbox provides the capability to solve a wide variety of optimization options, such as constrained and unconstrained problems, or linear and nonlinear problems. This section assumes previous familiarity with the basics of numerical optimization, and also with using MATLAB script files and functions (see Sec. 1.3). Before proceeding, please ensure that the MATLAB Optimization Toolbox is installed on the computer. To verify that the Optimization is indeed installed, open MATLAB and click the APPS tab (Fig. 1.1) at the top menu of the MATLAB window. The "Toolboxes" item should contain "Optimization" listed along with any other installed toolboxes.

Using the MATLAB Optimization Toolbox, it is possible to solve (i) nonlinear optimization and (ii) linear optimization problems. Chapter 5 provides the pertinent basic knowledge.

1.7 Popular Functions and Commands, and More

This section provides a list of functions and commands, as well as other general information. This list is presented in Tables (1.1 – 1.17) at the end of this chapter. This list will be useful for writing MATLAB programs. To know exactly how to use a given command or function, simply type help followed by the item in question. In addition, use the help command that is in the table title to obtain general related information.

1.8 Summary

In order to fully realize the power of computational Design Optimization, it is important to implement optimization methods through pertinent computer-based mathematical tools. MATLAB is one such mathematical tool that has gained notorious popularity in the science and engineering community over the last two decades, owing to the diverse set of mathematical functions that it provides and its ease of use (*i.e.*, user-friendliness). To facilitate effective use of MATLAB for implementing the modeling, analysis, and optimization techniques presented in this book, this chapter provided an important introduction to the MATLAB software. Specifically, it provided brief descriptions of the modular MATLAB interface, basic matrix handling and mathematical operations in MATLAB, and the GUI capabilities (*e.g.*, plotting functions) of MATLAB. The chapter ended with a comprehensive set of tables listing some of most useful classes of in-built functions available in MATLAB - from logical operators to equation solving functions. For those who are already proficient in MATLAB, this chapter may serve as a convenient reference.

1.9 Problems

Warm-up Problems

1.1   (a)   Create a folder on your computer and name it after your last name.

        (b)   Set the Current Directory of MATLAB to the directory created in Part (a).

        (c)   Create and save an empty M-file in this directory, and name it after your first name.

        (d)   Undock the Current Directory Window from the MATLAB Desktop and take a screen shot of this directory. Turn in this screen shot.

1.2   Start MATLAB.

        (a)   Define the array z=[1/6 3/4 2/9] in the Command Window.

        (b)   Using the help browser, find the command that can be used to change the (1) format, and (2) spacing of the numeric output of the above array in the Command Window.

        (c)   Find the commands that can be used to print (1) the complete contents in the Command Window, and (2) a selection of the contents in the Command Window.

        (d)   Give at least two examples each for different format and spacing commands available in MATLAB for the array z.

        (e)   For each example, print z and turn in the output displayed in the Command Window.

1.3   (a)   Create two arrays A and B (4x2 and 2x4, respectively) in the Command Window. You can randomly choose elements in A and B.

        (b)   From the Command History browser, create an M-file that includes the definitions of A and B.

        (c)   In this M-file, compute the following: (1) A*B, (2) B*A (3) Is A*B = B*A? What property of matrix multiplication can be recalled with the help of this example? Comment on these observations in about two lines.

        (d)   Change all the values of the array elements of A and B using the Variable Editor in the MATLAB Desktop environment to any values of your choice.

        (e)   Repeat Parts (a) through (c) for the new A and B.

        (f)   Submit the M files, the output in the Command Window, and screen shots of the Variable Editor before and after modifying the values.

1.4   (a)   In the Command Window, create an array, x, such that x ranges from 0 to 10.

        (b)   Compute four arrays: y1=sin(x), y2=exp(x), y3=x^2+2x+1, and y4=x^3+5 in the Command Window. You need to make sure that the sizes of x and y are the same.

        (c)   Save all the arrays computed in Parts (a) and (b) on your hard drive as .mat files.

(d) Write an M-file that loads these `.mat` files from your hard drive.

(e) In the same M-file, add a code to plot the arrays y1, y2 y3 and y4 against x, all in the same plot, with different line styles. Make the plots look professional. Add your name, title, axes legends, and labels to the plot.

(f) Identify which factors impact the smoothness of the above plotted curves. Create four new figures, each showing the plots of y1, y2, y3, and y4, respectively, as a function of x. In each figure, show at least three plots with increasing curve smoothness. Discuss your results in three to four lines. The objective here is to develop your understanding of what is sufficient to obtain a visually smooth curve in practice.

(g) Submit the M-file, plots, and the discussions.

1.5 Generate the following matrices: A = [2 4 6;3 5 1;7 5 9], B = [1 3 6], and C = [5;7;2;0].

(a) Generate a matrix D = [A;B].

(b) Now generate a new matrix E = [D C].

(c) Find the determinant of matrix E.

(d) Find the inverse of matrix E.

(e) Find the transpose of matrix E.

(f) Define a new matrix F = [3;17;12;-2].

(g) Define another matrix H = [5 7 4 -2;3 12 -6 14].

(h) Explain whether or not the following matrix multiplications are possible: (1) EE, (2) FF, (3) HH, (4) EF, (5) FE, (6) HE, (7) EH, (8) FH and (9) HF.

(i) In the cases for which multiplication is possible, perform it using MATLAB. Turn in a print out of your results (from the Command Window) of Parts (a) through (e), and (h).

1.6 For the matrix A = [2 4 6;3 5 1;7 5 9] given in the above problem, determine the eigenvalues and eigenvectors. Turn in a print out of your results (from the Command Window).

1.7 Generate the following matrices:
A = [12 14 16 40; 32 15 11 1; 7 25 19 10],
B = [9 1 36 4; 19 0 -31 2], and C = [7; 5; 7; 2; 0].

(a) Generate a matrix D = [A;B].

(b) Now generate a new matrix E = [D C].

(c) Find the determinant of matrix E.

(d) Find the inverse of matrix E.

(e) Find the transpose of matrix E.

(f) Define a new matrix F = [16;3;17;12;-2].

(g) Define another matrix H = [5 7 4 -2 -1;-9 3 12 -6 14].

(h) Give reasons: which of the following matrix multiplications are possible – (1) EE, (2) FF, (3) HH, (4) EF, (5) FE, (6) HE, (7) EH, (8) FH and (9) HF.

(i) In the cases for which multiplication is possible, perform it using MATLAB. Submit a printout of your results (from the Command Window) of Parts (a) through (e), and (h).

1.8 For the matrices A defined in Problem 1.6 and E defined in Problem 1.7, determine their eigenvalues and eigenvectors. Submit a print out of your results (from the Command Window).

1.9 Define the matrices: A = [12 16 4;23 1 21;9 10 1] and B = [2 7 14;3 11 2;-9 10 12].
(a) Perform matrix multiplication AB.
(b) Perform matrix multiplication BA.
(c) Are the answers from Parts (a) and (b) the same? If yes, explain why. If not, explain why.
(d) Find the inverse of A, and call it matrix X.
(e) Find the inverse of B, and call it matrix Y.
(f) Perform matrix multiplication A*X.
(g) Perform matrix multiplication B*Y.
(h) Are the answers from Parts (g) and (h) the same. If yes, explain why. If not, explain why.
(i) Submit a printout of your results (from the Command Window) for Parts (a) through (h).

1.10 Define the matrices A = [12 16 4;23 1 21;9 10 1], B = [2 7 14;3 11 2;-9 10 12], C = [43 12;13 12], and D = [1 2 3;4 5 6].
(a) Perform the following additions: (1) A+B, (2) A+C, (3) A+D, (4) B+C, (5) B+D, and (6) C+D.
(b) Are each of the above additions possible? If yes, explain why. If not, explain why.
(c) Perform the following operations: (1) A+B and (2) B+A. Is the answer to these two additions the same? If not, explain why. If yes, which matrix addition property is demonstrated using these two additions?
(d) Submit a print out your results (from the Command Window) for Parts (a) and (c), and discussions for the Part (b).

1.11 Define a matrix A = [12 16 4;23 1 21;9 10 1] and B = [2 7 14;3 11 2;-9 10 12].
(a) Perform matrix multiplication A*B, and call this D.
(b) Find the transpose of A, and call this E.
(c) Find the transpose of B, and call this F.
(d) Find the transpose of D, and call this G.
(e) Perform matrix multiplications: (1) E*F and (2) F*E.
(f) Which of the above two multiplications are the same as the matrix D^T?
(g) Which property of the matrix multiplication is demonstrated from Part (f)?
(h) Submit a printout of your results (from the Command Window) for Parts (a) through (g).

1.12 Solve the following systems of linear equations and turn in a print out of your results (from the Command Window).

(a) $3x + 4y = 12$ and $4x + 2y = 10$.

(b) $3x + 4y = 12$ and $4x = 10$.

(c) $-4x + y = 14$ and $4x + 3y = 10$.

(d) $13x + 12y = -6$, $-4x + 7y = -73$, and $11x - 13y = 157$.

(e) $2x + 3y - z = 8$, $4x - 2y + z = 5$, and $x + 5y - 2z = 9$.

(f) $4x - 8y + 3z = 16$, $-x + 2y - 5z = -21$, and $3x - 6y + z = 7$.

(g) $2a + 3b + c - 11d = 1$, $5a - 2b + 5c - 4d = 5$, $a - b + 3c - 3d = 3$, and $3a + 4b - 7c + 2d = -7$.

1.13 Write a simple MATLAB program in an M-file that generates a 1×25 array called A, where A = [1 2 3 4 ... 25]. Use for loop logic. Turn in a printout of your M-file and the command prompt output after running the M-file.

1.14 Write an M-file that tests whether two numbers are greater than zero. If both are greater than zero, then the program should print 'Both' on the screen. If neither of them is greater than zero, then the program should print 'None'. For any other case, the program should print 'Other'. Submit a printout of your M-file. Run your M-file for two cases, and submit a printout of the command prompt output.

1.15 This problem will test your MATLAB programming skills.
(a) Write an M-file that defines a row vector of 50 elements with all ones.
(b) Add a code that replaces every element that is in an even place (for example the 2nd, 4th, 6th,...) with the number 2.
(c) Add a code that replaces every element that is in a place divisible by three (for example the 3rd, 6th, 9th,...) with the number 3.
(d) Turn in a printout of the final M-file and the output after running Parts (a) through (c).

1.16 Write an M-file to generate three contour plots of $3x^2 + 4y^2 = C$, where $1 \le x \le 2$ and $3 \le y \le 5$. The first, second, and third plots should contain 2, 5, and 7 contours, respectively. Using the same M-file, also generate a scatter plot for $3x^2 + 4y^2 = 20$. Turn in the M-file and the plots.

1.17 Write an M-file to plot $y = 4x + 10$, and $y = 4x^2 + 10$ for $1 \le x \le 5$ on the same graph. Provide appropriate X and Y axis labels and the legend. All the labels and legends must be Times New Roman font and the font size must be 15. The font type and font size of these labels and legends must be changed through your M-file. Turn in your M-file and the plot.

1.18 Learn about the subplot command. You can refer to the MATLAB tutorial available on www.mathworks.com. Write an M-file to plot $y = 4x + 10$, $y = 4x^2 + 10$, and $y = \sin(x)$ for $0 \le x \le 3.5$ on the same plot. Now plot these three curves on different subplots using the subplot command. Every subplot should contain individual titles, axis labels, and legends. Turn in the M-file and the plots.

1.19 Learn about about the `peaks` command. You can refer to the MATLAB tutorial available on `www.mathworks.com`. Now write an M-file to generate a 3D plot of the `peaks` command using `mesh` and `surf`. The plot should contain a title, axis labels, and legends. Through your M-file, write your first name on the top-left corner of this plot, and your last name just below the peaks that appear in the plot. Turn in your M-file and the plots.

Intermediate Problems

1.20 This problem will test your ability to use M-files.
(a) Write a function file in MATLAB that takes in one input and returns a single output. Call the function `getSquared`. The function should output the square of the number that it takes as input. Remember that the function must actually RETURN the squared value, and not simply display it on screen. Save the function file in some directory that you created. Turn in a printout of the function file.
(b) Write a script M-file called `testSquared.m` that defines a variable called `xinput`. Give any numerical value to `xinput`. From within this M-file, call the function `getSquared`. Store the output from the function file in a new variable called `xoutput`. Write a statement in the M-file that will print both the `xinput` and `xoutput` on the screen. Save the script M-file in the same folder as the function file. Turn in a printout of the M-file.
(c) Run the file `testSquared.m` from the command prompt. Turn in a printout of the output you get.
(d) Run the function file `getSquared.m` from the command prompt to calculate the square of any variable in your workspace. If you do not have one, create it in the workspace. Turn in a printout of your command prompt output.
(e) Save your `getSquared.m` function as `getSquaredSpecial.m`. Modify this function. Add a test code in your new function to determine whether the input number is greater than or equal to zero. If yes, then the function does exactly what it did before (that is, squaring). If not, then the function simply returns the value -1. Turn in a printout of your `getSquaredSpecial.m` file.
(f) In a similar fashion as above, create the function `testSquaredSpecial.m` file. Run two cases: (i) `xinput` is a positive number, and (ii) `xinput` is a negative number. Turn in a printout of the command prompt output for both cases.

1.21 This problem will test your skills in structured programming using flow control statements.
(a) Write a function called `sumHundred` that takes an integer as the input. The function should determine whether the input is an integer between 1 and 100. If not, the function should display an appropriate error message. Turn in a printout of the function file.

(b) Test the function from the command prompt using (i) a number between 1 and 100, and (ii) a number greater than hundred. Turn in a printout of the command prompt output for both cases.

(c) You will now modify your function sumHundred. Add some logic to it using for or while loops, such that the function evaluates the sum of all integers from 1 to the input number. For instance, if the input number is 79, it should evaluate the sum of the first 79 integers. <u>Do not</u> use any built-in MATLAB functions. Turn in a printout of the modified function file. Run two test cases from the command prompt and turn in a printout of the results along with the M-file.

(d) Save your sumHundred function as sumEvenHundred. Add another logic code to the original function, such that now it evaluates the sum of all even integers from 1 to the input number. For instance, if the input number is 51, it should evaluate the sum of the even integers between 1 and 51. Turn in a printout of the function file and run it for two test cases from the command prompt. Turn in a printout of the results.

1.22 This problem will test your variable handling skills.

(a) Write a function called hiddenSum that takes two inputs and returns the sum of the two inputs. Turn in a printout of the function.

(b) Write a script M-file called testHiddenSum.m that defines two variables x and y. Assign some values to these variables. Call the function hiddenSum from within this script M-file and store the result in a local variable, z. Run the script M-file and turn in a printout with the value of z.

(c) Modify the function hiddenSum such that now it DOES NOT HAVE any output arguments. Now modify your script M-file such that it calls hiddenSum using x and y as inputs and assigns the result to the variable z. You will need to further modify your function file. Submit a printout of the new function and script files and command prompt results from running your script M-file.

(d) Now modify your function hiddenSum such that it DOES NOT HAVE any input OR output arguments. Modify your script M-file such that it calls hiddenSum using x and y as inputs, and stores the result in the variable z. You will need to further modify your function file. Submit a printout of the new function and script files, and the command prompt results from running your script M-file.

Table 1.1. *Arithmetic Operators* (help ops)

Command	Symbol	Description
plus	+	Addition
uplus	+	Unary addition
minus	-	Subtraction
uminus	-	Unary minus
mtimes	*	Matrix multiplication
times	.*	Array multiplication
mldivide	\	Left matrix divide
mrdivide	/	Right matrix divide
ldivide	.\	Left array divide
rdivide	./	Right array divide
mpower	∧	Matrix power
power	.∧	Array power

Table 1.2. *Logical Operators* (help ops)

Command	Symbol	Description
eq	==	Equal
ne	~=	Not equal
lt	<	Less than
gt	>	Greater than
le	<=	Less than or equal
ge	>=	Greater than or equal
and	&	Logical AND
or			Logical OR
not	~	Logical NOT
xor	N/A	Logical Exclusive OR
any	N/A	True if any element of vector is nonzero
all	N/A	True if all elements of vector are nonzero

Table 1.3. *Special Characters* (help ops)

Command	Symbol	Description
colon	:	Colon
punct	..	Parent directory
punct	%	Comment
punct	=	Assignment
transpose	'	Transpose

Table 1.4. *Program Control Flow Constructs* (help lang)

Command	Description
if	Conditionally execute statement
else	If statement condition
elseif	If statement condition
end	Terminate scope for while, switch, try, and if statement
for	Repeat statement for a specific number of times
while	Repeat statement for indefinite number of times
break	Terminate execution of for or while loop
continue	Pass control to the next iteration of for or while loop
switch	Switch among several cases based on an expression
case	switch statement case

Table 1.5. *Scripts, Functions and Variables* (help lang)

Command	Description
global	Define global variable
mfilename	Name of currently executing M-file
exist	Check if variable or function are defined
isglobal	True for global variable

Table 1.6. *Argument Handling* (help lang)

Command	Description
nargin	Number of function input argument
nargout	Number of function output argument

Table 1.7. *Message Display* (help lang)

Command	Description
error	Display error message and abort function
warning	Display warning
disp	Display and array
fprintf	Display formatted message
sprintf	Write format data to a string

Table 1.8. *Elementary Matrices* (help elmat)

Command	Description
zeros	Zeros array
ones	Ones array
eye	Identity matrix
rand	Uniformly distributed random numbers
randn	Normally distributed random numbers
linspace	Linearly spaced vector
logspace	Logarithmically spaced vector
meshgrid	x and y array for 3D plot
:	Regularly spaced vector and index into matrix

Table 1.9. *Basic Array/Matric Information* (help elmat)

Command	Description
size	Size of matrix
length	Length of vector
ndims	Number of dimensions
numel	Number of elements
isempty	True for empty matrix
isequal	True of arrays are identical
diag	Diagonal matrices
find	Find indices of nonzero elements
end	Last index

Table 1.10. *Special Variables and Constants* (help elmat)

Command	Description
ans	Most recent answer
eps	Floating-point relative accuracy
realmax	Largest positive floating-point number
realmin	Smallest positive floating-point number
pi	3.14159263
i, j	Imaginary unit
inf	Infinity
Nan	Not-a-number
isnan	True for not-a-number
isinf	True for infinite elements
isfinite	True for finite elements
why	Succinct number

Table 1.11. *Trigonometric Functions*
(`help elfun`)

Command	Description
sin	Sine
sinh	Hyperbolic sine
asin	Inverse sine
asinh	Inverse hyperbolic sine
cos	Cosine
cosh	Hyperbolic cosine
acos	Inverse cosine
acosh	Inverse hyperbolic cosine
tan	Tangent
tanh	Hyperbolic tangent
atan	Inverse tangent
atan2	Four quadrant inverse tangent
atanh	Inverse hyperbolic tangent
sec	Secant
sech	Hyperbolic secant
asec	Inverse secant
asech	Inverse hyperbolic secant
csc	Cosecant
csch	Hyperbolic cosecant
acsc	Inverse cosecant
acsch	Inverse hyperbolic cosecant
cot	Cotangent
coth	Hyperbolic cotangent
acot	Inverse cotangent
acoth	Inverse hyperbolic cotangenet

Table 1.12. *Exponential and Complex Functions* (`help elfun`)

Command	Description
exp	Exponential
log	Natural logorithm
log10	Common (base 10) logorithm
sqrt	Square root
abs	Absolute value
angle	Phase angle
complex	Construct complex data from real and imaginary part
imag	Complex imaginary part
real	Complex real part
isreal	True for real array

Table 1.13. *Rounding and Remainder (*help elfun*)*

Command	Description
fix	Round toward zero
floor	Round toward minus infinity
ceil	Round toward plus infinity
mod	Modulus (signed remainder after division)
rem	Remainder after division
sign	Signum

Table 1.14. *Specialized Math Functions (*help specfun*)*

Command	Description
cross	Vector cross product
dot	Vector dot product

Table 1.15. *Matrix Analysis (*help matfun*)*

Command	Description
det	Determinant
trace	Sum of diagonal elements
inv	Matrix inverse
eig	Eigenvalues and eigenvectors
svd	Singular value decomposition
expm	Matrix exponential
logm	Matrix logorithm
sqrtm	Matrix square root
fnum	Evaluate general matrix function

Table 1.16. *Basic Statistical Operations (*help datafun*)*

Command	Description
max	Largest component
min	Smallest component
mean	Average or mean value
median	Median value
std	Standard deviation
var	Variance
sort	Sort in ascending order
sortrows	Sort rows in ascending order
sum	Sum of elements
prod	Product of elements
hist	Histogram
histc	Histogram count
trapz	Trapezoidal numerical integration
cumsum	Cumulative sum of elements
cumprod	Cumulative product of elements

Table 1.17. *Optimization Functions*

Command	Description
bintprog	Binary integer programming problems
fgoalattain	Multiobjective goal attainment problems
fminbnd	Minimum of single-variable function on fixed interval
fmincon	Minimum of constrained nonlinear multivariable function
fminimax	Minimax constraint problem
fminsearch	Minimum of unconstrained multivariable function using derivative-free method
fminunc	Minimum of unconstrained multivariable function
fseminf	Minimum of semi-infinitely constrained multivariable nonlinear function
ktrlink	Minimum of constrained or unconstrained nonlinear multivariable function using KNITRO
linprog	Linear programming problems
quadprog	Quadratic programming problems

Table 1.18. *Equation Solving Functions*

Command	Description
fsolve	Solve system of nonlinear equations
fzero	Find root of continuous function of one variable

Table 1.19. *Least Squares (Curve Fitting)*

Command	Description
lsqcurvefit	Solve nonlinear curve-fitting problems in least-squares sense
lsqlin	Solve constrained linear least-squares problems
lsqnonlin	Solve nonlinear least-squares problems
lsqnonneg	Solve nonnegative least-squares constraint problem

BIBLIOGRAPHY OF CHAPTER 1

[1] M. S. Bazaraa, H. D. Sherali, and C. M. Shetty. *Nonlinear Programming: Theory and Algorithms*. John Wiley and Sons, 3rd edition, 2013.

[2] A. Ravindran, G. V. Reklaitis, and K. M. Ragsdell. *Engineering Optimization: Methods and Applications*. John Wiley and Sons, 2006.

[3] G. R. Lindfield and J. E. Penny. *Numerical Methods: Using MATLAB*. Academic Press, 2012.

[4] The MathWorks, Inc. www.mathworks.com.

[5] The MathWorks, Inc. *MATLAB Graphics*. 2014. www.mathworks.com/help/pdf_doc/matlab/graphg.pdf.

[6] T. A. Davis. *MATLAB Primer*. CRC Press, 8th edition, 2011.

[7] A. Knight. *Basics of MATLAB and Beyond*. CRC Press, 2000.

[8] R. Pratap. *Getting Started with MATLAB: A Quick Introduction for Scientists and Engineers*. Oxford University Press, 2009.

[9] J. J. Moré and S. J. Wright. *Optimization Software Guide*. Frontiers in Applied Mathematics - 14. SIAM, 1993.

[10] Vanderplaats Research and Development, Inc. GENESIS structural analysis and optimization software. www.vrand.com.

[11] MSC/Nastran. Structural optimization software. www.macsch.com.

[12] Altair Corp. Structural optimization software. www.altair.com.

[13] D. Xue and Y. Chen. *Solving applied mathematical problems with MATLAB*. CRC Press, 2011.

2

Mathematical Preliminaries

2.1 Overview

This chapter presents some of the basic mathematics needed for learning and using optimization. Specifically, this chapter reviews the basics of linear algebra (*e.g.*, vectors, matrices and eigenvalues) and calculus. A review of the material presented in this chapter may be helpful. For most, it will be well known material; for others, it will be a great opportunity for a quick review. For others still, it will be an opportunity to learn the minimally required mathematics for the material presented in this book.

2.2 Vectors and Geometry

A **scalar** is a quantity that is determined by its magnitude and sign. For example, length, temperature, and speed are scalars. A **vector** is a quantity that is determined by both its magnitude and its direction. It is a directed line segment. For example, velocity and force are examples of vectors.

A vector can be expressed in terms of its **components** in a Cartesian Coordinate system as $\mathbf{a} = a_1\mathbf{i} + a_2\mathbf{j} + a_3\mathbf{k}$, where $\mathbf{i}, \mathbf{j}$, and $\mathbf{k}$ denote the unit vectors along the X, Y and Z axes. The **magnitude** of this vector is given as $||\mathbf{a}|| = \sqrt{a_1^2 + a_2^2 + a_3^2}$.

The length of a vector $\mathbf{x}$ is also called the **norm (or Euclidean norm)** of the vector, denoted by $|\mathbf{x}|$. The **position vector** of a given point $A : \{x, y, z\}$ is the vector *from* the origin of the axes *to* the point A.

2.2.1 Dot Product

The dot product or Inner product of two vectors yields a scalar. Mathematically, the dot product of two vectors $\mathbf{a}$ and $\mathbf{b}$, denoted $\mathbf{a}\cdot\mathbf{b}$, is the product of their lengths times the cosine of the angle θ between them.

$$\mathbf{a} \cdot \mathbf{b} = |\mathbf{a}||\mathbf{b}| \cos\theta \qquad (2.1)$$

In terms of vector components, the dot product of two vectors $\mathbf{a} = a_1\mathbf{i} + a_2\mathbf{j} + a_3\mathbf{k}$ and $\mathbf{b} = b_1\mathbf{i} + b_2\mathbf{j} + b_3\mathbf{k}$ is given as

$$\mathbf{a} \cdot \mathbf{b} = a_1 b_1 + a_2 b_2 + a_3 b_3 \tag{2.2}$$

Two vectors are said to be **orthogonal** when their dot product is equal to zero ($\theta = 90^o$).

2.2.2 Equation of a Line

The **slope-intercept form** of the equation of a line that has a slope of m and y-intercept c is

$$y = mx + c \tag{2.3}$$

The **point-slope form** of the equation of a line that passes through a point P : (x_1, y_1) with a slope m is

$$(y - y_1) = m(x - x_1) \tag{2.4}$$

The **two-point form** of the equation of a line that passes through two points $P_1 : (x_1, y_1)$ and $P_2 : (x_2, y_2)$ is

$$(y - y_1)(x_2 - x_1) = (x - x_1)(y_2 - y_1) \tag{2.5}$$

The **two-intercept form** of a line with x intercept a and y intercept b is given as

$$\frac{x}{a} + \frac{y}{b} = 1 \tag{2.6}$$

2.2.3 Equation of a Plane

The equation of a plane with a normal vector $\mathbf{n} = [a, b, c]^T$ passing through the point (x_0, y_0, z_0) is given as

$$\mathbf{n} \cdot (\mathbf{x} - \mathbf{x}_0) = 0 \tag{2.7}$$

where $\mathbf{x} = (x, y, z)$ is any generic point on the plane. Substituting the value of $\mathbf{x}$ in Eq. 2.7, the general form of the equation of a plane is

$$ax + by + cz + d = 0 \tag{2.8}$$

where $d = -ax_0 - by_0 - cz_0$. The plane specified in this form has its X, Y, and Z intercepts at $p = \frac{-d}{a}$, $q = \frac{-d}{b}$, and $r = \frac{-d}{c}$.

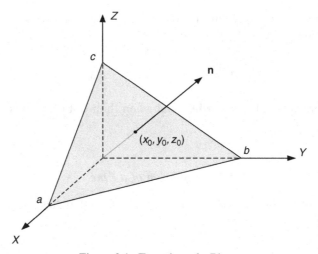

Figure 2.1. Equation of a Plane

In the intercept form, a plane passing through the points $(a,0,0)$, $(0,b,0)$, and $(0,0,c)$ can be given as

$$\frac{x}{a} + \frac{y}{b} + \frac{z}{c} = 1 \tag{2.9}$$

The plane passing through the two points (x_1,y_1,z_1) and (x_2,y_2,z_2), and parallel to the direction $[a,b,c]^T$ is given by the following equation.

$$\begin{vmatrix} x - x_1 & y - y_1 & z - z_1 \\ x_2 - x_1 & y_2 - y_1 & z_2 - z_1 \\ a & b & c \end{vmatrix} = 0 \tag{2.10}$$

The plane that passes through the three points (x_1,y_1,z_1), (x_2,y_2,z_2), and (x_3,y_3,z_3) is given by,

$$\begin{vmatrix} x - x_1 & y - y_1 & z - z_1 \\ x_2 - x_1 & y_2 - y_1 & z_2 - z_1 \\ x_3 - x_1 & y_3 - y_1 & z_3 - z_1 \end{vmatrix} = 0 \tag{2.11}$$

In the above two equations, we used the notation $|A| = det(A)$, for any given matrix A.

2.3 Basic Linear Algebra

Linear algebra can be defined as the study of the theory and application of linear systems of equations and linear transformations. Linear algebra uses matrices and their properties in a systematic way, often to solve physically meaningful problems (Ref. [1]).

2.3.1 Preliminary Definitions

A matrix is a rectangular array of numbers, symbols, or functions, which are called the elements of the matrix. The following matrices are presented to help discuss some basic matrix definitions.

$$\begin{bmatrix} 1 & 0 & 2 \\ 0 & 3 & -1 \end{bmatrix}_{2 \times 3} \tag{2.12}$$

$$\begin{bmatrix} 0 & 2 \\ 3 & -1 \end{bmatrix}_{2 \times 2} \tag{2.13}$$

$$\begin{bmatrix} 2 \\ -1 \\ 0.4 \\ 5 \end{bmatrix}_{4 \times 1} \tag{2.14}$$

$$\begin{bmatrix} 3 & -1 & 0 & 4 \end{bmatrix}_{1 \times 4} \tag{2.15}$$

$$\begin{bmatrix} \sin x & \cos x \\ \cos x & -\sin x \end{bmatrix}_{2 \times 2} \tag{2.16}$$

$$\begin{bmatrix} 1 & 0 & 0 \\ 0 & 1 & 0 \\ 0 & 0 & 1 \end{bmatrix}_{3 \times 3} \tag{2.17}$$

$$\begin{bmatrix} 1 & 2 & -1 & 6 \\ 0 & 2 & 3 & -5 \\ 0 & 0 & -2 & 2 \\ 0 & 0 & 0 & 7 \end{bmatrix}_{4 \times 4} \tag{2.18}$$

$$\begin{bmatrix} 2 & 0 & 0 & 0 \\ 3 & -5 & 0 & 0 \\ -1 & 2 & 3 & 0 \\ 3 & -7 & 8 & 6 \end{bmatrix}_{4 \times 4} \tag{2.19}$$

The first example provided in Eq. 2.12 has two *rows* (horizontal lines) and three *columns* (vertical lines). The **order** of this matrix is 2×3 since it has *two* rows and *three* columns. Matrices where the number of rows is not equal to the number of columns are called **Rectangular matrices**. The second example provided in Eq. 2.13 has two rows and two columns. Matrices where the number of rows is equal to the number of columns are called **Square matrices**. The third example found in Eq. 2.14 has only one column. These are called **Column matrices**. The fourth example in Eq. 2.15 has only one row. These are called **Row matrices**. A general notation for a matrix is given by

$$A = [a_{ij}] = \begin{bmatrix} a_{11} & a_{12} & \dots & a_{1n} \\ a_{21} & a_{22} & \dots & a_{2n} \\ \vdots & \vdots & \vdots & \vdots \\ a_{m1} & a_{m2} & \dots & a_{mn} \end{bmatrix}_{m \times n}$$

Any element in a matrix can be represented using the double subscript notation as a_{ij}, where i denotes the row and j the column in which the element is located. When $m = n$, the matrix is said to be **square**. The diagonal containing the elements a_{11}, $a_{22},\dots a_{nn}$ is called the **principal diagonal** of the matrix. The matrix given by Eq. 2.17, where all the elements in the principal diagonal are ones, and the rest of the elements are zeros, is called the **Identity Matrix**. The matrix in the form shown in Eq. 2.18, where all the elements below the principal diagonal are zeros, is called an **Upper Triangular matrix**. A **Lower triangular form** is provided in Eq. 2.19, where all the elements above the principal diagonal are zeros.

2.3.2 Matrix Operations

Addition

The addition of two matrices is possible only if they are of the same order. If two matrices A and B have the same order, then $A + B$ can be obtained by adding the corresponding elements of A and B. Consider the following example.

Example 1:

$$A = \begin{bmatrix} 1 & 0 & 2 \\ 0 & 3 & -1 \end{bmatrix}_{2 \times 3} \tag{2.20}$$

and

$$B = \begin{bmatrix} 5 & 6 & 1 \\ 3 & 0 & 0 \end{bmatrix}_{2 \times 3} \tag{2.21}$$

then,

$$A + B = \begin{bmatrix} 6 & 6 & 3 \\ 3 & 3 & -1 \end{bmatrix}_{2 \times 3} \qquad (2.22)$$

Subtraction

Similar to the addition operation, two matrices can be subtracted only if they have the same order. For the two matrices A and B given by Eqs. 2.20 and 2.21, $A - B$ can be obtained by subtracting the corresponding elements as,

Example 2:

$$A - B = \begin{bmatrix} -4 & -6 & 1 \\ -3 & 3 & -1 \end{bmatrix}_{2 \times 3}$$

Scalar Multiplication

The product of any matrix with a scalar can be obtained by multiplying each element of that matrix with that scalar. For example, for matrix A in Eq. 2.20, $4A$ can be obtained by simply multiplying each element of A by 4.

Example 3:

$$4A = \begin{bmatrix} 4 & 0 & 8 \\ 0 & 12 & -4 \end{bmatrix}$$

Matrix Multiplication

Unlike matrix addition, matrix multiplication is **not** obtained by multiplying the corresponding elements of two matrices. Consider two matrices C and D. These two matrices can be multiplied to yield CD only if the **number of columns in C is equal to the number of rows in** D. If C is an $m \times n$ matrix and D is an $n \times p$ matrix, then the product of C and D has an order of $m \times p$. For example, consider the following:

Example 4:

$$C = \begin{bmatrix} 2 & 1 & 2 \\ 1 & 1 & 4 \end{bmatrix}_{2 \times 3}$$

and

$$D = \begin{bmatrix} 4 & 0 \\ 0 & 12 \\ 3 & 2 \end{bmatrix}_{3 \times 2}$$

then

$$CD = \begin{bmatrix} 2 & 1 & 2 \\ 1 & 1 & 4 \end{bmatrix}_{2\times3} \begin{bmatrix} 4 & 0 \\ 0 & 12 \\ 3 & 2 \end{bmatrix}_{3\times2}$$

$$= \begin{bmatrix} 2\cdot4+1\cdot0+2\cdot3 & 2\cdot0+1\cdot12+2\cdot2 \\ 1\cdot4+1\cdot0+4\cdot3 & 1\cdot0+1\cdot12+4\cdot2 \end{bmatrix}_{2\times2}$$

$$= \begin{bmatrix} 14 & 16 \\ 16 & 20 \end{bmatrix}_{2\times2}$$

In addition, consider

$$DC = \begin{bmatrix} 4 & 0 \\ 0 & 12 \\ 3 & 2 \end{bmatrix}_{3\times2} \begin{bmatrix} 2 & 1 & 2 \\ 1 & 1 & 4 \end{bmatrix}_{2\times3}$$

$$= \begin{bmatrix} 4\cdot2+0\cdot1 & 4\cdot1+0\cdot1 & 4\cdot2+0\cdot4 \\ 0\cdot2+12\cdot1 & 0\cdot1+12\cdot1 & 0\cdot2+12\cdot4 \\ 3\cdot2+2\cdot1 & 3\cdot1+2\cdot1 & 3\cdot2+2\cdot4 \end{bmatrix}_{3\times3}$$

$$= \begin{bmatrix} 8 & 4 & 8 \\ 12 & 12 & 48 \\ 8 & 5 & 14 \end{bmatrix}_{3\times3}$$

Note that $CD \neq DC$. The following are some important properties of matrix multiplication.

1. $AB \neq BA$
2. $AI = IA = A$
3. $AB = 0$ does not necessarily imply $A = 0$ or $B = 0$
4. $k(AB)=(kA)B=A(kB)$, where k is a scalar
5. $A(BC)=(AB)C$
6. $(A + B)C=AC + BC$
7. $A(B + C)=AB + AC$

Transpose

The transpose of a matrix is obtained by interchanging its rows and columns. Consider the following example.

Example 5:

$$C = \begin{bmatrix} 2 & 1 & 2 \\ 1 & 1 & 4 \end{bmatrix}_{2 \times 3}$$

$$C^T = \begin{bmatrix} 2 & 1 \\ 1 & 1 \\ 2 & 4 \end{bmatrix}_{3 \times 2}$$

Some important properties of the transpose of a matrix follow:

1. If $A^T = A$, then A is called a **Symmetric Matrix**. If $A^T = -A$, then A is called a **Skew Symmetric Matrix**.
2. $(A^T)^T = A$
3. $(A + B)^T = A^T + B^T$
4. $(AB)^T = B^T A^T$

2.3.3 Determinants

The general form of a determinant of a matrix is given as

$$det(A) = \begin{vmatrix} a_{11} & a_{12} & \cdots & a_{1n} \\ a_{21} & a_{22} & \cdots & a_{2n} \\ \vdots & \vdots & \ddots & \vdots \\ a_{n1} & a_{n2} & \cdots & a_{nn} \end{vmatrix}$$

$$= \sum_{i=1}^{n} a_{ij} C_{ij}$$

$$= \sum_{i=1}^{n} a_{ij} (-1)^{i+j} M_{ij}$$

where j represents any row in A, C_{ij} is called the **Cofactor** of A, and M_{ij} is the corresponding **minor**. The minor determinant M_{ij} is the determinant of the sub-matrix obtained by deleting the ith row and the jth column of the matrix. A cofactor is a signed minor determinant. Specifically, $C_{ij} = M_{ij}$ when $i + j$ is even and $C_{ij} = -M_{ij}$ when $i + j$ is odd. For example, let us compute the determinant of a 2×2 matrix.

Example 6:

$$A = \begin{bmatrix} 2 & 1 \\ 3 & -1 \end{bmatrix}$$

$$det(A) = \begin{vmatrix} 2 & 1 \\ 3 & -1 \end{vmatrix}$$

$$= (2)(-1) - (3)(1)$$

$$= -5$$

The following example demonstrates computation of the determinant of a 3×3 matrix.

Example 7:

$$A = \begin{bmatrix} 2 & 1 & 1 \\ 3 & -1 & 2 \\ 1 & 0 & -1 \end{bmatrix}$$

$$det(A) = \begin{vmatrix} 2 & 1 & 1 \\ 3 & -1 & 2 \\ 1 & 0 & -1 \end{vmatrix}$$

$$= (-1)^{1+1}(2)\begin{vmatrix} -1 & 2 \\ 0 & -1 \end{vmatrix} + (-1)^{1+2}(1)\begin{vmatrix} 3 & 2 \\ 1 & -1 \end{vmatrix} + (-1)^{1+3}(1)\begin{vmatrix} 3 & -1 \\ 1 & 0 \end{vmatrix}$$

$$= 2(1-0) - 1(-3-2) + 1(0 - (-1))$$

$$= 8$$

The following are some important properties of determinants of (square) matrices.

1. $det(A^T) = det(A)$
2. $det(I) = 1$
3. If two rows or columns of a matrix are identical, then $det(A) = 0$.
4. If all entries of a row or column are all zeros, then $det(A) = 0$.
5. If B is obtained by interchanging any two rows or columns of A, then $det(B) = -det(A)$
6. If a row or column of a matrix is a linear combination of two or more rows or columns, respectively, then $det(A) = 0$.
7. $det(AB) = det(A)det(B)$
8. $det(cA) = c^n det(A)$, where c is a scalar.
9. If the determinant of a matrix is zero, the matrix is said to be **singular**. Otherwise, the matrix is said to be **non-singular**.
10. If a matrix is upper triangular or lower triangular, the determinant of that matrix is simply equal to the product of the elements in the principal diagonal.

2.3.4 Inverse

For a square matrix A, if there exists a matrix B, such that, $AB = BA = I$, then B is the inverse of A. A general definition of inverse is given as

$$A^{-1} = \frac{1}{det(A)}[Adjoint(A)] \tag{2.23}$$

where $Adjoint(A) = C^T$, and C is the matrix of the co-factors of A. For a 2×2 matrix, the above equation reduces to the following simple formula.

$$A = \begin{bmatrix} a & b \\ c & d \end{bmatrix}$$

$$A^{-1} = \frac{1}{ad - bc} \begin{bmatrix} d & -b \\ -c & a \end{bmatrix}$$

It follows from Eq. 2.23 that the inverse of a matrix exists if and only if the determinant is nonzero (*i.e.*, the matrix is non-singular). The inverse of a matrix can also be computed using the Gauss Jordan method. The following example illustrates how to compute the inverse of a 2×2 matrix using its determinant.

Example 8:

$$A = \begin{bmatrix} 3 & 1 \\ 2 & 4 \end{bmatrix}$$

$$A^{-1} = \frac{1}{3 * 4 - 2 * 1} \begin{bmatrix} 4 & -1 \\ -2 & 3 \end{bmatrix}$$

$$= \begin{bmatrix} 0.4 & -0.1 \\ -0.2 & 0.3 \end{bmatrix}$$

The following properties of matrix inversion (assuming the associated matrices are invertible) are also important to know.

1. $AA^{-1} = I$
2. $(AB)^{-1} = B^{-1}A^{-1}$
3. $(A^T)^{-1} = (A^{-1})^T$

2.3.5 Eigenvalues

The Eigenvalue problem is defined as follows:
For an $n \times n$ matrix, find all scalars λ such that the equation

$$Ax = \lambda x \tag{2.24}$$

has a nonzero solution x. Such a scalar λ is called the **eigenvalue** of A, and any nonzero $n \times 1$ vector, x, satisfying Eq. 2.24 is called the **eigenvector** corresponding to λ.

Given a matrix A, the eigenvalues of A, denoted by λ, can be computed by finding the roots of the equation $|A - \lambda I| = 0$. This equation $|A - \lambda I| = 0$ is called the **characteristic equation**. The following example illustrates how to find the eigenvalues of a 2×2 matrix.

Example 9

$$A = \begin{bmatrix} -5 & 2 \\ 2 & -2 \end{bmatrix}_{2 \times 2}$$

The characteristic equation of A can be written as $|A - \lambda I| = 0$, where λ is the eigenvalue we want to compute.

$$A - \lambda I = \begin{bmatrix} -5 & 2 \\ 2 & -2 \end{bmatrix} - \begin{bmatrix} \lambda & 0 \\ 0 & \lambda \end{bmatrix}$$

$$= \begin{bmatrix} -5 - \lambda & 2 \\ 2 & -2 - \lambda \end{bmatrix}$$

$$|A - \lambda I| = \begin{vmatrix} -5 - \lambda & 2 \\ 2 & -2 - \lambda \end{vmatrix}$$

$$= (-5 - \lambda)(-2 - \lambda) - 4$$

$$= \lambda^2 + 7\lambda + 6$$

$$= 0$$

The above quadratic equation is the characteristic equation of A. Recall that for a quadratic equation given by $ax^2 + bx + c = 0$, the roots are given as $\frac{-b \pm \sqrt{b^2 - 4ac}}{2a}$. The eigenvalues of A are the roots of $\lambda^2 + 7\lambda + 6 = 0$, which are found to be $\lambda_1 = -1$ and $\lambda_2 = -6$.

2.3.6 Eigenvectors

Given a matrix A, consider the equation $(A - \lambda I)\mathbf{x} = 0$, where λ is an eigenvalue of A. The nonzero values of $\mathbf{x}$ that satisfy the above equation for each eigenvalue are called the **Eigenvectors** of A. We will learn how to compute eigenvectors with the help of an example. Consider the matrix A in Example 9.

Example 10: Note that it is important to compute eigenvalues before we compute eigenvectors. Previously, in Example 9, we obtained the eigenvalues of A as

$\lambda_1 = -1$ and $\lambda_2 = -6$. For the first eigenvalue $\lambda_1 = -1$, we find values of $\mathbf{x}$ such that $(A - \lambda_1 I)\mathbf{x} = 0$, or $(A - (-1)I)\mathbf{x} = 0$.

$$(A - \lambda_1 I)\mathbf{x} = \begin{bmatrix} -5 & 2 \\ 2 & -2 \end{bmatrix} - (-1) \begin{bmatrix} 1 & 0 \\ 0 & 1 \end{bmatrix} \begin{bmatrix} \mathbf{x}_1 \\ \mathbf{x}_2 \end{bmatrix}$$

$$= \begin{bmatrix} -4 & 2 \\ 2 & -1 \end{bmatrix} \begin{bmatrix} \mathbf{x}_1 \\ \mathbf{x}_2 \end{bmatrix}$$

$$= 0$$

Expanding the above matrix, we obtain

$$-4\mathbf{x}_1 + 2\mathbf{x}_2 = 0$$

$$2\mathbf{x}_1 - \mathbf{x}_2 = 0$$

In order to find the eigenvector corresponding to $\lambda_1 = -1$, we need to find the values of $\mathbf{x}_1$ and $\mathbf{x}_2$ that satisfy the above two equations. By inspection, we find that $\mathbf{x}_1 = 1$ and $\mathbf{x}_2 = 2$ satisfy the above equations. Thus, the vector $[1, 2]^T$ is an eigenvector corresponding to the eigenvalue $\lambda_1 = -1$.

After finding a particular $\mathbf{x}$ as an eigenvector for a given eigenvalue, note that a multiple of $\mathbf{x}$, for example $k\mathbf{x}$, will also be an eigenvector for that given eigenvalue. For instance, for $\lambda_1 = -1$, $[\frac{1}{2}, 1]^T$ is also a valid eigenvector. Using the procedure outlined above, we can compute the eigenvector for the eigenvalue of $\lambda_2 = -6$ to be $[2, -1]^T$.

Note that there could be several linearly independent eigenvectors possible for a single eigenvalue. The maximum number of linearly independent eigenvectors corresponding to an eigenvalue of λ is called the **geometric multiplicity** of λ.

2.3.7 Positive Definiteness

A matrix is said to be positive definite if all its eigenvalues are positive. A matrix is said to positive semidefinite if all its eigenvalues are non-negative (that is, including zero).

A matrix is said to be negative definite if all its eigenvalues are negative. A matrix is said to negative semidefinite if all its eigenvalues are non-positive (that is, including zero).

2.4 Basic Calculus: Types of Functions, Derivative, Integration and Taylor Series

You can find a mathematical definition of function in numerous books. Instead of using mathematical language, we define function on a lighter note. Think about a vending machine that takes in money and gives you soda or coffee. Functions behaves similarly. They generally take in one or more numbers and output a number. For

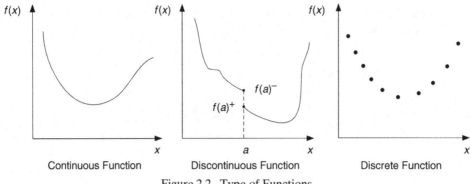

Figure 2.2. Type of Functions

example, consider a simple trigonometric function $f(x) = \sin(x)$. Now, when we input $x = \pi/2$, we obtain $f(x) = 1$. Thus, the sine function *maps* the input $\pi/2$ to the output, which is 1. Try typing `sin(pi/2)` in MATLAB. Most optimization books provide some useful mathematical background (Refs. [2, 3]). Note that our simplistic discussion here avaded many important issues, such as invertibility.

2.4.1 Types of Functions

In this section, we will introduce different types of functions commonly encountered in optimization.

Continuous Functions

A function $f(x)$ is continuous over an interval $[a, b]$ if the following conditions are satisfied for any point p $(a \leq p \leq b)$.

1. $f(x)$ is defined on the interval $[a, b]$.
2. The limit of $f(x)$, as x tends to p, should exist.
3. The limit of $f(x)$ as x tends to p is equal to $f(p)$. In other words, the limit should be equal to the value of the function evaluated at $x = p$.

The first plot in Fig. 2.2 illustrates a continuous function. For every value of x, there exists a unique value of $f(x)$. In simple terms, we note that if a function can be traced without lifting the hand, it is continuous.

Discontinuous Functions

Functions that do not satisfy the three conditions stated in the previous subsection are called discontinuous functions. As illustrated in the second plot in Fig. 2.2, at $x = a$, the function has two values, $f(a)^-$ and $f(a)^+$. As such, it violates the third condition. A discontinuous function cannot be traced from one end to the other without lifting the hand. Try it on the discontinuous function shown in Fig. 2.2.

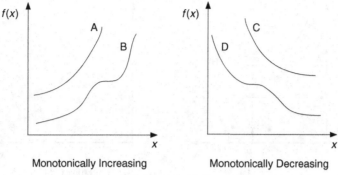

Figure 2.3. Monotonic Functions

Discrete Function

Discrete functions are not defined for all values of x within the given interval; rather, they are defined at particular values of x. As seen in the third plot in Fig. 2.2, function $f(x)$ is defined for discrete values of x. Some optimization algorithms approximate the discrete function using a continuous function.

Monotonically Increasing Function

A function $f(x)$ is said to be *monotonically increasing* if it satisfies the following property. For two points $x = a$ and $x = b$, such that $b > a$, if $f(b) \geq f(a)$, then the function is monotonically increasing. Monotonically increasing functions are shown by A and B in Fig. 2.3.

A function is called *Strictly increasing* if $f(b) > f(a)$, as represented by A in Fig. 2.3.

Monotonically Decreasing Function

A function $f(x)$ is monotonically decreasing if it satisfies the following property. For two points $x = a$ and $x = b$, such that $b > a$, if $f(b) \leq f(a)$, then the function is monotonically decreasing. Monotonically decreasing functions are shown by C and D in Fig. 2.3.

A function is called *Strictly decreasing* if, for any $b > a, f(b) < f(a)$; as represented by C in Fig. 2.3.

Unimodal Functions

Unimodal functions have a single optimum in a given interval (other than end-points), which may be either a maximum or a minimum. A unimodal function that has a minimum value at $x = p$ is represented by A in Fig 2.4. We note that $f(x)$ is strictly decreasing for $x < p$ and strictly increasing for $x > p$. In Fig. 2.4, B represents a special type of unimodal function. Unimodal functions need not be continuous. Function C in Fig. 2.4 is a discrete unimodal function.

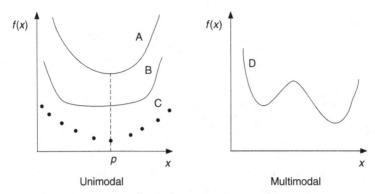

Unimodal   Multimodal

Figure 2.4. Unimodal and Multimodal Functions

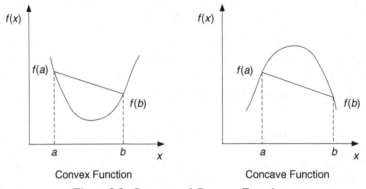

Convex Function   Concave Function

Figure 2.5. Convex and Concave Functions

Multimodal Functions

Unlike the previous functions, multimodal functions have multiple maxima or minima. One such multimodal function is represented by D in Fig. 2.4, which has two minima.

Convex Functions

A convex function is one for which the line joining any two points on the function lies entirely on or above the function defined between these two points. The left plot in Fig. 2.5 displays a convex function. As shown in Fig. 2.5, a and b are any two points in the domain of a function $f(x)$. The line joining their function values is entirely on or above the function defined between a and b.

If $f(x)$ is twice differentiable (that is, you can find its second derivative) in $[a, b]$, then a necessary and sufficient condition for it to be convex on that interval is that we have the second derivative $f''(x) > 0$ for all x in $[a, b]$.

Concave Functions

A function $f(x)$ is said to be concave on an interval $[a, b]$ if the function $-f(x)$ is convex in that interval. Thus, the definition of concave functions is exactly opposite that of convex functions. The right plot in Fig. 2.5 illustrates a concave function. As shown in Fig. 2.5, the line joining the function values of a and b lies entirely on or below the function defined between a and b.

2.4.2 Limits of Functions

Consider a function $f(x)$, and let the function $f(x)$ have a limit L when x tends to a value x_0. In simple language, the function $f(x)$ reaches a value L as x approaches the value x_0. Mathematically, it is denoted by

$$\lim_{x \to x_0} f(x) = L \tag{2.25}$$

Example 2.1 chap_math_limit Let the function be $f(x) = ((a+x)^2 - a^2)/x$. The limit of this function as x tends to 0 can be determined as

$$\lim_{x \to 0} \frac{(a+x)^2 - a^2}{x}$$

$$= \lim_{x \to 0} \frac{a^2 + 2xa + x^2 - a^2}{x}$$

$$= \lim_{x \to 0} \frac{2xa + x^2}{x}$$

$$= \lim_{x \to 0} 2a + x$$

To find the limit of $f(x)$ as x tends to 0, simply substitute $x = 0$ in the above equation, and obtain

$$\lim_{x \to 0} f(x) = 2a$$

2.4.3 Derivative

The derivative of a function $f(x)$ at a certain point x is a measure of the rate at which that function is changing as the variable x changes. That is, a derivative represents the rate of change of a function with respect to the input variable, or the derivative is the computation of the instantaneous slope of $f(x)$ at point x. Mathematically, the derivative is represented as

$$\frac{df(x)}{dx} = \lim_{h \to 0} \frac{f(x+h) - f(x)}{h} \tag{2.26}$$

where h is an infinitesimally small deviation in x.

2.4.4 Partial Derivative

For a function of several variables, it is often useful to examine the variation of the function with respect to one of the variables while all the other variables remain fixed. This is the purpose of a partial derivative. The partial derivative is obtained in the same way as ordinary differentiation with this constraint. Partial derivatives are defined as derivatives of a function of multiple variables when all but the variable of interest are held fixed during the differentiation. Consider a function $f(x, y)$, which is a function of two variables, x and y. The partial derivative of this function with respect to x is defined as the derivative of the function when y is held constant.

$$\frac{\partial f}{\partial x} = \lim_{h \to 0} \frac{f(x+h, y) - f(x, y)}{h} \tag{2.27}$$

Similarly, the partial derivative of the function $f(x, y)$ with respect to y is defined as the derivative of the function when x is held constant.

$$\frac{\partial f}{\partial y} = \lim_{h \to 0} \frac{f(x, y+h) - f(x, y)}{h} \tag{2.28}$$

2.4.5 Indefinite Integration

The previous subsection presented the derivative of a function. Let us assume that the derivative of a function $f(x)$ is $F(x)$. It is possible to recover $f(x)$, within a constant, by integrating $F(x)$, as follows

$$f(x) = \int F(x)dx + C \tag{2.29}$$

where, C is the integration constant. For example, if $F(x) = x^2 + 2$, then

$$f(x) = \int (x^2 + 2)dx + C = \frac{x^3}{3} + 2x + C \tag{2.30}$$

2.4.6 Definite Integration

Definite integration is the integration of a function, $F(x)$, over a particular range of the variable x. It is denoted by

$$\int_a^b F(x)dx \tag{2.31}$$

where a and b are the limits of integration. Let $f(x)$ be the indefinite integral of $F(X)$. The definite integral is determined as follows

$$\int_a^b F(x)dx = [f(x)]_a^b = f(b) - f(a) \tag{2.32}$$

For example, if $F(x) = x^2 + 2$, then from Eq. 2.30 we have $f(x) = \frac{x^3}{3} + 2x + C$. Now let us evaluate the definite integral over the interval from $x = 1$ to $x = 2$.

$$\int_1^2 (x^2 + 2)dx = \left[\frac{x^3}{3} + 2x + C \right]_1^2 \qquad (2.33)$$

$$= \left[\left(\frac{2^3}{3} + 2(2) + C \right) - \left(\frac{1^3}{3} + 2(1) + C \right) \right] \qquad (2.34)$$

$$= \left[\frac{8}{3} + 4 + C - \frac{1}{3} - 2 - C \right] = \frac{13}{3} \qquad (2.35)$$

2.4.7 Taylor Series

A Taylor series [4] provides an approach to approximate a function. Although it generally requires an infinite number of terms to obtain the exact value of the function, in practice only a small number of terms is required for an adequate approximation. The approximation takes the form of a polynomial. The approximation is accurate around a chosen point, and becomes progressively inaccurate as we move away from that point. For a single variable function, $f(x)$, the Taylor series expansion about a point $x = x_0$ is given by

$$f(x) = f(x_0) + f'(x_0)\frac{(x - x_0)}{1!}$$

$$+ f''(x_0)\frac{(x - x_0)^2}{2!} + f'''(x_0)\frac{(x - x_0)^3}{3!} + \dots \qquad (2.36)$$

where $'$ represents the derivative. To increase the accuracy of the function approximation, we increase the number of terms included in the Taylor series. At some point, including too many terms will become impractical. In many cases, the Taylor series progresses as far as the 2nd derivative term. This version of the approximation is referred to as a 2nd order Taylor series expansion. It is given as

$$f(x) = f(x_0) + f'(x_0)\frac{(x - x_0)}{1!} + f''(x_0)\frac{(x - x_0)^2}{2!} \qquad (2.37)$$

When higher order terms (HOT) are excluded, it is called a "Taylor series approximation." Let's evaluate the 2nd order Taylor series for $f(x) = \sin(x)$ at $x = \pi/2$. We have $f'(x) = \cos(x)$ and $f''(x) = -\sin(x)$. Therefore, the 2nd order Taylor series is given by

$$\sin(x) = \sin(\pi/2) + \cos(\pi/2)\frac{(x - \pi/2)}{1!} - \sin(\pi/2)\frac{(x - \pi/2)^2}{2!} \qquad (2.38)$$

Since, $\cos(\pi/2) = 0$ and $\sin(\pi/2) = 1$, we find

$$\sin(x) = 1 - \frac{(x - \pi/2)^2}{2!} \tag{2.39}$$

2.5 Optimization Basics: Single-Variable Optimality Conditions, Gradient, Hessian

A general optimization problem refers to finding the maximum or the minimum value of a function. Some examples of optimization problems include maximizing the mileage of a car or minimizing the failure rate of a product.

Consider a continuous single variable function $f(x)$. Optimization theory gives us the tools to find a "good" value of x that corresponds to a "good" value of $f(x)$. In many problems, the choice of the values of x is constrained in the sense that a candidate value of x must satisfy some conditions. Such optimization problems are called **Constrained Optimization** problems, which will be studied later. **Unconstrained Optimization** problems, on the other hand, involve no constraints on the values of x. In this section, we present some basic concepts of unconstrained optimization. We discuss the basic conditions of optimality for single variable functions. We also present the two basic entities that play a central role in multi-variable optimization: gradients and Hessians.

Definitions

Consider the function $f(x)$ in a set S. We define global and local minima as follows:

1. The function $f(x)$ is said to be at a point of **global minimum**, $x^* \in S$, if $f(x^*) \leq f(x)$ for all $x \in S$.
2. The function $f(x)$ is said to be at a point of **local minimum**, $x^* \in S$, if $f(x^*) \leq f(x)$ for all x within an infinitesimally small distance ϵ from x^*. That is, there exists $\epsilon > 0$ such that for all x satisfying $|x - x^*| < \epsilon$, $f(x^*) \leq f(x)$.

The concept of global and local minima is illustrated in Fig. 2.6. The word "optimum" refers to a maximum or a minimum.

2.5.1 Necessary Conditions for Local Optimum

Assuming that the first and second derivatives of the function $f(x)$ exist, the **necessary conditions** for x^* to be a **local minimum** of the function $f(x)$ on an interval (a, b) are:

1. $\left.\frac{df}{dx}\right|_{x=x^*} = 0$
2. $\left.\frac{d^2f}{dx^2}\right|_{x=x^*} \geq 0$

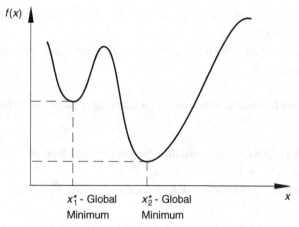

Figure 2.6. Local and Global Minima

The necessary conditions for x^* to be a **local maximum** of the function $f(x)$ on an interval (a, b) are:

1. $\left. \dfrac{df}{dx} \right|_{x=x^*} = 0$

2. $\left. \dfrac{d^2 f}{dx^2} \right|_{x=x^*} \leq 0$

It is important to understand that the above stated conditions are necessary, but not sufficient. This means that if the above conditions are not satisfied, x^* will not be a local minimum or maximum. On the other hand, if the above conditions are satisfied, it does not guarantee that x^* is the local minimum or maximum.

2.5.2 Stationary Points and Inflection Points

A **stationary point** is a point x^* that satisfies the following equation:

$$\left. \frac{df}{dx} \right|_{x=x^*} = 0$$

An **inflection point** may or may not be a stationary point. An inflection point is one where the curvature of the curve (second derivative) changes sign from positive to negative, or vice versa. That point is not necessarily a minimum or a maximum.

2.5.3 Sufficient Conditions for Local Optima

Consider a point x^* at which the first derivative of $f(x)$ is equal to zero, and the order of the first nonzero higher derivative is n. The following are the **sufficient conditions** for x^* to be a **local optimum**.

1. If n is odd, then x^* is an inflection point.
2. If n is even, then x^* is a local optimum. In addition,

(a) If the value of that derivative of $f(x)$ at x^* is positive, then the point x^* is a local minimum.
(b) If the value of that derivative of $f(x)$ at x^* is negative, then the point x^* is a local maximum.

The procedure to find the maximum of a function is illustated using the following example.

Example 10: In this example, we find the maximum value of a function given by $f(x) = -x^3 + 3x^2 + 9x + 10$ in the interval $-2 \le x \le 4$.
First, the stationary points are determined by solving $\frac{df}{dx} = 0$.

$$\frac{df}{dx} = -3x^2 + 6x + 9$$
$$= 0$$

Using the formula to compute roots of a quadratic equation, the roots of the first derivative are found by solving $x = \frac{-6 \pm \sqrt{6^2 - 4.(-3)(9)}}{2.(-3)}$. This process yields two solutions: $x = 3$ and $x = -1$ as the two stationary points, which are in the interval $-2 \le x \le 4$. The function values at these stationary points are evaluated to determine which of these points may correspond to a global maximum.
Evaluating $f(x)$ at $x = 3, -1, -2$ and 4 yields the function values as 37, 5, 12, and 30, respectively. Therefore, $x = 3$ corresponds to the maximum of the function in the interval $-2 \le x \le 4$.

2.5.4 Gradient and Hessian of a Function

In the previous subsection, we have only considered single variable functions; that is, $f(x)$, where x is a single variable. In practice, however, x could represent the two dimensions of a rectangular backyard. In this case, we could let x be a vector with two entries, a two-dimensional vector: $x = \{x_1, x_2\}$, where $x_1 = Length$ and $x_2 = Width$. Therefore, in general, we can consider multi-variable functions $f(x)$, where x is an n-dimensional vector. In these cases, the first and the second derivatives of the function are more complicated. They are respectively referred to as the *gradient* and the *Hessian* of the functions. The gradient is an $n \times 1$ vector and the Hessian is an $n \times n$ matrix.

Gradient of a Function:
Given a function $f(x)$, its gradient is given by

$$\nabla(f(x)) = \left\{ \begin{array}{c} \frac{\partial f}{\partial x_1} \\ \vdots \\ \frac{\partial f}{\partial x_n} \end{array} \right\} \tag{2.40}$$

For example, given $f(x) = 2x_1^4 + 4x_2^3 - 3x_1x_2^2$, the gradient is given by

$$\nabla(f(x)) = \left\{ \begin{array}{c} 8x_1^3 - 3x_2^2 \\ 12x_2^2 - 6x_1x_2 \end{array} \right\} \tag{2.41}$$

Recall that when the partial derivative of f is taken with respect to x_1, $\frac{\partial f}{\partial x_1}$, we treat x_1 as the variable and x_2 as a constant.

Hessian of a Function:

The Hessian of a function $f(x)$ is given by the following matrix

$$\nabla^2(f(x)) \equiv H(f) = \begin{bmatrix} \frac{\partial^2 f}{\partial x_1^2} & \frac{\partial^2 f}{\partial x_1 \partial x_2} & \cdots & \frac{\partial^2 f}{\partial x_1 \partial x_n} \\ \frac{\partial^2 f}{\partial x_2 \partial x_1} & \frac{\partial^2 f}{\partial x_2^2} & \cdots & \frac{\partial^2 f}{\partial x_2 \partial x_n} \\ \vdots & \vdots & \ddots & \vdots \\ \frac{\partial^2 f}{\partial x_n \partial x_1} & \frac{\partial^2 f}{\partial x_n \partial x_2} & \cdots & \frac{\partial^2 f}{\partial x_n^2} \end{bmatrix} \tag{2.42}$$

and when the above derivatives exist, the Hessian is symmetric. That is, the terms below the diagonal are the same as those above the diagonal. Therefore, we can either forgo evaluating the lower or upper triangular terms, or we can evaluate all the terms and verify that the resulting matrix is indeed symmetric.

For the above function, the Hessian is given by

$$\nabla^2(f(x)) = \begin{bmatrix} \frac{\partial(8x_1^3-3x_2^2)}{\partial x_1} & \frac{\partial(12x_2^2-6x_1x_2)}{\partial x_1} \\ \frac{\partial(8x_1^3-3x_2^2)}{\partial x_2} & \frac{\partial(12x_2^2-6x_1x_2)}{\partial x_2} \end{bmatrix} \tag{2.43}$$

leading to the symmetric matrix

$$\nabla^2(f(x)) = \begin{bmatrix} 24x_1^2 & -6x_2 \\ -6x_2 & 24x_2 - 6x_1 \end{bmatrix} \tag{2.44}$$

2.6 Summary

Quantitative optimization is founded on the understanding of important mathematical concepts, including calculus, geometry, and matrix algebra. This chapter hence provided a summary of the important mathematical concepts that are needed for learning and practicing optimization. Specifically, it provided a brief introduction to vectors, Euclidean geometry (*e.g.*, equation of a plane), matrix properties and operations, and differential and integral calculus (*e.g.*, function continuity, partial derivatives, and Taylor Series). The chapter ended with an introduction to single-variable optimality conditions. That is, how to estimate the gradient and the Hessian of a function and use them to determine (or search for) optimum points. These topics will be greater amplified in later chapters.

2.7 Problems

Warm-up Problems

2.1 A vector **a** starts from the point $P_1 : (0, -1, 1)$ and ends at the point $P_2 : (3, 4, -1)$. Find the length of **a**.

2.2 You are given three vectors: $\mathbf{a} = 2i + 3j + 4k$, $\mathbf{b} = -i + 2j - 3k$, and $\mathbf{c} = 2i + 3j - 4k$. Compute (1) $\mathbf{a} + \mathbf{b}$, (2) $\mathbf{b} + \mathbf{a}$, (3) $3\mathbf{a} + 3\mathbf{b}$, (4) $3(\mathbf{a} + \mathbf{b})$, (5) $(\mathbf{a} + \mathbf{b}) + \mathbf{c}$, and (6) $\mathbf{a} + (\mathbf{b} + \mathbf{c})$. Next, (7) What do you observe in the results that you obtain for Parts (1) and (2), Parts (3) and (4), and Parts (5) and (6)? (8) What are these properties called?

2.3 Given two vectors $\mathbf{a} = 2i + 3j + 4k$ and $\mathbf{b} = -i + 2j - 3k$, compute (a) $\mathbf{a} \cdot \mathbf{b}$ and (b) $\mathbf{b} \cdot \mathbf{a}$. Next, (c) What property of dot product of vectors is observed in this example? (d) What is the angle between **a** and **b**?

2.4 Given three vectors: $\mathbf{a} = -i + 2j - 5k$, $\mathbf{b} = 3i - j + 4k$, and $\mathbf{c} = 2i + 3j - 4k$. Compute (a) $\mathbf{a} \cdot (\mathbf{b} + \mathbf{c})$, (b) $\mathbf{a} \cdot (\mathbf{b} - \mathbf{c})$, (c) $\mathbf{a} \cdot \mathbf{b} + \mathbf{a} \cdot \mathbf{c}$, and (d) $\mathbf{a} \cdot \mathbf{b} - \mathbf{a} \cdot \mathbf{c}$. What properties of dot product of vectors are observed in the results of Parts (a) through (d)?

2.5 Let a force acting on a particle be given by the vector $\mathbf{f} = -2i + j$. As a result of this force, the particle moves from the point $A : (2, 4, -5)$ to $B : (2, 4, -7)$. (a) Compute the work done on the particle. (b) How can you explain the results you obtain? Interpret the results obtained in Part (a) from a Statics point of view.

2.6 You are given three vectors $\mathbf{a} = 2i + j$, $\mathbf{b} = 4k$, and $\mathbf{c} = -i + 2j$. Find the angles between (1) **a** and **b**, (2) **b** and **c**, and (3) **a** and **c**. Next, (4) what can you say about the vectors **a**, **b**, and **c** based on these results?

2.7 Find the equation of the line passing through the points $(-2, 2)$ and $(3, 4)$. What is the slope of the line joining these two points?

2.8 Find the equation of a plane passing through three points $P_1 : (2, 0, 0)$, $P_2 : (0, 2, 0)$ and $P_3 : (0, 0, 2)$ using the three point formula, and verify the equation you obtain using the intercept formula for the equation of the plane. Show your work for both methods.

2.9 Let A=[1 1;2 1], B=[0 1;3 2]. Do the following problems by hand and verify using MATLAB. (a) $(AB)^T$ and (b) $B^T A^T$. Turn in your hand written results, and a print out of the results at the MATLAB Command Window. What do you notice from your results?

2.10 Let A=[1 0 2;0 3 4;2 1 3] and B=[0 2 4;2 3 4;5 1 3]. Do the following problems by hand and verify using MATLAB. Turn in your hand written results and a print out of the results at the MATLAB Command Window. If you think any of the operations below cannot be performed, explain why.

(a) A^T, (b) $(A^T)^T$, (c) $(A+B)^T$, and (d) A^T+B^T (e) What matrix properties do you observe in these results?

2.11 Given A=[0 2 -1;-2 0 -4;1 4 0]. Find A^T by hand and MATLAB. Turn in your hand written results and a print out of the results at the MATLAB Command Window. What is the special property you observe about A? What are these kind of matrices called?

2.12 With the help of an example of your choice, prove that the det(A) where A is upper or lower triangular is equal to the product of elements in the principal diagonal. Take one example each of the upper and lower triangular kinds. Turn in your hand written results.

2.13 Find the determinant of A=[1 0 2;0 3 4;2 1 3] by hand, and verify your results using MATLAB.

2.14 Find the inverse of a 3×3 Identity matrix by hand. What do you observe?

2.15 Compute the eigenvalues and eigenvectors of A=[1 2;3 2] by hand. Find the command in MATLAB that can be used to compute eigenvalues and eigenvectors, and verify your results using MATLAB. Turn in printouts of the Command Window.

2.16 Find the eigenvalues of A=[3 2 1;2 2 1;1 1 1] by hand and using MATLAB. What can you say about the definiteness of the matrix? Give reasons. (You can use MATLAB to solve for the cubic equation.) Use MATLAB to compute the eigenvectors of A. Turn in your hand written calculations and the printouts of the Command Window.

2.17 Determine the following limits.
(a) $\lim\limits_{x\to 2} \frac{x^2-4}{x^2+3x-10}$
(b) $\lim\limits_{x\to 1} \frac{x-1}{x^3-1}$
(c) $\lim\limits_{x\to 2} \frac{1/x-1/2}{x-2}$

2.18 Determine if the function is convex or concave.
(a) $f(x) = e^{-x}$
(b) $f(x) = xlog(x)$
(c) $f(x) = 1/(x^2)$

2.19 Determine the derivatives of the following functions. (This is an opportunity to test or review your calculus).
(a) $f(x) = (x^2+x+2)(x^2+2)$
(b) $f(x) = x^2+x+x^{-1}+x^{-2}$
(c) $f(x) = xlog(x)$
(d) $f(x) = sin(x)$
(e) $f(x) = \frac{1}{x}cos(x)$
(f) $f(x) = tan(x)$

2.20 Determine the partial derivatives with respect to x of the following functions. (This is an opportunity to test or review your calculus).

(a) $f(x,y) = (y^2 + x + 2)(x^2 + 2)$

(b) $f(x,y) = x^2 + x + y^{-1} + y^{-2}$

(c) $f(x,y) = x log(y)$

(d) $f(x,y) = \frac{1}{x} cos(y)$

(e) $f(x,y) = \sqrt{2x - 3y}$

(f) $f(x,y) = e^x tan(x - y)$

2.21 Solve the following problems.

(a) Determine the 1st and 2nd order Taylor series for $f(x) = sin(x)$ about $x = 0$. Using MATLAB, plot the following functions for x ranging from -1 to 1: (i). $f(x) = sin(x)$, (ii) the 1st order Taylor series approximation, and (iii) the 2nd order Taylor series approximation. Comment on the accuracies of these two Taylor series approximations.

(b) Find $\lim\limits_{x \to 0} \frac{sin(x)}{x}$

(c) Find $\lim\limits_{x \to 0} cos(x)$

(d) Find $\lim\limits_{x \to 0} e^x$

2.22 X is a design variable vector. Find the Hessian of the following functions.

(a) $f(X) = \sum_{j=1}^{4} \sum_{i=1}^{4} c_{ij} x_i x_j$

(b) $f(X) = (sin x_1 + cos x_2)^N$, N is an integer

(c) $f(X) = x_1 \ln x_2 + x_2 \ln x_1$, at $(x_1, x_2) = (1, 1)$

2.23 Write a MATLAB program to generate a 5×5 matrix of random real numbers between A and B, where $[A, B] = [-5, 500]$. Show whether the matrix is positive definite or positive semi-definite or negative definite or negative semi-definite or indefinite. Print the matrix and the MATLAB M-file.

2.24 Consider the following function and do the following (by hand): $f(x) = 2x_1^2 - 3x_2^2 + 4x_1 x_2 + (x_3 + 2)^2 + 4x_1$

(a) What are the gradient and Hessian of $f(x)$?

(b) What are the stationary point(s) of $f(x)$?

2.25 Consider the following function and do the following (by hand): $f(x) = \frac{5}{2}x_1^2 + \frac{5}{2}x_2^2 + 2x_1 x_2 + 4x_1$

(a) What are the gradient and Hessian of $f(x)$?

(b) What are the stationary point(s) of $f(x)$?

Intermediate Problems

2.26 Let $f(x) = sin(x)$ be a function that you are interested in optimizing. Answer the following questions.

(a) What are the necessary conditions for a solution to be an optimum of $f(x)$?

(b) Using the necessary conditions obtained in (a), and considering the interval $0 \leq x \leq 2\pi$, obtain the stationary point(s).

(c) Confirm whether the above point(s) are inflection points, maxima, or minima. If they are maximum (or minimum) points, are they global maximum (or minimum) in the given interval?

(d) Plot the function $\sin(x)$ over the interval $0 \leq x \leq 2\pi$. Show all the stationary points on it, and label them appropriately (maximum, minimum, or inflection).

2.27 Consider the single variable function $f(x) = e^{-ax^2}$, where a is a constant. This function is often used as a "radial basis function" for function approximation.

(a) Is the point $x = 0$ a stationary point for (i) $a > 0$, and (ii) $a < 0$. What happens if $a = 0$? Is $x = 0$ still a stationary point?

(b) If $x = 0$ is a stationary point, classify it as a minimum, maximum, or an inflection point for (i) $a > 0$, (ii) $a < 0$, and (iii) $a = 0$.

(c) Prepare a plot of $f(x)$ for $a = 1$, $a = 2$, and $a = 3$. Plot all three curves on the same figure. By observing the plot, do you think $f(x) = e^{-ax^2}, a > 0$ has a global minimum? If so, what is the value of x and $f(x)$ at the minimum?

BIBLIOGRAPHY OF CHAPTER 2

[1] J. Snyman. *Practical Mathematical Optimization: An Introduction to Basic Optimization Theory and Classical and New Gradient-Based Algorithms.* Springer, 2005.

[2] E. K. Chong and S. H. Zak. *An Introduction to Optimization.* John Wiley and Sons, 2014.

[3] D. Z. Du, P. Pardalos, and W. Wu. *Mathematical Theory of Optimization.* Springer, 2010.

[4] A. Ravindran, G. V. Reklaitis, and K. M. Ragsdell. *Engineering Optimization: Methods and Applications.* John Wiley and Sons, 2006.

PART 2

USING OPTIMIZATION—THE ROAD MAP

In the first part of the book, we dealt with some material that prepared us to begin learning optimization. Specifically, we learned about MATLAB and about some basic mathematics. In this next portion of the book, we start to understand what optimization is all about. In particular, (i) we will be introduced to the fascinating world of optimization, (ii) we will differentiate optimization from other engineering activities (*e.g.*, analysis and modeling), and, finally, (iii) we will provide an important classification of optimization, which is important to know in practice.

Specifically, the topics presented, with the chapter numbers, are given below:

3

Welcome to the Fascinating
World of Optimization

3.1 Overview

Welcome to the fascinating world of optimization. Indeed, you will find optimization to be a powerful addition to your education or to your toolkit in the workplace. This brief initial chapter provides you with a clear perspective as to how the book will play a key role in your understanding of optimization. As discussed in the Preface, this book takes a squarely practical perspective. As such, your learning will involve practicing optimization using many problems. By the end of your study of this book, you should expect to be able to work with others to optimize practically any design or system and improve its performance. If the design you would like to optimize is in your technical area of expertise, you might be able to do all of the work yourself. If the design or system you would like to optimize involves modeling issues with which you are not familiar, you will simply need to collaborate with someone (or a team) who is able to provide you with the computational performance models. In this latter case, as it often is in practice, you will provide your knowledge of optimization, and someone else may model the system. In other words, one person might do the structural analysis, another might do the financial analysis, and you might use your knowledge of this book to optimize the combined system to make it competitive in the market place.

3.2 What Is Optimization? What Is Its Relation to Analysis and Design?

What is optimization? Optimization can be defined in different ways. Some define it as "the art of making things the *best*." Interestingly, many people do not like that definition as it may not be reasonable, or even possible, to do something in the very *best* possible way. In practice, doing something *as well as possible within practical constraints* is very desirable. Designing a product and doing all we can to increase profit as much as is practically possible is also very desirable. These comments may begin to give you an idea of what optimization is attempting to do in practice. It provides us with the means to make things happen in the best possible practical way.

"I've got too much work to do!
Get out of my way!"

Figure 3.1. Motivation for Optimization: Creating New Possibilities

Now, we may ask how this is different from what all engineers, all financial analysts, and most other professionals try to do? Well, the answer to this question is important indeed. Without optimization, we accomplish this by using experience, intuition, and just plain luck! With optimization, we do it in a systematic way, where we use the power of a computer to examine more possibilities than any human being could ever attempt. Furthermore, the optimization approach makes sure that the search is done as efficiently as possible.

Since you have made the decision to educate yourself in the art and science of optimization (see Refs. [1, 2, 3, 4, 5]), it is important to keep in mind that you are making a reasonable investment that is intended to bring tangible benefits. Figure 3.1 provides an interesting way to clarify this point. Some books that take a somewhat different but useful approach include (Refs. [6, 7, 8]).

In other settings, you may be exposed to a very mathematical view of optimization, which, in my opinion, sometimes obscures the basic beauty, simplicity, and practical power of optimization. To avoid that pitfall, let us immediately start getting an operational idea of the optimization process. We can do this without any equation or complex terminology. Let us simply examine Fig. 3.2. This figure illustrates the design process using (i) traditional design approaches, and (ii) using optimal design approaches.

- **Box A** displays the input, which includes two basic issues. The first defines our *dream design*: the desired performance levels (*e.g.*, maximize profit, minimize

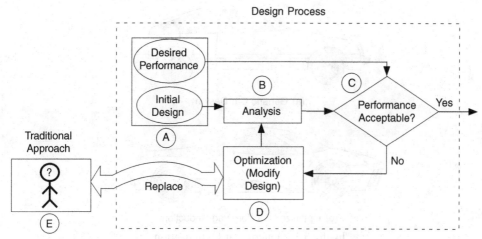

Figure 3.2. Traditional vs. Optimal Design Process

mass) and any constraints (*e.g.*, deformation less than 5 mm, cost less than \$7). The second item provides an initial design that we can obtain through any conventional means. That design is simply a starting point from which improvements can be made.

- **Box B** illustrates the analysis phase. Analysis is essentially what you do in almost all of your classes. Analysis usually tells you what the output result is for a given set of input conditions, whether you have a mechanical, electrical, or financial system under consideration. This is not design and this is not optimization; however, we need to be able to do analysis in order to design or optimize.

- **Box C** explains how the optimization process cycle begins. Using the initial design that is provided, it is very unlikely that it will satisfy all the constraints and maximize performance. Most likely, it will need to be improved by modifying it as intelligently as possible. This is where the power of optimization comes into play. This takes us to the next box.

- **Box D** is where the design is revised and improved in a very systematic way. This is not a trivial process. However, fortunately, the person who is mainly interested in applying optimization will not need to focus on most of what is in that box. In Parts I, II, and III of the book, we will learn what is needed to competently apply optimization, while Parts IV and V address more advanced topics as well as the details of **Box D**. In the process of optimization, **Box D** continually modifies the design with the expectation that the design will improve. Each modification is submitted to the analysis module. After the analysis is performed, the design performance is again evaluated to see whether it meets our objectives. If it does, we are done. If it does not, we go through the loop one more time. Please note that the actual process is more complex than this simplified explanation.

- **Box E** illustrates the human element involved in making the required improvements in a traditional way. In other words, **Box E** replaces **Box D**. The question that person is trying to answer is: How can I change the design to make it perform better? The options are: ask a friend, use intuition, use experience, or just

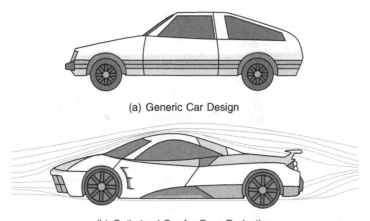

(a) Generic Car Design

(b) Optimized Car for Drag Reduction

Figure 3.3. Generic Car Optimization

hope to be lucky! As you might guess, as we move into this not-so-new world of computers and of extreme competition, this is not necessarily the best way to proceed. Please note that human design decision-making is critically needed, but not at the level of **Box D**.

Observation: A few observations regarding the conventional vs. the optimal approaches are in order. Using the traditional approach, we can only perform a small number of improvement loops. Using computational optimization approaches, we can move through the improvement loop more effectively and efficiently. In addition, the improvements are based on rigorous thinking using the power of optimization, while the traditional approach is usually ad hoc, based on intuition and experience that can fail us, particularly in complex and innovative designs.

Generic Car Optimization: Today, the car industry makes powerful use of optimization. Optimization takes place at many levels, from individual parts to crash-performance to save lives. Other applications range from fuel efficiency to pollution minimization or noise reduction. In Fig. 3.3, we see the possible impact of optimization in optimizing for drag reduction and various other important performance attributes. The examination of Figs. 3.3(a) and 3.3(b) yields endless questions regarding the process, the effectiveness and the potential impact of optimization. Many of these questions will be addressed in this book, and many will be discovered and addressed during the course of your career regarding real-world designs and systems.

Role of Optimization in the Revolutionary Transformation of the Airplane: Over two centuries ago, Sir George Cayley was reported to have advanced the concept of the modern airplane. A century later, in 1903, the Wright brothers are credited for the first sustained flight with a powered, controlled airplane. The past few decades have seen revolutionary transformations of the airplane, while its basic shape has not drastically changed. In Fig. 3.4, we observe how different attributes of interest can result in different layouts, sizes, fuel consumptions, cruise speeds for a business jet

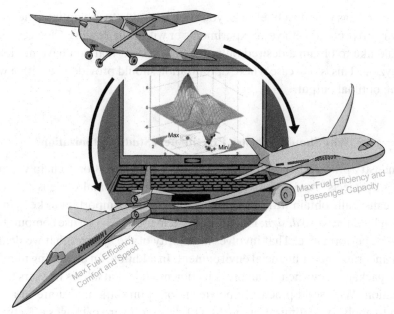

Figure 3.4. Evolutionary and Revolutionary Transformation of the Airplane

or a passenger jet. If you should gather a few experienced engineers in a room, you would have quite an engaging discussion about which of these transformation were or could have been influenced by computational optimization. The past three decades have experienced rapidly growing application of computational optimization in many critical aspects of airplane design, and the future is expected to bring us an acceleration of this trend, in part due to the exponential growth of computing power.

3.3 Why Should Junior and Senior College Students Study Optimization?

The next question that may come to mind is: as a Junior or Senior, why should I be studying optimization? Well, hopefully, what you have learned thus far should at least partially answer your question. In truth, this is a question that you will be able to answer in your own way after you will have had the opportunity to optimize designs yourself, and feel comfortable and confident that you could not have possibly gotten these optimal designs any other practical way.

In addition to these observations, you might also have the opportunity to be able to use optimization while you are still an undergraduate. There is no reason to wait. Applying optimization to your senior capstone design, for example, is one significant possibility that comes to mind.

3.4 Why Should Graduate Students Study Optimization?

To address this question, many of my previous comments to the undergraduates also apply. However, in your case, there are significantly more opportunities. If you are doing research, as you most likely are, you should be able to use optimization to find better ways to proceed with your experiments or with your designs. Chances are that you would like to obtain a desired output from a system that you have modeled or are analyzing. This is one case where optimization should provide you with a way to obtain an optimal output.

3.5 Why Should Industry Practitioners Study Optimization?

As an industry practitioner, you will be able to use optimization to help you in any number of projects. And when you do, you may find optimal designs that others cannot realistically obtain without optimization. It is also important to keep in mind that when I use the word *design*, I mean any system for which you use computation to evaluate its performance. That involves a majority of the systems that we deal with in an engineering and/or financial environment. In addition, an increasing number of software packages now include an optimization module that allows its users to apply optimization. With solid practical knowledge of optimization, you are in a strong position to apply the optimization portion of these software packages effectively.

3.6 Why Use this Book, and What Should I Expect from It?

There are several popular books on optimization in the market, which focus on either theories or applications. This book, however, provides the following key advantages, making it distinct.

1. This book serves as a practical guide to the application of optimization. This book uses a special way to teach optimization that requires sufficient practice. Like other books on optimization, this book also provides a mathematical background of optimization. In addition, it provides a large number of examples to help readers understand optimization. This book helps readers to quickly learn how to solve practical optimization problems. Readers can follow the examples to learn how to solve real engineering design problems. One unique aspect of this book is that numerical and modeling issues involved in practical optimization are also discussed.
2. This book covers a broad range of knowledge on the topic of optimization. Various aspects of optimization are covered in this book, such as linear optimization, nonlinear optimization, multiobjective optimization, global optimization, and discrete optimization. This book also teaches students how to solve contemporary complex engineering problems. The problems after each chapter are a good exercise to help readers master practical optimization problems.
3. This book is suitable for different classes of readers including undergraduate students, graduate students, and industry practitioners. Starting with the fundamentals of MATLAB and optimization, this book moves on to explore advanced and more recent topics in optimization.

3.7 How this Book Is Organized

This book contains five parts comprising 19 chapters. They are organized as follows:

Part I. Helpful Preliminaries: This part includes Chapters 1 and 2. These chapters provide an introduction to MATLAB, and the necessary mathematical preliminaries for optimization. This part includes the following chapters:

- MATLAB as a Computation Tool
- Mathematical Preliminaries

Part II. Using Optimization—The Road Map: This part includes Chapters 3 to 5. These chapters illustrate the benefits of optimization, the modeling of optimization problems, and the classification of optimization problems. This part includes the following chapters:

- Welcome to the Fascinating World of Optimization
- Analysis, Design, Optimization, and Modeling
- Introducing Linear and Nonlinear Programming

Part III. Using Optimization—Practical Essentials: This part includes Chapters 6 to 10. These chapters examine how to solve multiobjective optimization, global optimization, and discrete optimization problems. Important practical numerical issues of optimization are addressed. The links between optimization theories and applications are studied. Practical optimization examples are provided. This part includes the following chapters.

- Multiobjective Optimization
- Numerical Essentials
- Global Optimization Basics
- Discrete Optimization Basics
- Practicing Optimization - Larger Examples

Part IV. Going Deeper: Inside the Codes and Theoretical Aspects: This part includes Chapters 11 to 13. Theorems and optimization algorithms for linear and nonlinear optimization are presented in this part. This part includes the following chapters:

- Linear Programming
- Nonlinear Programming with No Constraints
- Nonlinear Programming with Constraints

Part V. More Advanced Topics in Optimization: This part includes Chapters 14 to 19. Advanced topics, including discrete optimization, design optimization under uncertainty, Pareto frontier generation, physical programming, and evolutionary algorithms, are investigated in this part. This part includes the following chapters:

- Discrete Optimization
- Modeling Complex Systems: Surrogate Modeling and Design Space Reduction
- Design Optimization Under Uncertainty

- Methods for Pareto Frontier Generation/Representation
- Physical Programming for Multiobjective Optimization
- Evolutionary Algorithms

3.8 How to Read and Use this Book

This book covers a broad range of knowledge of optimization. Undergraduate students, graduate students, and industry practitioners can use this book in different ways for different purposes.

Undergraduate students in their Junior or Senior year, who have learned calculus and linear algebra, have an opportunity to explore their application in the study of optimization. Part I of this book provides a brief review of the mathematical preliminaries for learning optimization. MATLAB serves as the programming language and computational tool for solving the optimization problems in this book. For those uninitiated in the use of MATLAB, it is necessary to carefully study Chapter 1. Parts II and III present the basics for modeling and solving optimization problems using MATLAB. The knowledge in these two parts is sufficient for those students who wish to apply optimization to practical engineering design problems.

To guide graduate students who would like to learn the theoretical aspects of the optimization algorithms behind the MATLAB functions, Part IV discusses the theorems and algorithms for linear and nonlinear programming. Advanced topics, including discrete optimization, optimization under uncertainty, Pareto frontier generation, physical programming, and evolutionary algorithms, are presented in Part V. These topics are important for research and development in engineering design. Doctorial students who conduct research on optimization are expected to learn these advanced topics.

This book teaches optimization with a practical approach, which favorably distinguishes it from other books on optimization. Industry practitioners, who are most concerned with how to apply optimization to practical problems, need only cover the chapters in Parts II and III. These chapters cover how to solve practical optimization problems, as well as the numerical and modeling issues encountered thereof.

3.9 Summary

This chapter provided a philosophical introduction to design optimization, its place in the world of science and engineering, and the importance of learning optimization to students, scholars, and industry practitioners. An illustration of how a generalized optimization process works is also provided. This chapter essentially serves as a gateway to the theory and practice of optimization taught in this book. To serve in that role, it provided an overview of what to expect from the upcoming chapters in this book, and how the overall content of this book is structured toward teaching optimization to undergraduate students, graduate students, and industry practitioners.

3.10 Problems

3.1 Describe a design problem of your interest (in 400-500 words) where optimization can be applied to enrich the design. It could be a problem you are currently working on (*e.g.*, Capstone design) or a problem you plan to work on. Clearly define the scope of applying optimization and the expected improvement in that context. Doctoral students are strongly recommended to identify a problem that is closely related to their principal area of research.

3.2 Conceive a modern real life product (*e.g.*, smartphone, solar PV, or PHEV) where optimization can be used to further improve its design. Describe the scope of applying optimization in that context (in 200-400 words). Specifically state (i) what objectives will need to be maximized and minimized, (ii) what features of the product could serve as design variables, and (iii) what practical constraints should be taken into consideration during optimization.

3.3 Compare and contrast (i) quantitative optimization-based design, and (ii) experience-based design.

3.4 From your own standpoint, explain the role of modern day computing (from portable ultrabooks to number-crunching supercomputers) in the application of optimization to real life design.

3.5 Expand on the discussion associated with Fig. 3.4 where the evolutionary transformation of the airplane is briefly discussed. Let your discussion be guided by your initial understanding of the role of computational optimization in modern engineering and related fields, as well as finance and other quantitative areas. There is not specific *good answer* to this question, while some may be more thoughtful and imaginative than others. A bit of research and cursory exploration of this book may be helpful. Limit your discussion to no more than one to two pages.

BIBLIOGRAPHY OF CHAPTER 3

[1] G. N. Vanderplaats. *Numerical Optimization Techniques for Engineering Design*. Vanderplaats Research and Development Inc., 3rd edition, 2001.
[2] P. Y. Papalambros and D. J. Wilde. *Principles of Optimal Design: Modeling and Computation*. Cambridge University Press, 2nd edition, 2000.
[3] M. Avriel, M. J. Rijckaert, and D. J. Wilde. *Optimization and Design*. Prentice Hall, 1973.
[4] J. N. Siddall. *Optimal Engineering Design: Principles and Applications*. CRC Press, 2nd edition, 1982.
[5] S. S. Rao. *Engineering Optimization: Theory and Practice*. John Wiley and Sons, 4th edition, 2009.
[6] A. D. Belegundu and T. R. Chandrupatla. *Optimization Concepts and Applications in Engineering*. Cambridge University Press, 2011.
[7] K. Lange. *Optimization*. Springer, 2013.
[8] R. A. Sarker and C. S. Newton. *Optimization Modeling: A Practical Approach*. Taylor and Francis, 2007.

4

Analysis, Design, Optimization and Modeling

4.1 Overview

In this chapter, we introduce the important activities of analysis, design, optimization, and modeling. While we are all generally familiar with these terms, it is important to understand them in the context of how they relate to our optimization activities. We also develop an important understanding of how modeling system behavior is a distinct activity from modeling an optimization problem. These involve two distinct lines of expertise. In most of your courses (*e.g.*, structures, dynamics and finance) you focused on the former. In this book, we focus on the latter.

4.2 Analysis, Design and Optimization

Analysis and *optimization* are two activities integral to the process of design (Refs. [1, 2, 3, 4, 5]). In this chapter, as well as in the remainder of the book, we will primarily focus on these two activities in the context of engineering or systems design. However, the mathematical concepts, approaches, algorithms, and software tools that you will learn in this book could be applicable to diverse fields beyond engineering (*e.g.*, optimization of market portfolios) [6]. Although, from a technical perspective, *analysis* and *optimization* can be considered to be steps within the process of design, they can also be performed as stand-alone activities toward other end goals within the scope of academic research and industrial R&D. In this section, you will learn the definition of these activities and how they are related to each other. At the same time, you will have the opportunity to understand and appreciate the roles and responsibilities of the users, researchers, or engineers who execute these activities, whether individually or as a part of a team.

In order to help you better understand the practical essence of these activities, the following simple design example will be used throughout this section. You are designing a table that can carry as much weight (of objects placed on it) as possible, while fitting within a particular corner of your room. You are allowed a limited budget to construct the table.

82

4.2.1 What Is Analysis?

Analysis by itself is a broad term and generally refers to the process of dissecting a complex system, topic, phenomena, incident, or substance into smaller (and likely more tractable) parts to acquire a better understanding of it. ***Engineering analysis** can be more specifically defined as the application of scientific principles and processes to reveal the properties and the state of a system, and also to understand the underlying physics driving the system behavior.* Now, in the case of the "table design problem," you are primarily required to analyze how the different design decisions (*e.g.*, the table geometry and material) impact the capacity of the table to carry weight without breaking down. This capacity will be referred to as the *weight-holding capacity* of the table in the remainder of this chapter. Considering the problem in more technical detail, we realize that there are multiple modes of breakdown (*e.g.*, buckling of the table legs, fracturing of the table top, or failure of one or more joints). Hence, analyses of the the different modes of breakdown or failure becomes necessary, which again illustrates that *analysis* generally involves decomposing a system/mechanism/problem into smaller parts to be studied. Now, if you are from a mechanical engineering or a related discipline, you might have already realized that the type of analysis needed for this problem is *structural analysis*. This realization brings us to another important aspect of *analysis*. The role of disciplinary knowledge in analysis and its implication in a research, industrial, or practical setting.

Analysis generally demands disciplinary knowledge pertinent to the system or mechanism being analyzed. More often than not, practical systems involve multiple disciplines. For example, designing an aircraft will require structural, aerodynamic, control, and propulsion analyses [7, 8]. In a practical setting, the design team generally involves experts from different disciplines. Since disciplinary understanding may have reached different levels of maturity (with scientific progress), mathematical tools might be readily available for conducting certain types of analyses, thus alleviating the necessity for dedicated disciplinary expertise. On the other hand, in the case of mechanisms or phenomena that are not yet well understood, in-depth and fundamental analyses might be required – thereby demanding the involvement of a disciplinary expert. Now that the importance of analysis and the role of disciplinary knowledge therein has been established, the next question is – what are the basic approaches to *engineering analysis*?

Analysis is sometimes misinterpreted as a purely mathematical or theoretical activity. Analysis could involve "experiments - testing - mathematical inferencing" as an iterative process, especially for the following commonly-occurring scenarios:

- the underlying physics is not well understood;
- the fundamental disciplinary principles or theory do not directly apply due to geometrical complexities and inherent uncertainties; or
- lack of knowledge of the material properties (*e.g.*, thermodynamic or structural properties).

For example, in the case of the "table design problem," ideally, the mathematical theory of solid mechanics (or mechanics or materials) can be used to fully analyze the system. However, in practice, you might not know the structural properties of the type of wood used for construction or the strength of the fasteners to be used at the joints. Appropriate experiments could be conducted in that case to fill this knowledge gap. Such scenarios are expected to be more common in designing new and innovative systems and, as such, a comprehensive understanding of the scope of "analysis" is important (for researchers/engineers) to effectively contribute to technological innovation.

4.2.2 What Is Design?

Design, in general terms, can be defined as the creation of a plan and/or strategy for constructing a physical system or process. *Engineering design* itself could be readily classified into multiple (often overlapping) categories based on the "object of design," such as *product design*, *systems design*, *industrial design*, and *process design*. An entire dedicated book (or even sets of books) would be necessary to fully explain and demonstrate one of the "design" categories (*e.g.*, Ref. [9]). This book will primarily focus on engineering design in the context of mathematical optimization. From this standpoint, "engineering design" itself can be perceived as a multi-stage process, an example of which is presented in Fig. 4.1, which includes stages up to

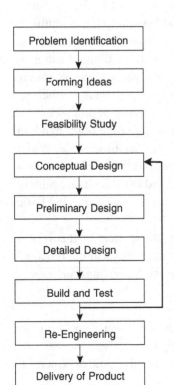

Figure 4.1. Multi-stage Design Process

product delivery. In practice, the design process might not include all the stages shown in Fig. 4.1 or might include some additional unique stages.

Considering the example of the "table design problem," *conceptual design* will involve (i) planning the overall shape or configuration of the table (*e.g.*, round top or rectangular top; rigid or collapsible design), and (ii) choosing the class of material to be used (*e.g.*, metal, glass top, or wood construction). *Preliminary design* in this case will primarily involve determining the optimum dimensions of the different parts of the table and the material to be used for these parts. The objective of maximum weight-holding capacity and the constraints imposed by the budget and the size of the room will guide this preliminary design process. Detail design will finally involve determinimg the necessary modularity of the table and the joint mechanisms from the perspective of manufacturing.

It is important to note that, in practice, these design stages may not be distinctly defined in a linear fashion; significant overlap is common. Additionally, iterations among these stages could also become necessary. For example, if a feasible design or satisfactory value of the objectives could not be obtained in the *preliminary design* stage, you might need to go back and re-think your conceptual design.

4.2.3 What Is Optimization?

In the previous chapter, you were introduced to a philosophical and practical understanding of the role of optimization in academic research and industrial systems development. You were also presented the opportunity to appreciate the importance of learning optimization as an undergraduate student, a graduate student, or as a professional. In this section, we will take a step further into the world of optimization by looking at it from a quantitative design perspective without getting into the mathematical intricacies (which will be described in later chapters).

From the general standpoint of searching for the best available design, *optimization* can be defined as follows. *Mathematical optimization is the process of maximizing and/or minimizing one or more objectives without violating specified design constraints, by regulating a set of variable parameters that influence both the objectives and the design constraints.* It is important to realize that in order to apply mathematical optimization, you need to express the objective(s) and the design constraint(s) as quantitative functions of the variable parameters. These variable parameters are also known as design variables or decision variables.

To better understand the definition of optimization, we consider the preliminary design stage for the "table design problem." For this problem, optimization of the table design can be articulated as the following process:

- maximize the weight-holding capacity of the table (*i.e.*, the total weight of objects that can be placed on the table), while
- satisfying (i) the geometrical constraints imposed by the size and shape of the room (where the table will be located) and (ii) the cost constraints imposed by the allowed budget for building the table,

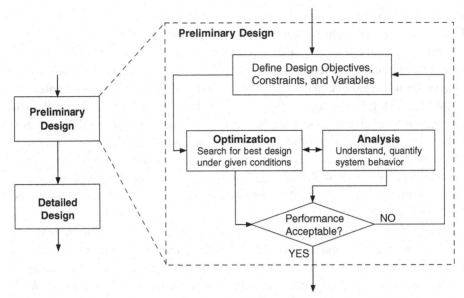

Figure 4.2. Relationship Between Design, Analysis and Optimization (A Representative Example)

- by varying the geometrical configuration (*e.g.*, length, breadth, and thickness of the table top) and the material used for constructing the table.

4.2.4 Interdependence of Analysis, Design and Optimization

Whether in academic research or in an industrial R&D setting, design, analysis, and optimization are generally undertaken as strongly interrelated activities toward developing better products and technologies. However, is no unique structure as to how they are related – the relational structure generally depends on the available human, computational, and physical resources and on the choices of decision-makers (*e.g.*, the design team leader). We use, as an example, one of the common relational structures to particularly focus on understanding how the activities interact with each other on a one-on-one basis. This structure is shown in Fig. 4.2.

Design, in general, is the enveloping process that includes *analysis* and *optimization* as sub-processes. Figure 4.2 expands on the preliminary design stage. The primary steps within this stage are (i) defining the design objectives, constraints, and variables, (ii) performing or using analysis, and (iii) performing optimization. In this case, *optimization* is the main driver for improving the preliminary design. However, in order to yield meaningful results in a time-efficient manner, *optimization* generally depends on a well-thought definition of the design objectives, constraints, and variables, which is also known as *effective problem formulation*. For the "table design problem," this step calls for a clear definition of the following:

1. Which geometrical parameters will be considered as variables and what material choices are available for constructing the table?

2. What is the total budget/cost constraint, and what are the geometrical constraints imposed by the room shape and size?
3. What is the minimum needed weight-holding capacity of the table, and what is the desired maximum weight-holding capacity (if any)?

With a clear problem formulation, you are ready to perform optimization where the objective is to maximize the weight-holding capacity of the table. Optimization is a methodical process of changing the variable parameters (design variables) to determine better values of the objectives within the feasible design space (defined by the constraints). Thus, in order to implement optimization, you need a quantitative understanding of how the weight-holding capacity of the table is related to the table geometry and the table construction material. This necessity brings us to the relationship between *optimization* and *analysis*.

Analysis provides you with the knowledge or, more specifically, a mathematical model that accepts the variable values (defining a candidate design) and outputs the corresponding values of the objectives of interest and the values of constraint violations (if any). This knowledge is used by the optimization process in searching through the design space for the optimum results. In the case of the "table design problem," a solid mechanics-based analysis of the table structure is required to provide a quantitative understanding of the stress distribution of the table as a function of the table geometry, table material, and the force acting on the table (attributed to the weight of the objects placed on it). This knowledge will, in turn, provide a strategy for quantifying the maximum weight-holding capacity for a table of a given geometry and material. Additionally, a cost analysis is necessary to estimate the total cost of constructing a table of any given geometry and material, which can include material costs, tool costs, utility costs, and labor costs. *Analysis* also provides the opportunity to investigate the performance of the final optimum design. Hence, in determining whether the optimum design satisfies the desired goals, input from both the optimization process and analysis is needed, as shown in Fig. 4.2.

On the other hand, optimization could provide food for further analysis, especially in the context of practical optimization where the design process rarely stops at a single optimization run (instead generally requiring several iterations). For example, optimization could provide insight into which region of the design space (*i.e.*, a more focused range of designs than that initially allowed) show better promise in terms of objective values, thereby inciting more in-depth analysis of the system over that region of the design space. As a result, *optimization* and *analysis* are considered to be mutually contributive elements of the design process, and are linked with a bidirectional arrow in Fig. 4.2.

Although *analysis* and *optimization* are two central elements of quantitative design, engineering design in itself could also involve qualitative elements that are often beyond the scope of *analysis* and *optimization*. Two typical qualitative elements include:

1. *Creativity- and aesthetics-driven design decisions*: For example, the overall configuration (*e.g.*, round top) and the color of the table surface could be conceived

simply based on aesthetics and/or prior experience with tables. Although qualitative in nature, these decisions have important quantitative implications for the later stages of design where *analysis* and *optimization* are involved (*e.g.*, regulating the material options).

2. *Market-driven design decisions*: Design decisions can also be driven by an understanding of the market, especially in the case industrial and product design. Although, quantitative market analysis might be available in certain cases, such availability is not necessarily generic (*e.g.*, imagine the first Iphone or major changes to popular automobile models). A qualitative understanding of customer preferences or simply a clear vision for the product (*i.e.*, generating new customer preferences) is necessary to make design decisions in such cases.

4.3 Modeling System Behavior and Modeling the Optimization Problem

In the previous section, you learned about how *analysis* and *optimization* provides the necessary tools for performing engineering design. Now, in order to pursue such a quantitative design process, you require a tractable mathematical representation of the system analysis and of the optimization problem. This requirement brings us to the role of *modeling* in performing design through analysis and optimization. An introduction to this role is provided in this section. Further mathematical description of this role will be provided in later chapters of this book.

4.3.1 Modeling System Behavior

There are different definitions of a model in the context of systems design:

- Traditional definition: A model is a scaled fabricated version of a physical system.
- Simulation-oriented definition: A model is a symbolic device built to simulate and predict characteristics of the behavior of a system.

Modeling can be defined as *a process by which an engineer or a scientist translates the actual physical system under study into a mathematical model of the system.*

From the standpoint of design, one is generally concerned with quantifying certain parameters of interest through analysis and modeling. These parameters can be collectively termed as criteria functions. Thus, modeling system behavior boils down to developing a set of functions that represent the parameters of interest as functions of the variable parameters that can be controlled through design. Mathematically, modeling can be represented as:

$$P = f(X) \tag{4.1}$$

where the P is a parameter of interest that can be represented as a function of the design variables defined by the vector X. Therefore, the process of modeling system behavior is basically the process of determining the function f. In practice, f may not be a simple analytical function. It could be a collection of functions or a computational simulation [10].

Depending on the approach used to develop system behavior models, they can be classified into the following major categories:

1. *Physics-based Analytical Models*: These models are developed based on the physics of the system. If the physics of the system is defined by a set of differential equations, the analytical models represent the functional solution to those differential equations.

2. *Simulation-based Models*: These models generally leverage a discretized representation of the system in translating the system behavior to a set of algebraic equations that are solved using numerical techniques (by harnessing the number-crunching power of computers). Depending on modeling assumptions and the resolution of the discretization, the fidelity of these models can vary significantly. High-fidelity simulations, especially for complex systems, generally tend to be computationally expensive and more often than not require dedicated software for generating 3D geometries and performing the simulations (with limited portability). Examples of simulation-based models include finite element models, finite volume models, and spectral analysis models.

3. *Surrogate Models*: Surrogate models are purely mathematical and/or statistical models with certain generic functional forms and coefficients that can be tuned. These models are trained (*i.e.*, the coefficients are tuned) using a set of input-output data (*i.e.*, $[P, X]$ data) generated from a high fidelity source. The high-fidelity source could be comprised of experimental or simulations-based analysis. As a result, surrogate models by themselves lack any direct physical information of the system; however, they provide the advantage of being tractable, fast, and highly portable (generally not requiring any specialized software) (see Ref. [11]).

With the exception of surrogate models, the development of other types of models (*i.e.*, physics-based models) generally necessitates disciplinary knowledge, and that's where your disciplinary courses come in handy in the design process. For example, in the case of the "table design problem," your "Solid Mechanics," "Mechanics of Materials," or "Structures" course will prove helpful in developing a model of the maximum weight-holding capacity of the table as a function of the table geometry and material.

At this point, you must be wondering about the challenges involved in designing real-life engineering systems (*e.g.*, an aircraft) where knowledge from multiple disciplines is required at a level which is unlikely to come from a single expert. Practical engineering design generally requires a team effort. Working with others to develop or use physics-based models that are outside of your field of expertise is a pervasive practice in industrial settings, where the expertise of one person is generally insufficient to model the global system. As a team contributor, you need to feel comfortable with the idea of understanding only part of the analysis.

4.3.2 Modeling the Optimization Problem

Modeling the optimization problem is also called problem formulation, a process that you will learn in more detail in later chapters. Essentially, it involves developing a clear definition of the design variables, design objectives, and design constraints. In this context, design variables and design constraints could be of different types (*e.g.*, continuous and discrete variables, and equality and inequality constraints). Problem formulation also involves defining the upper and lower bounds of the design variables, which are sometimes perceived as linear constraints.

Modeling the optimization problem is also strongly correlated with the choice of optimization algorithms. In other words, the class of optimization algorithms available to solve a problem depends on how that problem is formulated. This relationship often drives researchers to make important approximations in their problem formulation (*e.g.*, converting equality constraints to inequality constraints using a tolerance value) in order to leverage powerful algorithms that perform well in the absence of equality constraints.

4.3.3 Interdependence of System Behavior Modeling and Optimization Modeling

It is important to ensure that optimization problem formulation is coherent with the system behavior model. From Fig. 4.2 you can recall that analysis and optimization are interrelated. Therefore, if the optimization process demands a set of output parameters (criteria functions) to be estimated by the analysis model for a given set of input parameters (design variables), the analysis model should be able to provide the right outputs. Any discrepancy in this information exchange will crash the optimization process. In other words, the choice of objective and constraint functions and the choice of design variables should be made in view of the capabilities of the analysis model when accounting for the associated relationships. Alternatively, if the analysis model cannot meet the needs of the optimization formulation, new analysis models will need to be developed to represent the necessary functional relationships. When you put these issues in the context of practical design, where the analysis models are often developed by disciplinary experts and the optimization problem is formulated by a design expert (who may not have in-depth knowledge of the multiple disciplines involved), you will realize that there is often significant room for discrepancies. Effective communication is a necessary component of engineering design - essentially a collaborative effort.

There are also other practical considerations in harmonizing the optimization modeling and systems behavior modeling. For example, if you choose an optimization algorithm that requires a relatively high number of system evaluations, you would most likely require a fast (computationally-efficient) model of the system behavior to complete the optimization in a reasonable amount of time. Similarly, if the system behavior model is inherently highly nonlinear, you will need to formulate the optimization problem such that a nonlinear optimization algorithm can be used to solve the problem. To summarize, the characteristics of the optimization problem formulation and the system behavior model should be aligned with each other and with the overall objectives of the design effort.

4.4 Summary

This chapter introduced the key components of design optimization, namely analysis, design, modeling, and optimization. A holistic view of design, including the major activities involved in a design process (*e.g.*, preliminary design and detailed design), is provided. The importance of analysis, modeling, and optimization is then described in the context of engineering design. In doing so, this chapter also provided important insights into the relationship between these different components of design. The chapter ended with a bi-level perspective to modeling in the context of design. That is, modeling the behavior of the system being optimized, and modeling the optimization problem itself. This bi-level perspective essentially shows how modeling decisions in these two steps are distinct but strongly correlated, such as is terms of pertinent computational consequences. These thoughts will become ever clearer as we move along.

4.5 Problems

4.1   Consider the table design problem discussed in this chapter and do the following:

1. Provide a hand-drawn sketch of a representative table (four-legged with rectangular table top) and clearly label and list the geometrical design variables. Be as comprehensive as possible.
2. Identify three differen types of analysis models (analytical, simulation-based, and surrogate models) that can be used to represent the weight-holding capacity of the table as a function of the geometrical variables. If you are not from the mechanical engineering, aerospace engineering, structural engineering or other related disciplines, feel free to discuss the problem with your peers who are from those disciplines in order to identify the analysis models.

4.2   Compare and contrast the process of (i) conceptual design, (ii) preliminary design, and (iii) detailed design in general and in the development of an automobile. Feel free to refer to the appropriate literature for this purpose.

4.3   Give an example of an original engineering design problem and clearly outline the objectives, the constraints, and the design variables.

4.4   Outline two advantages and two limitations (each) of experiment-based analysis and simulation-based analysis.

4.5   From your current understanding of design, analysis, and optimization, elaborate on the relationship of *analysis* and *optimization* in the context of *computational expense* (in 200-300 words).

4.6   In 300-400 words, discuss the role of simulation software (*e.g.*, ANSYS, Abaqus, and Simulink) in engineering analysis and optimization, and compare and contrast the impact of licensed software and open-source software (*e.g.*, free codes) in performing engineering design in the 21st century.

BIBLIOGRAPHY OF CHAPTER 4

[1] M. N. Horenstein. *Design Concepts for Engineers*. Prentice Hall, 2010.

[2] M. Beaney. Conceptions of analysis in early analytic philosophy. *Acta Analytica*, pages 97–115, 2000.

[3] F. P. Brooks. *The Design of Design: Essays from a Computer Scientist*. Pearson Education, 2010.

[4] N. Cross, K. Dorst, and N. Roozenburg. Research in design thinking. Technical report, Delft, Netherlands: Delft University Press, 1992.

[5] A. Ertas and J. C. Jones. *The Engineering Design Process*. John Wiley and Sons, 1996.

[6] A. K. Dixit. *Optimization in Economic Theory*. Oxford University Press, 1990.

[7] I. M. Kroo, S. Altus, R. Braun, P. Gage, and I. Sobieski. Multidisciplinary optimization methods for aircraft preliminary design. Technical report, NASA Langley Technical Report Server, 1994.

[8] J. Sobieszczanski-Sobieski and R. T. Haftka. Multidisciplinary aerospace design optimization: Survey of recent developments. *Structural optimization*, 14(1):1–23, 1997.

[9] C. L. Dym, P. Little, and E. Orwin. *Engineering Design: A Project-Based Introduction*. John Wiley and Sons, 4th edition, 2013.

[10] J. T. Oden, T. Belytschko, T. J. R. Hughes, C. Johnson, D. Keyes, A. Laub, L. Petzold, D. Srolovitz, and S. Yip. Revolutionizing engineering science through simulation: A report of the national science foundation blue ribbon panel on simulation-based engineering science. *Arlington, VA: National Science Foundation*, 2006.

[11] G. J. P. Gigch. *System Design Modeling and Metamodeling*. Springer, 2014.

5

Introducing Linear and
Nonlinear Programming

5.1 Overview

This chapter introduces important fundamental principles for the optimization of any system whose performance can be quantitatively defined in term of quantities that can be controllably changed (i.e., design variables). In Sec. 5.2, we discuss the various classes of optimization problems. Specifically, we discuss optimization in terms of (i) linearity vs. nonlinearity, (ii) single vs. multiple objectives, (iii) discrete vs. continuous design variables, and (v) constrained vs. unconstrained optimization. In Sec. 5.3, we discuss the important realization that in most cases, problems are treated as if they involved a single objective. In Sec. 5.4, we present the solution approaches to optimization: analytical, numerical, graphical, and experimental. In the last section, Sec. 5.5, we introduce several of the software packages available for solving different classes of optimization problems.

5.2 Problem Classes

As we deal with a given optimization problem, it is important to know what class (or type) of problem it represents. This is because each different class of optimization problem generally calls for a different type of solution approach, as well as a different class of software or algorithm. In this section, we define the major classes of optimization problems, part of which is presented in Fig. 5.1. A general optimization problem can be formulated using the following set of equations (Refs. [1, 2].

$$\min_{x} \quad J(x) \tag{5.1}$$

subject to

$$g(x) \leq 0 \tag{5.2}$$

$$h(x) = 0 \tag{5.3}$$

$$x_l \leq x \leq x_u \tag{5.4}$$

The function $J(x)$ represents the *objective function*, which we would like to minimize. The idea being that as we minimize the objective function, the system or design

93

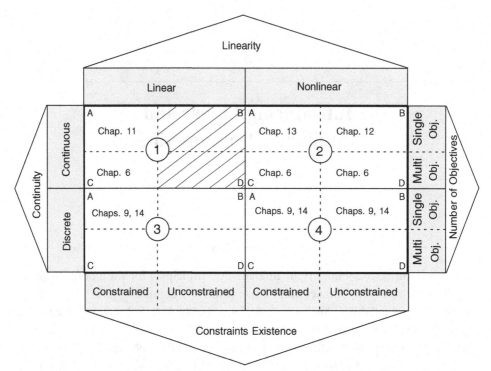

Figure 5.1. Classification of Optimization Methods

will behave better (*e.g.*, cheaper, stronger, lighter, or faster). The function $g(x)$ represents a vector of inequality constraints and the function $h(x)$ represents a vector of equality constraints. These constraints make the design *feasible* (*i.e.*, not unacceptable). For example, they might ensure that the mass is not negative or that a process is not prohibitively expensive. They are called *behavioral constraints*. The vector x represents the vector of design variables. These are the quantities that we can change in the design to improve its behavior. The constraints on the design variables are called *side constraints*. A set of design variables that fully satisfies all the constraints is called a feasible solution (even if it does not minimize the objective function). However, thus far, we have provided no information regarding the important properties of the above quantities. These properties define the *class of the optimization problem*. We will now be more specific.

Broadly speaking, optimization problems can be classified along seven major categories. Knowledge of, and appreciation for, these categories is important as they help us understand whether the problem at hand is simple or involved, and whether a given algorithm or software applies to our problem. Figure 5.1 provides a helpful presentation of the interaction between the first four. The categories are as follows:

1. Linear vs. Nonlinear
2. Constrained vs. Unconstrained
3. Discrete vs. Continuous
4. Single vs. Multiobjective

5. Single vs. Multiple Minima
6. Deterministic vs. Nondeterministic
7. Simple vs. Complex

These categories are discussed below.

(1) Linear vs. Nonlinear: Are the functions $J(x)$, $g(x)$, and $h(x)$ linear or nonlinear functions of x? When the objective function and the constraints are all linear, the problem is called a *linear programming* (LP) problem. (As mentioned earlier, linear programming and linear optimization mean the same thing. The former is often used for historical reasons). When either the objective function or any of the constraints is a nonlinear function of the design variables, the problem is called a *nonlinear programming* (NP) problem (which means the same as a nonlinear optimization problem). Generally, LP problems are much easier to solve than NP problems, particularly when the dimension of the problem is large (*i.e.*, large number of design variables). In addition, the more nonlinear the problem, the more difficult it may be to optimize. It is also more likely to present numerical difficulties. Fortunately, there is much that we can do to minimize these difficulties. As a general rule, solving a large LP problem is a more reliable process than solving a large NP problem. There are many different ways to pose a Linear Programming problem, all of which are practically equivalent. One possible option is as follows.

$$\min_{x} \quad c^T x \tag{5.5}$$

subject to

$$Ax \leq b \tag{5.6}$$

$$A_{eq} x = b_{eq} \tag{5.7}$$

$$x_l \leq x \leq x_u \tag{5.8}$$

In the above equations, the quantities A and A_{eq} represent matrices of constant coefficients. The quantites c, b, b_{eq}, x_l, and x_u represent constant vectors. The important observation to make here is that the objective function, as well as the constraints, are all linear functions of the design variable x, which is what makes this a linear programming problem. As a final comparison, we examine the equality constraints in the linear and nonlinear programming formulations. Specifically, let us compare Eq. 5.3 and Eq. 5.7. In the former, h is a function of x, which may be linear or nonlinear. In the latter, we have a multiplication between a matrix, A_{eq}, and a vector, x, which is a linear operation (see Refs. [3, 4]).

(2) Constrained vs. Unconstrained: Does the optimization problem have constraints? When the optimization problem does have constraints, we call it a *constrained optimization problem*. When we do not have any constraints at all, the problem is an *unconstrained optimization problem*. Most practical problems involve constraints. Thus, unconstrained problems generally have more theoretical than practical value. Specifically, in the case of nonlinear programming, the problem statement would only involve Eq. 5.1 – the relations 5.2 and 5.3 would not be part of the (unconstrained) optimization problem. However, for reasons that will be explained later, we must have constraints in the case of linear programming. Otherwise, the solution will not generally be finite. (We invite you to ponder why this is the case.) In other words, in the case of linear programming, at least one of the relations 5.6, 5.7, and 5.8 must be part of the linear programming problem. This discussion applies in the case of continuous design variables (why this limitation?). Continuity or discreteness of design variables is discussed next.

(3) Discrete vs. Continuous: Are any of the design variables discrete, or are they all continuous? If any design variable is discrete, we no longer have a continuous optimization problem. We note the following typical cases. (i) In the first, the design variables could be restricted to take only the values of 0 or 1. This case is called *0-1 programming* or *binary programming*. (ii)In the second case, the design variables are restricted to take on only integer values. This case is called *integer programming* (see Refs. [5, 6]). (iii) In the third case, the design variables are restricted to take on only a given prescribed set of real values. This case is generally called *discrete optimization*. Often, we have a mixture of the above cases, where some variables may be 0-1 type, some may be integers, some may take on given/prescribed real numbers, while others might be continuous. Different methods are available to handle these different cases. Continuous optimization problems are usually much easier to deal with. It is worth noting that in some cases, the integer programming problem might allow for the design variable to take on any integer value, while in other cases, it might call for a restricted given set of integer values. We invite you to think of practical optimization cases that may involve the above different cases. One such example is the case of optimizing the diameter of a pipe that is available in ten different sizes. Another case may involve optimizing profit in the case where a loan package must be chosen among ten possibilities, each of which involves a different interest rate and a different origination fee (see Ref. [7]).

(4) Single vs. Multiobjective: In general, the design or optimization of practical systems involves tradeoffs among competing objectives. For example, to improve a car's crashworthiness, we make it heavier; but, in the process, the fuel efficiency is worsened. Therefore, these two objectives (crashworthiness and fuel efficiency) compete against each other and require a compromise where we tradeoff one for the other as needed for an optimal situation. As we think of the above situation and of most practical problems, we realize that most optimization problems involve more than a single objective. Most are multiobjective in nature. There are many methods available to deal with both single objective and multiobjective problems. This is an important area of study for the practical application of optimization. In fact, we devote a whole chapter (Chapter 6) to this topic. As such, we will not discuss it here any further. We only note that this is an important classification of optimization problems.

(5) **Single vs. Multiple Minima:** Does the optimization problem have a single min-imum/maximum (optimum), or multiple optimum values (*i.e.*, unimodal or multimodal). Solving optimization problems that have several optima is referred to as *global optimization*. This optimization case is much more difficult to han-dle than *single optimum optimization*, as we will see later. Furthermore, the algorithms that are used to handle unimodal or multimodal problems are quite different. A significant class of engineering problems can be managed using single-optimum optimization algorithms, which is the main focus of this book. Finally, we note that single minimum search is referred to as *local optimization*, while multi-minima search is referred to as *global optimization*.

(6) **Deterministic vs. Nondeterministic:** In recent years, designers have realized that information is almost never exact. They began to understand that there is a high cost associated with low/tight tolerances. The more precisely we manufacture a part, the more costly it will be. We also started to understand that it is not neces-sary for every part in a design to have the same tolerance. In addition, there are certain aspects of the product over which we do not have direct control. Demand for a product, for example, is something that we can only estimate. The net result is that there are some aspects of the design that can be represented by design variables that are deterministic, while other design variables might have to be treated as as nondeterministic (*e.g.*, probabilistic or stochastic). In this book, we will explicitly focus on deterministic optimization, although many of the tech-niques discussed in this book can be used for nondeterministic optimization (see Ref. [8]).

(7) **Simple vs. Complex Problem:** Perhaps the most critical aspect of a problem at hand is to understand whether it will be *Simple* or *Complex* endeavor. Let us explain what we mean. A *simple* problem can be viewed as one that can be solved relatively easily by virtue of certain characteristics. This can be the case (i) because the model of the system is provided or readily created, (ii) because it only involves continuous variables, (iii) because it is not strongly nonlinear, (iv) because it is expected that local optimization will be sufficient, (v) because the computational model of the system behavior can run on a computer in seconds or minutes (not hours), (vi) because the number of design variables is *not large*, (vii) because all the models needed to describe the system behavior can run on a single computer, or (viii) because all the design variables are deterministic. An assessment of the above items gives us a sense of how simple or complex it will be to optimize the design or system. In practice, however, each of the above items only provides us with indications, and not absolutes, in terms of how easy a problem will be to solve.

In Fig. 5.1, we observe the interaction of the first four categories discussed above. The shaded area represents a case that should not be considered. Specifically, this case represents continuous unconstrained linear programming, which leads to a solu-tion at ∞ or $-\infty$. The linear programming chapter (Chapter 11) presents this scenario in detail. The chapters in which various cases are discussed in the book are also provided. Boxes 1 and 2 are covered by the bulk of this book and most introductory texts. Parts of Boxes 3 and 4 are also covered to offer the possibility of a more advanced treatment of practical optimization.

5.3 Single Objective Optimization—An Inclusive Notion

Most practical problems in design or in other technical areas involve conflicting considerations. For example, when you make a part as light as possible, it cannot also become as strong as possible. In this case, we are dealing with two conflicting objectives: minimize mass and maximize strength. In practice, we must settle for a compromise between the two objectives. One way to deal with this compromise (*i.e.*, tradeoff) is through so-called multiobjective optimization methods. We present this important topic in greater detail in Chapter 6.

However, there are important reasons to learn about single objective optimization:

(i) As you will see, in practice, a multiobjective problem can be transformed to make it appear to be single objective. This seemingly single objective is actually some form of aggregation of the multiple objectives involved. This point will become clearer later.

(ii) In line with the above comment, we often treat computational optimization as single objective.

(iii) Finally, we note that some problems are truly single objective. The most glaring example is in the case where one wishes to simply maximize one quantity – profit.

5.4 Solution Approaches: Analytical, Numerical, Experimental and Graphical

In the above section, we discussed the various categories, or classes, of optimization problems. We also discussed their associated implications in terms of overall difficulty. In this section, we discuss the various approaches that are available to us to solve optimization problems. Generally speaking, we can identify four broad solution approaches: analytical, algorithmic, experimental, and graphical. We briefly discuss the first three, followed by a more detailed presentation of the fourth (graphical) approach.

5.4.1 Analytical Optimization

The *analytical approach* is one that most high school and first year engineering students are actually already familiar with. To use this approach, we must have a mathematical function available to us that we wish to minimize. This mathematical function is supposed to represent the performance of our design or system. By finding the conditions that maximize this function, we will have identified the conditions that maximize the performance of the design. We learned in calculus that if we differentiate this function, and we identify the point where the obtained derivative is zero, we will also have identified an optimum point. We can then take the second derivative. If it is positive, we have a minimum. If it is negative, we have a maximum. In the case where we also have constraints in the problem, the situation is a bit more involved. It requires us to use *lagrange multiplyers*, which we present in Sec. 13.5.

This approach was the most viable in the pre-computer age, when differential calculus rather than numerical approaches was *king*. Unfortunately, for most practical problems, this approach is too complicated to be viable. However, while this approach is not a viable practical approach, it does help us understand how to develop more useful numerical algorithms upon which the whole modern field of computational optimization is based.

5.4.2 Numerical (or Algorithmic) Optimization

The numerical optimization approach is indeed the modern way to optimize. It fully exploits the power of a computer as a computational workhorse. It helps us explore endless possibilities that we would not be able to identify in any manual way. Basically, this approach uses *algorithms* to help us *search* the possible options. Stated differently, we start with one possible design (generally a bad one!). Based on some logic in an algorithm, we estimate a next design (which we hope will be better). We then keep iterating until we reach what we hope will be the optimal design. The better the algorithm, the more likely we are to reach the optimal design and reach it faster.

As we can see, this approach involves an iterative process. The advantage and power of this approach is two-fold. First, it uses an algorithm to determine what the next improved design should be. (There is no guess-work). Second, it uses the computer to perform the iterations. Since the computer does not get tired, we can perform a thorough exploration and generally obtain an optimal design in a reasonable amount of time.

Here, we state a key distinctive approach of this book to our learning optimization. Traditionally, when we study optimization, we spend most of our time learning how these search algorithms work. Unfortunately, we might be extremely knowledgeable as to how they work, while they may play little role in helping us become better designers in practice. The mere knowledge of the inner workings of these algorithms may or may not play a critical role in our design activities. Instead, we need to know how to use them in a computational design infrastructure. This is what is presented in the first 10 chapters of this book. The knowledge of these algorithms is, however, a useful addition to our advanced understanding of the topic of optimization. As such, we do provide a presentation of these topics in Chapters 11 through 13. In addition, we provide some advanced practical topics in Chapters 14 through 19.

Describing the optimization process can be accomplished more vividly using numerical and graphical illustrations. The following two examples illustrate how the minimization procedure is executed to obtain optimal solutions.

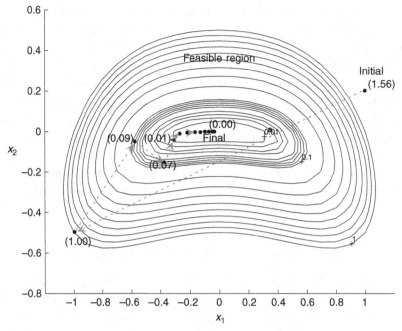

Figure 5.2. Numerical Illustration of Optimization

For our first example, let us consider the following optimization problem:

$$\min_{x} \; f(x) = (x_1^2 + x_2)^2 + 3x_2^2 \tag{5.9}$$

subject to

$$-1 \leq x_1 \leq 1 \tag{5.10}$$

$$-0.5 \leq x_2 \leq 0.5 \tag{5.11}$$

Figure 5.2 helps to illustrate how optimization is carried out numerically on an objective function, with two design variables satisfying the constraints in (5.10) and (5.11).

The function, $f(x)$, generates the contour plot shown in the figure. Each point on the diagram represents an iteration, and the number in the brackets represents the value of the objective function for the given design variables. The optimal solution is defined as *the smallest possible value of the objective function satisfying the specified constraints*. As indicated on the diagram, the contour plot corresponding to $f(x) = 1.00$ forms the feasible boundary of the space possessing possible solutions (feasible region). The initial point, $x = [1, 0.2]$, yields the initial design objective value of $f(x) = 1.56$, which is outside of the feasible region. The arrows show the step by step procedure carried out by the optimizer to obtain the minimized solution. The iterative process indicates that, even though the value of the objective function is being decreased, it is still possible to achieve lower objective function values satisfying the constraints. The iterations will continue until the objective function reaches its minimum value (the optimal solution). Executing the `fmincon` function in MATLAB to minimize this objective function requires 13 iterations. From examining this function, we observe that the optimal solution lies at $x^* = [0, 0]$, yielding $f(x)^* = 0$. In practice, MATLAB, or any other optimization code, will generate the optimum with a finite number of accurate digits. The default will sometimes be 4 to 6. For the above example, MATLAB generated the optimal design values of $x^* = [-0.0341, -0.002]$, yielding $f(x)^* = 1.0337 \times 10^{-6}$. Obtaining more accurate values for the design variables would require either scaling the objective function or changing some of the MATLAB default accuracy parameters, as discussed in the Numerical Essentials Chapter (Chapter 7).

5.4.3 Experimental Optimization

Experimental optimization, as it is called, can be viewed as a traditional trial and error approach to design. It essentially involves the following steps: (i) build a version of the physical system, (ii) evaluate its performance, and (iii) if we are satisfied with the current performance, we STOP; if not, we review what changes might be helpful, and go back to the first step, hopefully yielding an improved version.

This approach is practically obsolete. It can be very costly and time consuming, and it might not converge easily to a good solution. It might lead to a very sub-optimal solution. It may also require quite a bit of experience in order to work. This approach makes innovative new designs less likely. Fortunately, in this new highly competitive world where high performance computing is available, we can do much better using other approaches.

5.4.4 Graphical Optimization

It is also possible to explain this phenomena graphically. We now consider the following optimization problem:

$$\min_{x} \quad f(x) = x_1 + 2x_2 \tag{5.12}$$

subject to

$$x_1^2 + x_2^2 - 1 \leq 0 \tag{5.13}$$

$$-1 \leq x_1, x_2 \leq 1 \tag{5.14}$$

In this example, the given constraint is a circle whose perimeter forms the boundary of the feasible region. Figure 5.3 illustrates this problem.

In the diagram, each line corresponds to the value of the objective function during the minimization process. If the initial point, P, where $x = [1, 1]$ is chosen, then $f(x) = 3.00$. It is clear that this line does not intersect the circle at any point, making all the points lying on this line infeasible. Going through a similar iterative process as in the first example, the objective function is sequentially reduced to find the optimal solution. Therefore, for the case when $f(x) = 2.00$, we can obtain possible solutions

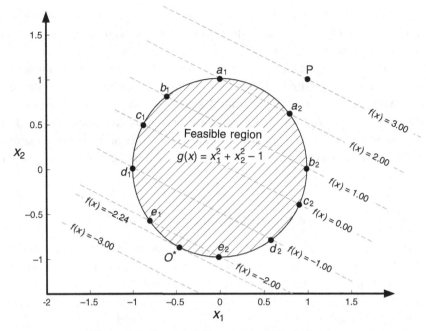

Figure 5.3. Graphical Illustration of Optimization

that fall between points a_1 and a_2. Even though these points are feasible solutions, it is still possible to obtain a smaller value for $f(x)$ satisfying the constraints (recall the definition of the optimal solution). The minimization process will result in smaller and smaller values of the objective function, with feasible solutions bounded between the points $a_1a_2, ..., e_1e_2$. As indicated in the figure, a point, $O*$, will be reached, such that the value of the design variables at this point will result in the smallest possible value of the objective function, which also satisfes the constraint equations Eqs. 5.13 and 5.14. This is the optimal solution. MATLAB generated $x^* = [-0.4472, -0.8944]$, yielding $f(x)^* = -2.2361 \approx -2.4$ as optimal values for this problem. Note that the optimal point falls on the line which is a tangent to the circle. Further minimization is no longer required, as smaller values for the objective function will yield points that fall outside the feasible region.

5.5 Software Options for Optimization

In this section, we discuss the various options available for optimizing problems using a computer software. As presented in Table 5.1, we define three main classes of optimization software. **The first** class involves *stand-alone* optimization software, where the primary focus is to solve various types of prescribed optimization problems. **The second** class involves design and/or analyses *integration frameworks*, where analyses codes from different engineering disciplines can be conveniently integrated and designs can be optimized. **The third** class of optimization software involves large scale *analyses codes that have optimization capabilities* as one of their offered features – typically added in recent years with the growing popularity of optimization.

Table 5.1. *Broad Classification of Software for Optimization*

Software for Optimization (SO)			
Stand-Alone **(SO-SA)**	**Within Design Framework** **(SO-WDF)**	**Within Analysis Package** **(SO-WAP)**	**Discrete Integer or Mixed**
MATLAB Toolbox	iSIGHT	GENESIS	XPRESS
NEOS Server	PHX ModelCenter	NASTRAN	CPLEX
DOT-VisualDOC	modeFRONTIER	ABAQUS	Excel and Quattro
NAG	XPRESS	Altair	NEOS Server
NPSOL	LINDO/LINGO	ANSYS	MINLP
GRG2	GAMS	COMSOL	GAMS WORLD
LSSOL	Boss Quattro	MS Excel	
CPLEX		What'sBest!	
BTN		RISKOptimizer	
PhysPro		Busi. Spreadsh.	

Details of the above options are discussed next. For convenience, we respectively refer to these classes as (i) Software for Optimization as Stand-Alone (SO-SA), (ii) Software for Optimization Within Design Framework (SO-WDF), and (iii) Software for Optimization Within Analysis Package (SO-WAP).

Within this section, we first provide a table (Table 5.1) that lists many of the popular software for each of these classes in the first three columns. The last column lists those software that perform optimization for problems with discrete, integer, or mixed design variables. In Section 5.5.1, we discuss the MATLAB Optimization Toolbox. In Sections 5.5.2, 5.5.3 and 5.5.4, we discuss each of the three classes in sequence. The last column is discussed in Section 14.3.7, where the topic of discrete optimization is presented.

We also note that it may be useful to classify optimization software as being (i) small scale or large scale, (ii) easy or difficult to use, or (iii) particularly effective or less so. The choice of a particular optimization software will generally depend on pertinent user experience and on the problem under consideration.

5.5.1 *MATLAB Optimization Code*—fmincon *and* linprog

Nonlinear Optimization/Programming

Here, we describe the procedure for solving a constrained nonlinear optimization problem using MATLAB. The first step is to carefully examine how MATLAB defines the optimization problem, which is as follows

$$\min_{x} \quad f(x) \tag{5.15}$$

subject to

$$c(x) \le 0 \tag{5.16}$$

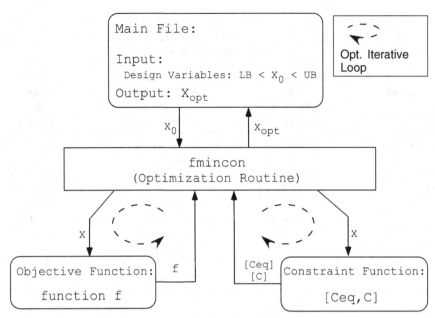

Figure 5.4. Optimization Flow Diagram

$$ceq(x) = 0 \tag{5.17}$$

$$A\,x \leq b \tag{5.18}$$

$$Aeq\,x = beq \tag{5.19}$$

$$LB \leq x \leq UB \tag{5.20}$$

Each of the above variables is defined within the context of the following example. Figure 5.4 provides a top level view of the optimization algorithm. A more detailed presentation is provided in Sec. 7.3, and illustrated in Fig. 7.1. Note that if we wish to *maximize* a function $f(x)$ and use the `min` MATLAB function, we simply need to perform the operation $\min_x$ `-f(x)`.

Let's use the following example to further explain the optimization problem definition given by Eqs. 5.15–5.20

Example 5.1

$$\min_{x} f(x) = x_1^2 + 3x_2^2 - 2x_1x_2 - 15 \tag{5.21}$$

subject to

$$g_1(x) \equiv -2x_1 - 2x_2 + 8 \leq 0 \tag{5.22}$$

$$-5 \leq x_1, x_2 \leq 5 \tag{5.23}$$

We would like to obtain the optimum values of the design variables x_1 and x_2 that minimize the objective function $f(x)$ in Eq. 5.21, subject to the constraints in Eqs. 5.22 and 5.23 being satisfied.

The `fmincon` **function**

To solve the above problem, we use the `fmincon` function provided in MATLAB. The syntax of the function is as follows.

```
[xopt, fopt] = fmincon('fun', x0, A, b, Aeq, beq, LB, UB,
                'nonlcon')
```

where `x0, A, b, Aeq, beq, LB,` and `UB` are the input variables that need to be defined before calling `fmincon`. `'fun'` is the name of the function file containing the definition of $f(x)$, and `'nonlcon'` is the name of the function file containing the nonlinear constraints. The variables `xopt` and `fopt` are the outputs of `fmincon`, where `xopt` is the optimum vector of variables $[x_1, x_2]$ and `fopt` is the minimum value of the objective function f.

The Calling Function: `main.m`

Define the input variables: `x0`: This is the initial guess provided to `fmincon`. It is a vector of size equal to the number of variables in the optimization problem. In this case, we have two variables, x_1 and x_2. Let us define `x0 = [1;1]`.

`A, b, Aeq,` and `beq`: These variables need to be defined only if the problem has linear constraints. In many cases, all constraints (*linear* and nonlinear) can be defined in the `nonlcon.m` file, so these variables can simply be defined as empty matrices. For example, `A=[ ]`. However, there are great numerical advantages to defining the linear constraints within Eqs. 5.18 and 5.19, rather than within Eqs. 5.16 and 5.17, respectively.

`LB, UB`: These are the vectors that contain the lower and upper bounds on the variables x_1 and x_2, respectively. From the above problem definition (Eq. 5.23), we define: `LB = [-5;-5]` and `UB = [5;5]`. If a problem does not have bounds, these variables can simply be declared as empty vectors. We recommend creating a `main.m` file that defines all the above variables and calls `fmincon` for the optimization.

After defining the above quantities, we call the function `fmincon.m`. The function `fmincon.m` calls (i) `nonlcon.m` to evaluate the constraints and (ii) `fun.m` to evaluate the objective function as needed.

Objective Function file: `fun.m`

This file should be saved in the same folder as the above main M-file. This function returns the value of the objective function at any given point x. The contents of `fun.m` are as follows:

```
function f = fun(x)
f = x(1)^2 + 3*x(2)^2 - 2*x(1)*x(2) - 15;
```

Nonlinear constraints file: `nonlcon.m`

This file should be saved in the same folder as the main and the function files. It returns the values of the inequality and equality constraints at any given point x.

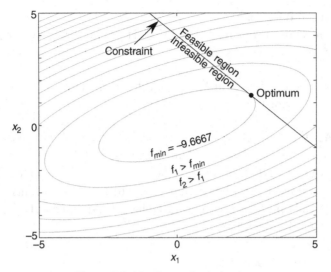

Figure 5.5. Nonlinear Optimization

Note that all inequality constraints should first be written in the form $g(x) \leq 0$, as shown in Eq. 5.22. The expression on the left-hand-side of the inequality is then defined in the constraint file `nonlcon.m` as shown below.

```
function [C,Ceq] = nonlcon(x)
C(1) = -2*x(1) - 2*x(2) + 8;
Ceq = [ ];
```

Note that `C(1)` is the first and only nonlinear inequality constraint in this problem. If the problem had more inequality constraints (say, n), they would be defined as `C(2)`, `C(3)`,...,`C(n)`. Ceq is the vector of all equality constraints, which is empty in this problem.

Calling `fmincon`

Once all the variables and function files are defined, we can call the `fmincon` function as shown above, to obtain the output

```
xopt =
    2.6667
    1.3333
fopt =
    -9.6667
```

Figure 5.5 provides a plot of the objective function and the feasible region. `fmincon` also allows the user to set a number of different optimization parameters, such as the maximum number of iterations and other termination criteria. It also allows the user to pass problem parameters to the `fun.m` and the `nonlcon.m` files. More help on these features can be obtained by typing 'help fmincon' at the MATLAB command prompt.

Options

Optimization options can be set for `fmincon` using the command `optimset`. Some options apply to all algorithms, while others are relevant to particular algorithms. You can use `optimset` to set or change the values of the options arguments. The options arguments include algorithms selection, information display settings and gradient estimation in series or parallel. Please see Optimization Options help in MATLAB for detailed information.

Display

Depending on the option selected for display, MATLAB can show different outputs in the Command Window. The information includes outputs at each iteration and provides technical exit messages. The exit messages can include hyperlinks. These hyperlinks bring up a window containing further information about the terms used in the exit messages.

Linear Optimization/Programming

Here, we describe the procedure for solving a linear programming problem using the MATLAB function `linprog`. The first step is to carefully examine the way MATLAB defines the optimization problem, which is as follows.

$$\min_{x} \quad f^T x \tag{5.24}$$

subject to

$$A x \leq b \tag{5.25}$$

$$Aeq\, x = beq \tag{5.26}$$

$$LB \leq x \leq UB \tag{5.27}$$

where f is a so-called cost coefficient vector, A and Aeq are constant matrices, and b and beq are constant vectors. The quantities LB and UB are the lower and upper bounds on the design variables. This optimization statement is further explained within the context of the following example.

Example 5.2

$$\min_{x} 4x_1 + 4x_2 \tag{5.28}$$

subject to

$$-5x_1 - 3x_2 \leq -30 \tag{5.29}$$

$$-3x_1 - 5x_2 \leq -15 \tag{5.30}$$

$$0 \leq x_1, x_2 \leq 3 \tag{5.31}$$

The `linprog` function

The basic syntax of the `linprog` function is as follows

```
[xopt,fopt] = linprog(f, A, b)
```

where xopt and fopt are the optimum values of the design variables and objective function, respectively. Before solving the linear program, we need to ensure that the objective function is of a minimization type, and that all the inequality constraints are written in the form of $a_1x_1 + a_2x_2 \leq b_1$, where a_1, a_2, and b_1 are constants (see Eqs. 5.29 and 5.30).

Define the input variables

f: This is a row vector corresponding to the coefficients of the design variables in the objective function f. Thus, in the current problem,

```
f = [4;4]
```

A: This is a matrix in which every row corresponds to the coefficients of the design variables in each "≤" constraint. Therefore, A will have as many rows as we have inequality constraints, and as many columns as we have design variables.

```
A = [-5 -3; -3 -5]
```

Aeq and beq: These matrices do not exist in this example, so we write

```
Aeq = [ ]
beq = [ ]
```

b: This is a vector in which each element is the number that appears on the right-hand-side of each "≤" constraint. We need to make sure that the order of the constraints used to define A and b is the same.

```
b = [-30; -15]
```

Calling linprog

Calling linprog using the above syntax yields

```
xopt =
    1.8750
    1.8750
fopt =
   15.0000
```

Figure 5.6 illustrates the objective function (dashed line) being minimized and the feasible region defined by the constraints in Eqs. 5.29 and 5.30. The dash lines represent constant values of the objective function. As this line moves downward, to decrease the value of the objective function, it reaches the optimum point. If it decreased any further, there would be no part of the line that would remain in the feasible region. Type "help linprog" at the MATLAB command prompt to explore other features and options.

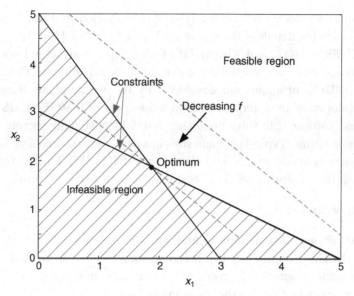

Figure 5.6. Linear Optimization

5.5.2 *Software for Optimization as Stand-Alone (SO-SA)*

Stand-alone optimization software refers here to those software that are created primarily for optimization purposes. We note that the following is only a representative sample of the software options available. A more complete listing of available software packages is provided in [9]. As shown below, there are software packages that have the capability to solve different types of optimization problems. A representative list is provided below with associated descriptions.

Please note that the websites provided for some of the software options listed below, while active as of the writing of this chapter, are subject to change in the future. The purpose of the following discussion is to inform the readers of the broad set of software options available in the marketplace to perform optimization. In practice, the latest information can be readily obtained through an internet search.

1. **Multipurpose Stand-Alone Optimization Software**

   a) **MATLAB optimization toolboxes:** MATLAB has two optimization toolboxes that can solve various types of optimization problems, such as linear, nonlinear, and multiobjective problems. General information regarding the MATLAB toolboxes were presented in Chapter 1.

   b) **NEOS server:** The NEOS server is an environment to solve optimization problems over the internet. It is available at the website www.neos-server.org/neos, and can be used to solve a variety of optimization problems, such as linear, nonlinear, discrete, integer programming, and combinatorial problems. The NEOS server also offers global optimization algorithms. The user needs to provide the optimization problem in a specific format, which depends on the type of optimization problem being solved. The software details of the solver can be found in Ref. [10].

c) **GENESIS, DOT, and VisualDOC:** GENESIS is a structural optimization software that also has finite element analysis capabilities. The DOT and VisualDOC programs are developed by the Vanderplaats Research and Development Inc., and are distributed as part of the GENESIS package. These software can solve nonlinear, multiobjective, and discrete optimization problems. Typical problem sizes are well over 100 design variables with thousands of constraints. Details are provided in [11]. Further information about these software tools can be found at the following website.

   www.vrand.com

d) **NAG library:** Several implementations, such as C and Fortran, are available for the NAG library [12, 13]. It can be used to solve linear, quadratic, and nonlinear programming problems. Further information regarding the NAG Library can be found at the following website.

   www.nag.com/optimization/availnaglib.asp

2. **Packages for Specific Classes of Problems**

   a) **Nonlinear constrained problems:**

      i. **NPSOL:** NPSOL is a software used to solve constrained optimization problems. It is especially effective for nonlinear problems whose functions and gradients are expensive to evaluate. These algorithms are claimed to be numerically stable, and to provide global convergence. Details are available at the following website.

         www.sbsi-sol-optimize.com/asp/sol_product_npsol.htm

      ii. **GRG2:** The GRG2 software [14] is used for nonlinear programming problems. It can be used as a stand-alone system or as a subroutine. It can handle up to two hundred active constraints. Another implementation of GRG2, known as LSGRG2 [15], can solve large scale optimization problems with over 500 constraints.

   b) **Linear programming problems:**

      i. **LSSOL:** It is a software package for solving constrained linear programs. The software is claimed to be numerically stable; and is used to solve problems in Statistics, Economics, and Finance. Details are available at the following website.

         www.sbsi-sol-optimize.com/asp/sol_product_lssol.htm

      ii. **CPLEX:** CPLEX is used to solve linear programming and integer programming problems. It employs both Simplex and interior point algorithms for linear programming. It can be used directly or as a subroutine.

The software is claimed to be robust and reliable with a user-friendly interface. Details are available at the website:

`www.ibm.com/software/commerce/optimization/cplex-optimizer`

e) **Unconstrained nonlinear problems:**
The **BTN** software can be used to solve unconstrained nonlinear optimization problems, and is implemented in a parallel computing environment. This software is particularly suitable for large scale problems. Details are in [16].

d) **Multiobjective problems:**

   i. **NIMBUS:** This software can be used to solve multiobjective problems. In the NIMBUS method, the user examines the values of the objective functions calculated at a current solution, and divides the objective functions into as many as five classes. The classes are functions whose values: (1) should be improved, (2) should be improved until some aspiration level is reached, (3) are satisfactory at the moment, (4) are allowed to worsen up to some bound, or (5) are allowed to change freely. A new subproblem is constructed and solved according to the classification and the associated information. Details are provided in [17]. Further information regarding NIMBUS can be found at the following website.

   `https://wwwnimbus.it.jyu.fi`

   ii. **PhysPro:** This software offers a unique and comprehensive formulation and solution to multiobjective problems. It embodies the Physical Programming Method, which is briefly presented in Chapter 18, and fully in Refs. [18, 19]. For more information regarding **PhysPro**, please visit `www.physpro.com` (Ref. [20]).

Next, we discuss some examples of integration frameworks where multidisciplinary analysis and optimization can be performed.

5.5.3 *Software for Optimization Within Design Framework (SO-WDF)*

The following is a representative list of various integration frameworks and their optimization features.

1. **iSIGHT:** This software helps engineers manage the computer software required to execute simulation-based design processes, including commercial computer aided design/analysis software and Excel spreadsheets. iSIGHT enables the rapid integration of these programs and automates their execution to accelerate the evaluation of several design alternatives. iSIGHT has multiobjective optimization capabilities, and contains state-of-the-art multiobjective genetic algorithm routines. A toolkit is available in iSIGHT for users to integrate their own proprietary or obtained optimization algorithms. Through a standard procedure, these user algorithms can be made available through the iSIGHT

optimization interface, including algorithm specific tuning parameters. Details are available at the following website.

www.simulia.com

2. **PHX ModelCenter:** Phoenix Integration has developed the PHX ModelCenter, which can be used for process integration and design optimization. PHX Model-Center is an environment for process integration to support the design team. It also enables visualizing multidimensional design spaces to seek the best designs. PHX ModelCenter allows the user to construct the design process as a series of linked applications with a simple interface. Once completed, the user will have organized analyses models that facilitate optimization and rapid design space exploration. Further details are available at the following website.

www.phoenix-int.com

3. **LMS Samtech Boss Quattro:** This is an applications manager that works in a multidisciplinary environment and allows the user to explore the design space using optimization. It allows interfacing with major computer aided design and analysis software, with in-house codes, and with management of data sheets such as MS-EXCEL. Boss Quattro is one of the software packages provided by SIEMENS. SIEMENS provides various software solutions that can be adapted to the applications at hand and to the industrial environment. Details are available at the following website.

www.plm.automation.siemens.com

4. **modeFRONTIER:** modeFRONTIER is a user-friendly and full-featured design environment that facilitates easy integration of several computer aided engineering (CAE) tools, whether commercial or in-house. It has the capability to integrate finite element structural analysis and computational fluid dynamics software. It also provides multiple optimization features, which include gradient based optimization, genetic algorithms, and multiobjective optimization. The modeFRONTIER software can be viewed as a wrapper around the CAE tool that performs optimization by modifying the input variables and monitoring the output variables. Details can be found at the following website.

www.esteco.com

Next, we discuss the third category of software that provides optimization features.

5.5.4 *Software for Optimization Within Analysis Package (SO-WAP)*

The following is a representative list of various analysis codes from different disciplines that have optimization routines.

1. **Altair Optistruct:** Altair OptiStruct is a commercially available finite element-based software program for both structural analysis and design optimization.

Altair OptiStruct is used to design, evaluate, and improve performance of mechanical structures. Users can define a target and constraints such as maximum allowable deflection and stress, modal response, member sizing, and method of manufacture. Altair OptiStruct then solves and visualizes the structurally optimal and manufacturable design proposal based on the specifications. Details are available at the following website.

`www.altairhyperworks.com`

2. **NASTRAN**: MSC NASTRAN allows the user to perform the optimization of designs where the objectives/constraints can be responses from the finite element analysis, such as weight, stress, and displacement. Large scale optimization with hundreds of design variables and millions of responses can be handled in NASTRAN. The optimization module can also handle numerical issues, such as design variable linking (see Sec. 15.3.1). Advanced capabilities, such as approximation methods and robust optimization, are also available. For further details, see `www.mscsoftware.com`.

3. **ANSYS:** ANSYS offers design optimization features in its ANSYS Mechanical suite of products. ANSYS has integrated optimization tools, such as topological optimization and probabilistic design. The DesignXplorer software, a subset of ANSYS, allows users to study and quantify various structural and thermal analysis responses on parts and assemblies, and to subsequently perform optimization. It also offers design for six sigma capabilities. More details are available at the following website.

`www.ansys.com`

4. **ABAQUS:** ABAQUS provides optimization capabilities through relationships with independent software vendors. More details are available at the following website.

`www.simulia.com`

The disciplinary solution is directly provided by the respective software vendor.

5. **COMSOL:** The optimization module in COMSOL provides optimization codes that are suitable for computationally intensive finite element analyses and multiphysics problems. The disciplines of the problems range from traditional engineering disciplines, such as structural mechanics and chemical engineering, to emerging technologies, such as bioengineering. The user can input objectives and constraints, which could be simple algebraic expressions, or an analysis model of any physical phenomena. The optimization module has solvers for linear and nonlinear constrained problems. Details are available at the following website.

`www.comsol.com`

6. **Microsoft Excel Add-ins for optimization:** Microsoft Excel has an optimization solver that can be included as an add-in [21]. The hybrid evolutionary solver,

which is a combination of genetic algorithms and classical nonlinear optimization methods, enables the user to optimize models with any Excel functions. Details are available at the website: `www.solver.com`. In addition to the above solver, additional add-ins are available from various companies. A representative list follows.

a) **What'sBEST!** This add-in from LINDO Systems lets the user solve large scale optimization models by using Microsoft Excel. It provides solvers for linear, nonlinear, integer, and global optimization. It is claimed to be fast, reliable, and easy to use. Details are available at the following website.

   `www.lindo.com`

b) **RISKOptimizer:** RISKOptimizer is an optimization add-in for Microsoft Excel. RISKOptimizer allows the optimization of Excel spreadsheet models that contain uncertain values. RISKOptimizer is claimed to find solutions quickly and is easy to use. Details are in [22].

c) **Business spreadsheets:** Business Spreadsheets (formerly Excel Business Tools) provides purpose-built Microsoft Excel templates that can be applied to a range of financial analysis and business decision-making scenarios. The Portfolio Optimization, which is one of the templates of Business Spreadsheets, evaluates the optimal capital weightings for a basket of investments that gives the highest return for the least risk. Details are available at the following website.

   `www.business-spreadsheets.com`

5.6 Summary

The previous chapters provided the perquisite knowledge for learning the theory of optimization, and for implementing optimization through computational tools, followed by a philosophical introduction to the essence and the principal components of design optimization. This chapter built on that foundation by delving directly into the mathematical theory of optimization. More specifically, this chapter provided an introduction to the fundamental concepts of linear and nonlinear programming. An insightful classification of optimization problems was first provided, based on the function (non)linearity, the number of objectives, the presence of constraints, and the nature of design variables. This was followed by an introduction to the major classes of approaches used to solve such optimization problems; namely, analytical optimization, numerical/algorithmic optimization, experimental optimization, and graphical optimization. The capabilities of MATLAB in solving optimization problems was discussed next, specifically focusing on the `fmincon` and the `linprog` functions. The chapter concluded with short descriptions of different currently-available optimization-related software, including commercial software, open-source codes, multi-purpose packages, optimization frameworks, and simulation software that include optimization modules.

5.7 Problems

Warm-up Problems

5.1 Formulate the following problem (by hand) to represent the generic optimization format shown in Eq. (5.15) to (5.20) of the textbook.

$$\min_x f(x) = 5x_1^2 - 23x_1x_2 - 9x_2^2 + 9x_3^2 + 15 \qquad (5.32)$$

subject to

$$x_1^2 + x_2^2 = 10 \qquad (5.33)$$

$$x_1x_3 \le 100 \qquad (5.34)$$

$$4x_1 - 19x_2 = 50 \qquad (5.35)$$

$$3x_1 + 4x_3 = 100 \qquad (5.36)$$

$$x_1 + x_2 + x_3 \ge -10 \qquad (5.37)$$

$$-500 \le x_1, x_2, x_3 \le 500 \qquad (5.38)$$

5.2 Solve the following constrained optimization problem using MATLAB. Use the following as your starting points: (i) [1;1] and (ii) [1;6].

$$\min_x f(x) = x_1^2 - 5x_2 \qquad (5.39)$$

subject to

$$x_1 + x_2 \le 5 \qquad (5.40)$$

$$-\infty \le x_1 \le 10 \qquad (5.41)$$

Answer the following questions:
(a) Solve the above problem using MATLAB. Report the optimum value of x_1 and x_2, and the corresponding minimum value of $f(x)$ for both starting points.
(b) Does the minimum objective function value change when the starting point is changed?
(c) What is the effect of the starting point on the optimum value of x_2?
(d) Explain in your own words any interesting features of this problem in view of the above questions.

5.3 Consider the following problem.

$$\min_x f(x) = x_1^2 + 10x_2^2 - 3x_1x_2 \qquad (5.42)$$

subject to

$$2x_1 + x_2 \ge 4 \qquad (5.43)$$

$$x_1 + x_2 \geq -5 \qquad\qquad\qquad (5.44)$$

$$-5 \leq x_1, x_2 \leq 5 \qquad\qquad\qquad (5.45)$$

(a) Solve the above problem using MATLAB. Report the optimum value of x_1 and x_2, and the corresponding minimum value of $f(x)$.

(b) Solve the above problem by removing the first constraint $2x_1 + x_2 \geq 4$. Report the optimum value of x_1 and x_2, and the corresponding minimum value of $f(x)$.

(c) Now, solve the same problem by removing **all** the constraints. Report the optimum value of x_1 and x_2, and the corresponding minimum value of $f(x)$.

(d) Create a contour plot similar to Fig. 1.11 (a) showing the contours of the objective function. Show on it the locations of the optima obtained in Parts (a) - (c).

5.4 The following is a linear programming problem.

$$\min_{x} f(x) = 20x_1 + 64x_2 \qquad\qquad\qquad (5.46)$$

subject to

$$25x_1 + 70x_2 \geq 2,100 \qquad\qquad\qquad (5.47)$$

$$0 \leq x_1 \leq 70 \qquad\qquad\qquad (5.48)$$

$$0 \leq x_2 \leq 50 \qquad\qquad\qquad (5.49)$$

(a) Solve the above linear optimization problem using the `linprog` command in MATLAB.

(b) Solve the above problem using the `fmincon` command in MATLAB. Compare the results with the results obtained in (a).

Intermediate Problems

5.5 Discuss the nature of the solution of a continuous linear programming problem that does not have constraints. Next, discuss the nature of the solution of a discrete linear programming problem that does not have constraints.

5.6 Prepare a one to two page review of the application of design optimization in a field of your choice (*e.g.*, aerospace industry, energy industry, biomedical/biotechnology industry, operations research (OR), and finance). The review should summarize the contribution of design optimization to the concerned field in the last few decades. A brief survey of some key literature and real life implementation of optimization in that field should also be included. Clearly state all the references that you use in the review.

5.7 The water tank shown in Fig. 5.7 is supported by a vertical column and subjected to a wind load, w [23]. As the engineer, you are required to determine

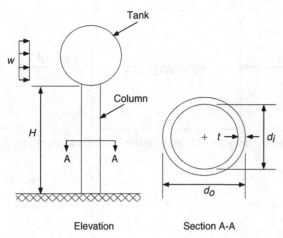

Figure 5.7. Structure of Water Tank

the best value for the thickness, t, and inner diameter, d_i, of the column in an effort to minimize the total mass for the design (total mass is given as the mass of the column plus the mass of the tank).

The following constants are given: the height of the column, $H = 35$m, diameter of the tank, $D = 15$m, wind pressure, $w = 700$N/m^2, gravity, $g = 9.81$m/s^2, average thickness of the tank wall, $t_{avg} = 0.015$m, unit weight of steel, $\gamma_s = 80$KN/m^3, allowable bending stress, $\sigma_b = 165$MPa, and allowable deflection, $\Delta = 0.2$m.

You are presented with a simple function `wtower.m` in the book website (`www.cambridge.org/Messac`) that describes the loading condition shown in Fig. 5.7. This function allows you to enter values for the inner diameter and thickness (inputs); giving, in return, the corresponding values of the following (outputs):

1. Outer diameter of the column, d_o,
2. Allowable axial stress, σ_a (calculated using the critical buckling load with a factor of safety $\frac{23}{12}$),
3. Surface area of the tank, A_s,
4. Deflection of the tank, δ,
5. Bending stress, f_b, and
6. Axial stress, f_a

Figure 5.8 illustrates the function `wtower.m`.

The following constraints should be imposed on the water tank design:

1. $\frac{do}{t} \leq 92$,
2. $\delta \leq \Delta$,
3. $\frac{f_a}{\sigma_a} + \frac{f_b}{\sigma_b} \leq 1$,
4. $0.7 \leq d_i \leq 2.0$m and
5. $0.01 \leq t \leq 0.2$m

Figure 5.8. Input and Output for the Function `wtower.m`

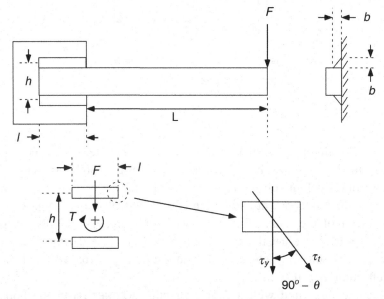

Figure 5.9. Schematic of Cantilever Beam

You are required to complete the following:

1. Enter the constants given above as a row vector *y*. That is

$$y = [H \quad D \quad w \quad g \quad t_{avg} \quad \gamma_s \quad \sigma_b \quad \Delta]$$

2. Using the function `wtower.m`, with initial values for the inner diameter and thickness of the column being 1m and 0.1m, respectively, determine the optimal values for the design variables and design objective. (Note: Your objective function and constraints should be functions of the design variables, *x*, and the constants, *y*).

3. If you minimize only the mass of the column, will the optimal values for the design variables and design objective change? Discuss the reasons for your results.

5.8   A welded cantilever beam subjected to a tip force is shown in Fig. 5.9. The volume of the weld holding the cantilever should be as small as possible while maintaining the applied tip force. As shown in Fig. 5.9, the weld has two segments, each of length l and height b. The cantilever beam, of length L, is subject to a tip force of F. This force, F, induces shear stresses in the welds indicated by the components, τ_t and τ_y. T corresponds to the turning effect (torque) of F.

1. Minimize the weld volume subject to a maximum shear stress limit in the weld. Find the optimal solution using MATLAB.
2. Make a plot of the design variable space, showing the optimal variable values. On the same plot, draw (i) the objective function contour that passes through the optimal design variables, and (ii) an objective function contour that is not feasible.
3. Indicate on your plot: the constraint functions, and the direction for which the objective function contours worsen.

Given the parameter values:

$$\tau^u = 30,000 \, psi \text{ (upper limit for shear stress)}$$

$$F = 6,000 \, lb$$

$$h = 3 \, in$$

$$L = 14 \, in$$

$$0.125 \, in \le b \le 0.75 \, in$$

$$1 \, in \le l \le 6 \, in$$

The expression for shear stress is given as:

$$\tau_y = F/(bl\sqrt{2})$$

$$\tau_t = [6F(L + 0.5l)\sqrt{h^2 + l^2}]/[\sqrt{2}bl(l^2 + 3h^2)]$$

$$\tau = \sqrt{\tau_y^2 + 2\tau_y\tau_t \cos\theta + \tau_t^2}$$

$$\cos\theta = l/\sqrt{h^2 + l^2}$$

BIBLIOGRAPHY OF CHAPTER 5

[1] H. W. Kuhn. *Nonlinear Programming: A Historical View*. Springer, 2014.
[2] L. T. Biegler. *Nonlinear Programming: Concepts, Algorithms, and Applications to Chemical processes*. SIAM, 2010.
[3] R. Bronson, G. B. Costa, and J .T. Saccoman. *Linear Algebra: Algorithms, Applications, and Techniques*. Elsevier Science, 2013.
[4] D. G. Luenberger and Y. Ye. *Linear and Nonlinear Programming*. Springer, 2008.
[5] J. K. Karlof. *Integer Programming: Theory and Practice*. CRC Press, 2005.
[6] D. S. Chen, R. G. Batson, and Y. Dang. *Applied Integer Programming: Modeling and Solution*. John Wiley and Sons, 2010.
[7] J. Hromkovic. *Algorithmics for Hard Problems: Introduction to Combinatorial Optimization, Randomization, Approximation, and Heuristics*. Springer, 2010.
[8] P. M. Pardalos, Q. P. Zheng, and A. Arulselvan. Deterministic global optimization. *Wiley Encyclopedia of Operations Research and Management Science*, 2010.

[9] J. J. Moré and S. J. Wright. *Optimization Software Guide*. Society for Industrial and Applied Mathematics, Philadelphia, 1993.

[10] M. C. Ferris, M. P. Mesnier, and J. J. Moré. NEOS and CONDOR: Solving optimization problems over the internet. *ACM Transactions on Mathematical Software*, 26(1):1–18, 2000.

[11] G. N. Vanderplaats. *Multidiscipline Design Optimization*. Vanderplaats Research and Development, Inc., 2007. www.vrand.com/BookOnOptimization.html.

[12] NAG Ltd. NAG C library manual. www.nag.com/numeric/CL/CLdescription.asp.

[13] NAG Ltd. NAG fortran library manual. www.nag.com/numeric/fl/fldocumentation.asp.

[14] L. S. Lasdon, A. D. Waren, A. Jain, and M. Ratner. Design and testing of a generalized reduced gradient code for nonlinear programming. *ACM Transactions on Mathematical Software*, 4:34–50, 1978.

[15] S. Smith and L. S. Lasdon. Solving large sparse nonlinear programs using GRG. *ORSA J. Comput.*, 4:1–15, 1992.

[16] S. G. Nash and A. Sofer. BTN: Software for parallel and unconstrained optimization. *ACM Transactions on Mathematical Software*, 18:414–448, 1992.

[17] K. Miettinen and M. M. Mäkelä. Synchronous approach in interactive multiobjective optimization. *European Journal of Operational Research*, 170(3):909–922, 2006.

[18] A. Messac. Physical Programming: Effective optimization for computational design. *AIAA Journal*, 34(1):149–158, 1996.

[19] A. Messac, S. Gupta, and B. Akbulut. Linear Physical Programming: A new approach to multiple objective optimization. *Transactions on Operational Research*, 8:39–59, 1996.

[20] PhysPro. www.physpro.com.

[21] D. Fylstra, L. S. Lasdon, J. Watson, and A. D. Waren. Design and use of the Microsoft Excel Solver. *Interfaces*, 28(5):29–55, 1998.

[22] W. L. Winston. *Decision Making Under Uncertainty With RISKOptimizer : A Step-To-Step Guide Using Palisade's RISKOptimizer for EXCEL*. Palisade Corporation, 1999.

[23] J. Arora. *Introduction to Optimum Design*. Academic Press, 3rd edition, 2011.

PART 3

USING OPTIMIZATION—PRACTICAL ESSENTIALS

In Part I, we presented some prerequisite material for our study of optimization: and introduction to MATLAB, and some elementary mathematics. In Part II, we learned what we needed to explore what optimization is all about, and the important role it can play in your life as and engineer, scientist, business person, or anyone dealing with improving things quantitatively (*e.g.*, performance and cost). In Part III, we get into the thick of things! We learn what we need to know to credibly address optimization problems in practice.

Specifically, the topics presented, with the chapter numbers, are given below:

6. Multiobjective Optimization

7. Numerical Essentials

8. Global Optimization Basics

9. Discrete Optimization Basics

10. Practicing Optimization—Larger Examples

6

Multiobjective Optimization

6.1 Overview

In Part I of this book, we reviewed preparatory knowledge needed to start learning optimization: MATLAB, and elementary mathematics. In Part II, we were exposed to the world of optimization and its potentially powerful role in our lives as engineers or professionals in quantitative fields. We then delved into the specific activities of analysis, design, and optimization, their links and distinctions. In the previous chapter, we then began addressing the fundamental aspects of computational optimization. These included single objective optimization, and the different approaches to optimization (analytical, numerical, experimental, and graphical). This was followed by a discussion of software options.

In this chapter, we study one of the most important aspects of optimization in practice, the notion of multiobjective optimization. Stated simply, Multiobjective optimization is the art and science of formulating how to optimize a set of competing objectives, which is almost always the case in practice. A detailed presentation of the pertinent methods is provided in such a way as to allow you to be readily productive and effective in practical design.

6.2 The Multiobjective Problem Definition

The identification of the right design objectives plays a crucial role in the design of any system. More often than not, in real-life design, you will find that your optimization problem contains more than one design objective. For example, wouldn't it be nice if your car dealer would tell you that the car you like happens to feature *more* miles per gallon *and* also *costs less* than a competitor's car? As an engineer, you would think twice before making a decision based on his interesting comments. If he were right, the other car company would probably not stay in business for very long. The point here is that while designing any product or system, you will almost always have to consider several competing design objectives. As a car designer, you would like the car to provide the most miles per gallon possible, while taking care that the car does not cost a million dollars. This is where multiobjective optimization comes to the rescue. It is a methodical approach to solving problems involving several

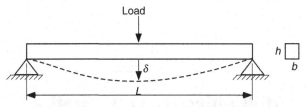

Figure 6.1. A Simple Beam Design Problem

competing design objectives simultaneously. The fundamental message is that you will almost always have to *compromise* between your various objectives and find a way to prioritize them somehow (see Refs. [1, 2, 3, 4]).

6.2.1 Example Problem

Let us illustrate a typical multiobjective optimization problem through a design example. As a member of a design team in an automobile company, your first simple task is to design a pinned-pinned beam shown in Fig. 6.1 [5]. Several of the beam's features are already fixed by the design team, including its length L and width b, the material, and the load acting on it. The only parameter that you are allowed to change is the height, h. As part of the design, you have been asked to minimize the bending stress in the beam and, at the same time, minimize the deflection at the mid-span.

Using this information, we can now set up our multiobjective design problem as follows. The design objectives are:

$$\mu_1 = \text{Bending stress} \tag{6.1}$$

$$\mu_2 = \text{Deflection} \tag{6.2}$$

The design variable is

$$x_1 = \text{Height} \tag{6.3}$$

6.2.2 Multiobjective Optimization Problem Statement

We explore the structure of a generic multiobjective optimization problem statement. A mathematical definition is given below.

$$\min_x [\mu_1(x) \quad \mu_2(x) \quad ... \quad \mu_n(x)]^T \tag{6.4}$$

subject to

$$g(x) \le 0 \tag{6.5}$$

$$h(x) = 0 \tag{6.6}$$

$$x_l \le x \le x_u \tag{6.7}$$

Our example problem had only two objectives. Note that, in the above problem statement, the number of design objectives μ_i is equal to n. When $n = 2$, we call it a *bi-objective problem*. Recall that x is the design variable vector. Each element x_i is a design variable. In the above problem statement, $g(x)$ represents the vector of inequality constraints, and $h(x)$ is the vector of equality constraints. That is, $g_1(x)$ is the first inequality constraint, and $h_3(x)$ is the third equality constraint. The vectors x_l and x_u are the lower and upper bounds on the design variable vector x, respectively.

6.3 Pareto Optimal Solution
6.3.1 Introducing the Pareto Solution

Thus far, we have discussed how to formulate a multiobjective optimization design problem. The next important question is: how do we solve this optimization problem? It indeed does not resemble the various single objective problems that we have seen thus far. Does it have a single optimum solution or multiple solutions? What is the value (or values) of x that seeks to minimize μ_1 and μ_2 simultaneously, while providing a desirable tradeoff between the two? One of the interesting features of multiobjective optimization is that the solution to the problem is generally not unique as different tradeoff levels may be desirable (with each tradeoff level yielding a different solution). A set of solutions called *Pareto Optimal Solutions* form the complete solution set of the optimization problem.

To understand the concept of Pareto optimality, we review the folloing example. For simplicity, we consider an unconstrained problem involving two design objectives, μ_1 and μ_2, which are functions of a single design variable x. We are interested in minimizing both design objectives simultaneously. Figure 6.2 provides the plot of each objective function on the same vertical axis and the design variable x on the horizontal axis.

If we minimize each objective function independently, ignoring the other objective, we will obtain the point that corresponds to the minimum of the objective being

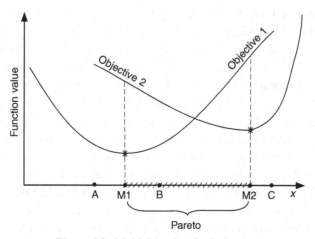

Figure 6.2. Multiobjective Optimization

minimized. These two points are indicated by stars in Fig. 6.2. Suppose you are at the minimum of Objective 1 (*i.e.* Point M1), and you decide that you want a design that has a lower value for Objective 2 than what you have at Point M1. To achieve this, you will have to move to the right of Point M1, say to a point B. In doing so, Objective 2 has decreased in value, but what happened to the value of Objective 1? It increased as we moved from Point M1 to Point B. Remember, we are trying to minimize both objectives. Thus, to improve the value of μ_2, we had to compromise on the performance of μ_1. In fact, this is true for all the points between M1 and M2. We call these points *Pareto optimal solutions* or *non-dominated* solutions.

Definition: Pareto optimal solutions are those for which any improvement in one objective will result in the worsening of at least one other objective [6]. That is, a tradeoff will take place.

Thus, if a point is Pareto optimal, we can be assured that there cannot be simultaneous improvement in all the objectives. In Fig. 6.2, if we move to the right of Point A, both objectives decrease simultaneously. Therefore, we call points, such as A, *non-Pareto* or *dominated* solutions. The same it true with other points such as C. The hatched region in the central portion is the set of all design variable values that are Pareto optimal. As an optimization engineer, if you are solving a multiobjective optimization problem, Pareto points are what you should be looking out for.

6.3.2 The Pareto Frontier

In the previous subsection, we determined how to identify Pareto solutions in the *design variable* space; that is, x space. We introduce the concept of *design objective* space. This is a plot with a design objective plotted on each axis. We are particularly interested in what happens if we plot the design objective values of the Pareto solutions in this objective space. The pattern of points that you see in the objective space is called the *Pareto frontier*. The name might sound highly technical, but it is simply a plot of all Pareto solutions in the objective space. Figure 6.3 is a plot of all the Pareto points that we identified in Fig. 6.2.

All the points (M1, M2, A, B, and C) illuatrated in Fig. 6.2 are plotted with their respective objective function values in Fig. 6.3. M1 and M2, in particular, form the end points of the Pareto frontier, also known as anchor points. Point M1 is where Objective 1 has the least value, while M2 is where Objective 2 has the least value. Points A and C, as dominated points, do not lie on the Pareto frontier.

You can appreciate why such a plot can be of immense use to a designer who is trying to optimize a particular system. By looking at the Pareto frontier, one can clearly see the tradeoffs associated with each Pareto point. For example, if Objective 1 and Objective 2, respectively, denote stress and deflection, then one can immediately identify regions of the Pareto frontier corresponding to the low values of stress, or regions with low values of deflection, whatever the preference of the designer. The designer can then select a particular Pareto point (say Point B), and map it back to the design variable space (Fig. 6.2) to determine what values of the design variables

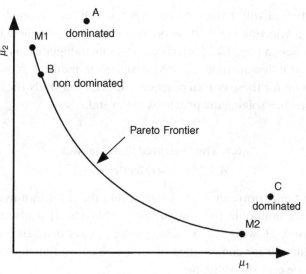

Figure 6.3. Pareto Frontier

(x) yield Point B. Thus, the concept of Pareto frontier is central to the understanding and application of multiobjective decision making.

6.3.3 Obtaining Pareto Solutions

From the above discussion, it is clear that to solve a multiobjective problem, we need to obtain a Pareto optimal point or a set of points on the Pareto frontier. We cannot always plot the design objectives as shown in Fig. 6.2. In most real-life design problems, the design objectives are functions of several variables, and we need more methodical techniques to obtain Pareto solutions. Note that most optimization algorithms are developed for single objective problems. As such, one intuitive way to solve a multiobjective problem is to combine all the objectives into a single *Aggregate Objective Function (AOF)* in such a way that, when the AOF is optimized, a Pareto solution is obtained.

6.3.4 Aggregate Objective Function

Definition: An Aggregate Objective Function (AOF), generally denoted by J, is a function that combines the design objectives into a scalar function.

The AOF typically contains parameters to be selected by the designer. These parameters reflect the relative importance of each design objective. An objective with higher importance will be given priority during the optimization process. That is a form of inter-criteria preference. In addition, as effective AOF should provide the ability to express the relative preference for different values of a given objective (intra-criteria preference). Note that the most commonly used AOFs do not effectively provide both of these crucial attributes. Keep in mind that the final solution that can be achieved will depend on the type of AOF that is used. An important consideration in selecting the AOF to be used is its ability to allow the designer to impose his/her design objective preferences in an unambiguous manner. Below are some popular AOF formulations used by designers in industry. You will develop a deep appreciation for these crucial practical issues as you study the material in this chapter, and practice solving the problems at the end.

6.4 The Weighted Sum Method
6.4.1 Two-Objective Case

The Weighted Sum approach is the simplest and the most intuitively meaningful means of solving multiobjective optimization problems. It is also the one that is most widely used. However, it possesses some serious deficiencies. The AOF is simply a weighted linear combination of all the objective functions. Let's consider the two-objective case shown below.

$$J(x) = w_1\mu_1(x) + w_2\mu_2(x) \tag{6.8}$$

The above equation is a linear function of the two objective functions. Therefore, if we plot the constant value curves (contour plot) of J in the *objective* space (see Fig. 6.3), these contours will appear as straight lines. Setting the weights in front of each objective function, and minimizing J yields a Pareto point. By changing the weights ($w \geq 0$) uniformly, say from 0 to 1, we can obtain a series of Pareto points on the Pareto frontier. The mathematical formulation of the optimization problem is given below.

$$\min_{x} J(x) = w_1\mu_1(x) + w_2\mu_2(x) \tag{6.9}$$

subject to

$$g(x) \leq 0 \tag{6.10}$$

$$h(x) = 0 \tag{6.11}$$

$$x_l \leq x \leq x_u \tag{6.12}$$

Note that we are considering the generic constrained optimization problem in Eq. 6.9. The constraints $g(x)$ and $h(x)$ define the feasible region, which is shown in Fig. 6.4 as the shaded region. A pictorial representation of what goes on behind the scenes during the optimization process is also provided. The dashed lines are the constant value curves of J.

As the optimization process decreases the value of J, the solution moves closer and closer to the Pareto frontier. The optimization will terminate when it is no longer possible to decrease both μ_1 and μ_2 simultaneously while remaining in the feasible region. Notice that if you set $w_1 = 0$, the constant value curves of J are parallel to μ_1; and the optimum point obtained will be M2, the minimum of μ_2. Similarly, if you set $w_2 = 0$, you will obtain M1 as the minimum. For all other combinations of

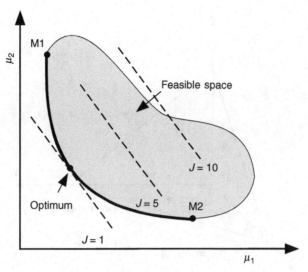

Figure 6.4. Weighted Sum Method

w_1 and w_2, different points on the Pareto frontier can be obtained. A typical way to set the weights is to vary them between 0 and 1 such that their sum is equal to 1. One of the deficiencies of the weighted sum method is that there is no easy way to know what values of w to use. They are generally chosen by trial and error.

6.4.2 Addressing More than Two Objectives

For two-objective (*i.e.*, bi-objective) optimization problems, the two weights in Eq. 6.9 can be made to vary between 0 and 1 in an attempt to generate Pareto solutions. Here, we address the case of m objectives, where $m > 2$ and the Aggregate Objective Function can be expressed as

$$\tilde{J}(x) = h_1\mu_1(x) + h_2\mu_2(x) + \cdots + h_m\mu_m(x) \qquad (6.13)$$

where h_i is greater than or equal to 0, and represents the weight for the ith objective function. Note that h_i cannot be equal to 0 for all i. We now make the following observations. The solution obtained using $\tilde{J}(x)$ is the same as that obtained using $J(x)$, where $J(x) = \frac{1}{\alpha}\tilde{J}(x)$ with α as a positive constant. If we let $\alpha = h_1 + h_2 + \cdots + h_m$, then the new objective function can be written as

$$J(x) = w_1\mu_1(x) + w_2\mu_2(x) + \cdots + w_m\mu_m(x) \qquad (6.14)$$

where

$$w_1 = \frac{h_1}{h_1 + h_2 + \cdots + h_m} \qquad (6.15)$$

$$w_2 = \frac{h_2}{h_1 + h_2 + \cdots + h_m} \qquad (6.16)$$

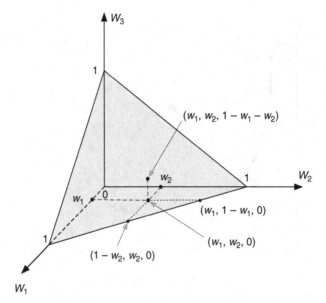

Figure 6.5. The Plane of Weights for a Three-Objective Optimization Problem

$$\vdots \tag{6.17}$$

$$w_m = \frac{h_m}{h_1 + h_2 + \cdots + h_m} \tag{6.18}$$

With the above construction, we note that

$$\sum_{1}^{m} w_i = \frac{h_1}{h_1 + \cdots + h_m} + \frac{h_2}{h_1 + \cdots + h_n} + \cdots + \frac{h_m}{h_1 + \cdots + h_m}$$

$$= \frac{h_1 + h_2 + \cdots + h_m}{h_1 + h_2 + \cdots + h_m} \tag{6.19}$$

$$= 1$$

From Eq. 6.19, we observed that the weights, $(w_1, w_2, \cdots, w_m)$ lie on the hyperplane defined by $w_1 + w_2 + \cdots + w_m = 1$, where

$$0 \le w_i \le 1 \tag{6.20}$$

For three-objective optimization problems, the points belonging to the plane $w_1 + w_2 + w_3 = 1$ $(0 \le w_1, w_2, w_3 \le 1)$ define the ranges of appropriate weights (see Fig. 6.5). To examine how the weights may change relative to each other, we recall that the sum of the weights is equal to 1, and we make the following observations: (i) We begin by allowing w_1 to vary from 0 to 1. (ii) For a given w_1, w_2 must vary between 0 and $(1 - w_1)$, in order to have $w_1 + w_2 = 1$. (iii) For a given w_1 and w_2,

w_3 must vary between 0 and $(1 - w_1 - w_2)$ in order to have $w_1 + w_2 + w_3 = 1$.
(iv) (***For four objectives, we would have:***) For a given w_1, w_2 and w_3, w_4 must vary
between 0 and $(1 - w_1 - w_2 - w_3)$ in order to have $w_1 + w_2 + w_3 + w_4 = 1$. And the
process continues.

Figure 6.5 describes the range of weights for the three-objective case, where the
horizontal dashed line describes the generic range for w_2 and the vertical dashed line
describes the generic range for w_3.

The associated MATLAB code for the three-objective case is presented below. If
we use an increment of 0.01 for the weights iterations, the range of weights can be
evaluated using the following nested loop code.

```
for w1=0:0.01:1
    for w2=0:0.01:1-w1
        w3=1-w1-w2;
%       (evaluate the objectives u1, u2, and u3)
        J = w1*u1 + w2*u2 + w3*u3;
    end
end
```

6.5 Compromise Programming

An important question is whether the weighted sum method has the ability to generate all correct results. To answer this question, consider Fig. 6.6. It displays, what
we call in multiobjective optimization terminology, a *non-convex* Pareto frontier.
(Recall that the Pareto frontier in Fig. 6.4 was part of a convex set.)

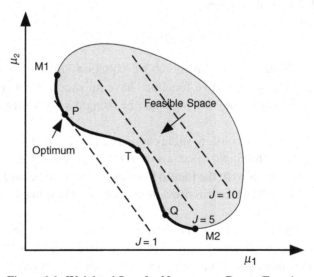

Figure 6.6. Weighted Sum for Non-convex Pareto Frontier

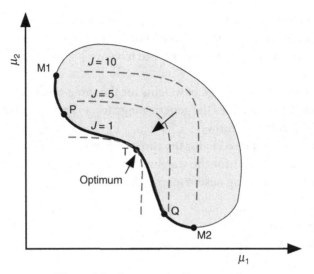

Figure 6.7. Compromise Programming

If we try to use the weighted sum method for this type of problem, the Pareto points that lie in the non-convex regions (all points between P and Q) will be unobtainable. A pictorial representation of the optimization is shown in Fig. 6.6. As the optimization progresses, you might think that it will stop when the dashed line passes through point T. In fact, it continues beyond point T and yields point P as the optimum solution. No matter what set of weights you use, points such as T (and any point between P and Q) are unobtainable, because they lie in a *non-convex* portion of the Pareto frontier. This is a serious drawback of the weighted sum method, and one that calls for particular caution.

Is there a way out? Yes, one option is called Compromise Programming. It is a simple extension of the weighted sum strategy. The mathematical form of the pertinent AOF is provided below.

$$J(x) = w_1\mu_1^n(x) + w_2\mu_2^n(x) \tag{6.21}$$

Notice the difference between the above equation and the weighted sum in Eq. 6.8. In Eq. 6.21, the objective functions have an exponent, n. If we plot the constant value curves of J, they will no longer be straight lines. Instead, they will be n-th order curves.

Figure 6.7 presents the constant value curves of J as dashed lines as the optimization progresses. Note that if we select a sufficiently high value for n, we can *reach into* the non-convex portions of the Pareto frontier. In this specific case, the optimum point obtained is point T, which was previously unobtainable using the weighted sum method.

As a general guideline, always choose n to be an even integer less than or equal to 8. In most cases, using $n = 2$ or $n = 4$ should yield satisfactory results. When $n = 2$, the compromise programming approach is sometimes called the weighted square sum method.

6.6 Generating the Pareto Frontier—with MATLAB

Previously in Sec. 5.5.1, we discussed how to use `fmincon` to solve a single nonlinear optimization problem with one objective. In this section, we explain how to develop a code in MATLAB to generate several Pareto solutions, or the Pareto frontier, for a multiobjective optimization problem.

Consider the following bi-objective optimization problem for which we wish to generate the Pareto frontier.

$$\min_{x_1, x_2} \quad \{\mu_1 = x_1^2 + 4x_2, \mu_2 = x_2^2 + 2\} \qquad (6.22)$$

subject to

$$2x_1 + 3x_2^2 - 8 \leq 0 \qquad (6.23)$$

$$x_1 + x_2 - \frac{7}{2} = 0 \qquad (6.24)$$

$$0 \leq x_1 \leq 10 \qquad (6.25)$$

$$0 \leq x_2 \leq 5 \qquad (6.26)$$

Figure 6.8 provides the MATLAB code for Pareto frontier generation using the weighted sum method. The procedure is summarized below:

1. **Program Structure:** The main program is called `main.m`, which upon execution, generates and plots a set of Pareto points. The file `main.m` calls the two functions `objfun.m` and `confun.m`, which generate the objective functions and the constraints, respectively.

2. **Initialization:** The file `main.m` begins by providing the initial guess (`x0`) and specifying lower and upper bounds (`LB` and `UB`) for the design variables. The quantities `x0`, `LB`, and `UB` are arguments in `fmincon`. A counter `j = 1` is defined. This counter is used to increment the indices of the arrays in which the Pareto points are stored.

3. `for` **loop and Details:** After initialization, a `for` loop is defined. Like most optimization codes, the MATLAB solver `fmincon` can only solve single objective optimization problems. Therefore, we combine the two objectives in the problem at hand, μ_1 and μ_2, to form a single AOF. In order to obtain the Pareto frontier of the bi-objective problem, we solve a series of single objective problems. Each iteration of the `for` loop solves one single objective optimization problem. This approach is further discussed below.

   a) Each single objective problem has a different AOF that can be formed using the weighted sum method or the compromise programming method, using different sets of weights. In order to automate this AOF generation for each single objective optimization, we use the `for` loop to generate the sets of weights.

Main file: **main.m**

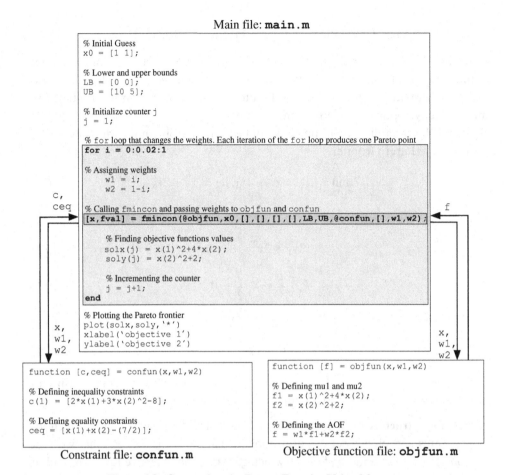

Figure 6.8. Generating the Pareto Frontier Using MATLAB

b) As shown in Fig. 6.8, each iteration of the for loop generates one set of weights that defines an AOF for that iteration. The corresponding single objective problem is solved by calling fmincon. The command fmincon, in turn, calls the two function files objfun.m and confun.m.

c) Figure 6.8 describes the values that are exchanged among the files main.m, objfun.m, and confun.m. The file main.m provides the design variable and weight values to objfun.m and confun.m. Note that although the weights are not used in confun.m, they still need to be passed. All problem parameters in fmincon (in this problem, w1 and w2) must be provided to both objfun.m and confun.m; otherwise, a MATLAB error will occur.

d) Based on the weights that are passed into objfun.m from main.m, an AOF is formed in objfun.m. Every time fmincon calls the objective function file, objfun.m returns the value of f, which is the AOF value for the corresponding set of weights. The file confun.m returns to main.m the values of the inequality and equality constraints [c, ceq] every time fmincon calls confun.m.

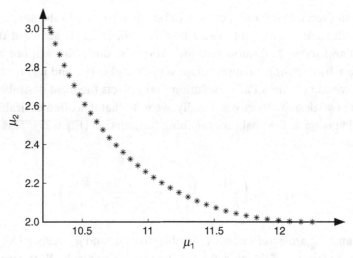

Figure 6.9. Pareto Frontier for the Example

e) The optimum values of the corresponding design variables obtained by calling
    `fmincon` are stored in the variable x. Next, the values of the corresponding
    objective functions are computed and stored in two arrays, `solx` and `soly`.
    After each iteration of the `for` loop, the counter j is incremented.

4. **Plotting:** After the execution of the `for` loop is completed, the objective values
   in the two arrays, `solx` and `soly`, are plotted to obtain the Pareto frontier.

The number of Pareto points to be generated can be set using the `for` loop
arguments. Different sets of weights used to form the AOF generally yield different
Pareto points. The above statement, however, might not be true in problems with
numerical issues.

The code in Fig. 6.8 generates the Pareto frontier shown in Fig. 6.9. Readers are
encouraged to use this example to reproduce the Pareto frontier generation code
and the results shown in Fig. 6.9.

6.7 Reaching a Target—Goal Programming

In some problems, instead of simply minimizing the design objectives, we may
wish to reach a given target value for each objective. For example, say the stress
design objective (μ_1) is required to be as close to 1,500 MPa as possible, while the
value of deflection (μ_2) is required to be as close to 2 inches as possible. Such a
design statement can be easily formulated as a compromise programming AOF, as
shown below.

$$J(x) = w_1 (\mu_1 - 1,500)^n + w_2 (\mu_2 - 2)^n \qquad (6.27)$$

where n is an even integer, often chosen to be equal to two. In the above equation, notice that the smallest possible value of J is zero, which is achieved only when $\mu_1 = 1500$ and $\mu_2 = 2$. In most real-life problems, this point will not be achievable. Instead, by choosing an appropriate set of weights (w_1 and w_2) for the design objectives, we can obtain a Pareto solution that reflects the most desirable tradeoff between the two design objectives. Finally, we note that it is often desirable to use a normalized version of the goal programming formulation (Eq. 6.27). This version is expressed as

$$ J(x) = w_1 \left(\frac{\mu_1 - \overline{\mu}_{1g}}{\overline{\mu}_{1b} - \overline{\mu}_{1g}} \right)^n + w_2 \left(\frac{\mu_2 - \overline{\mu}_{2g}}{\overline{\mu}_{2b} - \overline{\mu}_{2g}} \right)^n \tag{6.28} $$

where $\overline{\mu}_{1g}$ and $\overline{\mu}_{2g}$ are *good* values of the objectives μ_1 and μ_2, respectively; and $\overline{\mu}_{1b}$ and $\overline{\mu}_{2b}$ are *bad* values of the objectives μ_1 and μ_2, respectively. With respect to Eq. 6.27, we would have $\overline{\mu}_{1g} = 1,500$ and $\overline{\mu}_{2g} = 2$; and we might have $\overline{\mu}_{1b} = 8,000$ and $\overline{\mu}_{2b} = 10$. The formulation in Eq. 6.28 will have more stable numerical behavior than that in Eq. 6.27. Such numerical conditioning and scaling strategies will be discussed in more detail in Chapter 7.

We conclude by noting that, in the above methods, we have described formulations from the point of view of two objectives. While the bi-objective case is indeed more computationally tractable, we find that, at least conceptually, these formulations can be readily extended to cases of more than two objectives by simply adding appropriate terms in the AOF.

6.8 Expressing a Preference—Physical Programming

In the case of Goal Programming (GP), we try to reach a given target for each objective. However, in practice, we generally have a more complex set of preferences. The Physical Programming method provides the means for a more realistic expression of preference. Instead of saying that the goal is for the mass to be close to single fixed target, we can say that we wish to minimize the mass with different ranges of differing desirability. Specifically, we might say that (i) it is *Highly Desirable* for the mass to be less than 2 kg, (ii) *Desirable* between 2 and 3 kg, (iii) *Tolerable* between 3 and 5 kg, (iv) *Undesirable* between 5 and 7 kg, and (v) *Highly Undesirable* between 7 and 8 kg. Beyond 8 kg is *Unacceptable*. This expression of preference through physical programming is depicted in Fig. 6.10 for the mass objective. A similar expression of preference for other objectives – deflection, sig1, and sig2 – are shown for minimization, and sig3 for maximization. This added flexibility is a distinguishing feature of the Physical Programming (PP) method. A more detailed presentation of the PP method is provided in Chapter 18. Figure 6.10 is part PhysPro (Ref. [7]), which is a software implementation of the Physical Programming method. (Note that this figure is color coded in the actual code Graphical User Interface).

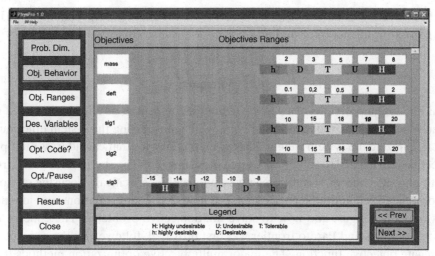

Figure 6.10. Expressing Preference Using Physical Programming

Brief contrast of the two methods:

Goal Programming Key Characteristics: (i) GP is somewhat easier to implement than PP. (ii) GP is much easier to code. (iii) GP can provide useful answers in many practical cases. (iv) GP can be overly dependent on the target/goal values chosen. (v) GP requires the assignment of weights for each goal, which can be extremely difficult in cases of more than two objectives.

Physical Programming Key Characteristics: (i) PP is more amenable to leading the designer to the most preferred solution.(ii) PP is shown not to be highly sensitive on the desirability values chosen. (iii) PP does not require the assignment of weights, which are notoriously difficult to correctly determine in practical cases of more than two objecives. (iv) PP has been shown to work well in diverse multiobjective practical cases. (v) PP brings optimization within the reach of users with minimal technical knowledge thereof.

6.9 Multiobjective Optimization Using MATLAB Optimization Toolbox

The MATLAB Optimization Toolbox provides two functions for two distinct formulations of multiobjective optimization problems: (i) goal attainment and (ii) minimax.

The goal attainment problem involves reducing the value of a linear or nonlinear vector function in order to attain the goal values given in a goal vector. A weight vector is used, which is intended to indicate the relative importance of the goals. The goal attainment problem may also be subject to linear and nonlinear constraints. The function used to solve the goal attainment problem is `fgoalattain`. Please use the command `help fgoalattain` for further information.

The minimax problem involves minimizing the worst case value of a set of multivariate functions, possibly subject to linear and nonlinear constraints. The function used to solve the minimax problem is `fminimax`. Please use the command `help` `fminimax` for further information.

6.10 Summary

Real-life design problems often involve multiple objectives or criteria that need to be achieved; such as, reducing system cost, while maximizing the system efficiency, and minimizing the system weight. Special methodologies that can address multiple objectives are needed to solve such design problems. This chapter provided an introduction to the concept of multiobjective optimization, specifically focusing on how to formulate such multiobjective problems, how to compare solutions and identify best tradeoff designs, and how to solve multiobjective problems. The concept of Pareto solutions and Pareto frontier was first provided, followed by the concept of aggregate objective function. The major multiobjective optimization methods that were presented in this chapter included traditional methods such as the weighted sum method, compromise programming, and goal programming, as well as the contemporary method of physical programming. The chapter ended with a brief overview of the provisions in MATLAB for solving multiobjective optimization problems.

6.11 Problems

Warm-up Problems

6.1   Consider the following bi-objective optimization problem. This is a standard single-variable bi-objective problem often used to test multiobjective optimizers.

$$\mu_1 = x^2 \tag{6.29}$$

$$\mu_2 = (x - 2)^2 \tag{6.30}$$

$$-5 \leq x \leq 5 \tag{6.31}$$

(a) Obtain several optimal points on the Pareto frontier using the weighted sum method. Use the MATLAB function `fmincon` for optimization. Plot each design objective as a function of x on the same figure (as shown in Fig. 6.2). Identify on this plot, the Pareto solutions that you just obtained. Turn in your M-files. (b) Plot the Pareto optimal points in the μ_1-μ_2 space. Turn in your M-files and the plot.

6.2   Consider the following bi-objective optimization problem.

$$\mu_1 = \sin \theta \tag{6.32}$$

$$\mu_2 = 1 - \sin^7 \theta \tag{6.33}$$

$$0.5326 \leq \theta \leq 1.2532 \tag{6.34}$$

(a) Obtain several optimal points on the Pareto frontier using the weighted sum method. Use the MATLAB function `fmincon` for optimization. Plot the points in the μ_1-μ_2 space. Turn in your M-files and the plot. (b) Do you think that the weighted sum method performs satisfactorily in obtaining points on the Pareto frontier? (c) Use the compromise programming approach with an appropriate value for the exponent to obtain the Pareto frontier. Turn in the plot and the M-file that you think yields the most satisfactory results.

6.3 Design a single-support water tower of maximum height, h, and maximum water storage capacity. For this exercise, the only mode of failure to protect against is support column buckling. The basic tower design is shown in Fig. 6.11.

Assumptions:

(1) The steel water tank is a spherical pressure vessel of thickness t.
(2) The support column has a circular cross section and is made of steel.
(3) The weight of the full tank acts vertically at Point B.

Design Variables:

(1) Column height: h
(2) Tank radius: r

Side constraints:

(1) Height: $10 \leq h \leq 16$ (meters)
(2) Radius: $2.13 \leq r \leq 14$ (meters)

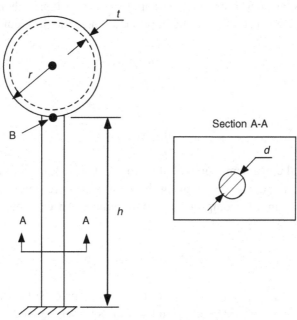

Figure 6.11. Problem 6.3—Water Tower

Numerical Constants:

(1) Modulus of steel, $E = 206$ GPa

(2) Diameter of support column, $d = 0.3$ m

(3) Thickness of tank, $t = 0.0127$ m

(4) Buckling safety factor, SF $= 2$

(5) Gravitational constant, $g = 9.8$ m/s^2

(6) Density of steel, $\rho_s = 7{,}800$ kg/m^3

(7) Density of water, $\rho_w = 1{,}000$ kg/m^3

The questions are:

(a) Find the tallest water tower design, h.

(b) Find the largest storage capacity design.

(c) Find four other Pareto solutions that are significantly different from any other design you have obtained. How do you compare your designs to decide the extent to which they are different?

(d) Plot all your designs from the previous parts in the μ_1-μ_2 space. Turn in your plot and your M-files.

Intermediate Problems

6.4   You are a new member of an optimization team at an automobile company. As your first task, you are asked to optimize a particular automotive component. You have been asked to simultaneously minimize the mass (μ_1) and the cost (μ_2) of the component. The other team members give you a hint, saying that if the mass and cost are modeled as optimization variables, then the following optimization constraints can be imposed to solve the problem.

$$\mu_1 = x_1 \tag{6.35}$$

$$\mu_2 = x_2 \tag{6.36}$$

$$5e^{-x_1} + 2e^{-0.5(x_1-2.1)^2} \le x_2 \tag{6.37}$$

$$0 \le x_1, x_2 \le 6 \tag{6.38}$$

To solve the problem, you will need to construct an appropriate Aggregate Objective Function (AOF) using the approaches presented in this chapter: weighted sum and compromise programming. An example of an AOF is shown below:

$$J = w_1 \mu_1^m + w_2 \mu_2^m \tag{6.39}$$

(a) Your task is to duplicate the optimal values provided in the table below. To do so, you will need to choose values for $w_1, w_2,$ and m that yield the given μ_1, μ_2 when J is minimized. For each case in the table, you will have

Table 6.1. *Problem 6.4—Results*

μ_1	μ_2	w_1	w_2	m
1.22	2.84			
1.99	2.67			
2.51	2.25			
5.04	0.059			

to solve a separate optimization problem. You can use weighted sum (m = 1) or compromise programming (m = 2, 4, 6,..., even numbers). Fill in the following table after you are confident that you have the correct values for the three parameters. Provide values of the weights correct to three decimal places. In each case, use the smallest value of m possible. Use the MATLAB function `fmincon` for the optimization. Provide your M-files.

(b) For the above problem, generate a number of optimal solutions on the Pareto frontier. Use the following AOF formulations: (1) Weighted sum, (2) Weighted square sum, (3) Compromise programming with $m = 4$, and (4) Compromise programming with $m = 10$. Use the MATLAB function `fmincon` for the optimization. Generate about 60-80 points on the Pareto frontier. Plot your results in the μ_1-μ_2 space. Label your plots. Provide your M-files.

(c) Comment on the performance of the weighted sum and compromise programming approaches. Which approach would you recommend for this problem?

6.5

$$\min_{x} \{\mu_1, \mu_2, \mu_3\} \tag{6.40}$$

subject to

$$(\mu_1 - 1)^2 + (\mu_2 - 1)^2 + (\mu_3 - 1)^2 \leq 1 \tag{6.41}$$

$$\mu_1 = x_1 \tag{6.42}$$

$$\mu_2 = x_2 \tag{6.43}$$

$$\mu_3 = x_3 \tag{6.44}$$

Using the weighted sum method, plot the Pareto frontier for the above problem. (Hint: You will need three weights, each corresponding to a design objective. Allow each of these weights to vary between 0 and 1 when executing your code.)

6.6 You have a good friend who is about to buy two items. For some strange reason, the costs of these two items depend on a variable x which varies from

Table 6.2. *Problem 6.6—Results*

Case	x	w_1	w_2	m	μ_1	μ_2
A						
B						
C						
D						
E						
F						

-4 to 4. The cost for the first and second items are μ_1 and μ_2, respectively. After much market research, you determine that the following relationships apply.

$$\mu_1 = e^x + R \tag{6.45}$$

$$\mu_2 = e^{-x} + R \tag{6.46}$$

where

$$R = 2\left(\frac{x^2 + 1}{x^4 + 1}\right) \tag{6.47}$$

Being a good friend yourself, you would like to save him money. You decide to minimize the function

$$J = w_1\mu_1^m + w_2\mu_2^m; m = 2, 4, 6, ...; w_1, w_2 \geq 0 \tag{6.48}$$

Your friend has various requests. For each request, you need to solve an optimization problem. Therefore, you need to find the correct scalar weights, w_1 and w_2, in addition to the appropriate power m. For each case, use the smallest m possible. (a) For Cases A through F below, your friend would like to have the price of Item 1 to be very near 1.5, 2, 2.9, 3.5, 4, and 5.1, respectively. For each case, fill in the missing information in following table. Turn in your M-files. (b) Explain the relationship between the AOF that you use and the Pareto frontier in geometrical terms (*e.g.*, convex or concave).

6.7   Consider the following multiobjective problem:

$$\min_x \{\mu_1, \mu_2\}$$

subject to

$$\mu_1 = x_1$$

$$\mu_2 = x_2$$

$$x_1^2 + \frac{x_2^2}{9} \geq 1$$

$$x_1^4 + x_2^4 \geq 16$$

$$\frac{x_1^3}{27} + x_2^3 \geq 1$$

$$0 \leq x_1 \leq 2.9$$

$$0 \leq x_2 \leq 2.9$$

1. Generate the Pareto frontier using the weighted sum method. Do you obtain a good representation of the Pareto frontier? Explain why or why not.
2. Use the compromise programming method to obtain the Pareto frontier. What value of the exponent gives an adequate representation of the Pareto frontier?
3. Discuss your ability to generate the Pareto frontier in terms of the exponents you used in the compromise programming formulation. Which exponents worked satisfactorily for the complete Pareto frontier generation?

6.8 Consider the following multiobjective problem:

$$\min_x \{\mu_1(x), \mu_2(x)\}$$

subject to

$$\mu_1(x) = 1 - \exp\left[-\sum_{i=1}^{3}\left(x_i - \frac{1}{\sqrt{2}}\right)^2\right]$$

$$\mu_2(x) = 1 - \exp\left[-\sum_{i=1}^{3}\left(x_i + \frac{1}{\sqrt{2}}\right)^2\right]$$

$$-4 \leq x_1, x_2, x_3 \leq 4$$

1. Plot the Pareto frontier for this problem using the weighted sum method. Do you obtain a good representation of the Pareto frontier?
2. Give one possible reason why the weighted sum method does not give all the Pareto points for this problem. Suggest another method that you think will achieve a good representation of the Pareto frontier.
3. Solve the above problem using the method you suggested in response to the previous question. Plot at least 50 Pareto points.
4. Explain why the method you suggested is more suitable than the weighted sum method for solving similar problems.

6.9 SPEED REDUCER: Figure 6.12 shows an illustration of the Golinski speed reducer [8]. The design of the speed reducer involves the design of a simple gear box. This mechanism can be used in a light airplane between the engine and the propeller to allow optimum rotating speed for each.

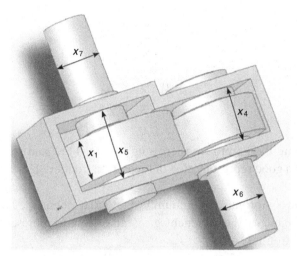

Figure 6.12. Golinski Speed Reducer

There are seven design variables in the problem: (1) gear face width, x_1 (cm), (2) teeth module, x_2 (cm), (3) number of teeth of the pinion, x_3 (we will treat this as a continuous variable), (4) distance between the bearing set 1, x_4 (cm), (5) distance between the bearing set 2, x_5 (cm), (6) diameter of shaft 1, x_6 (cm), and (7) diameter of shaft 2, x_7 (cm).

We are interested in minimizing two objectives: (1) the volume of the speed reducer and (2) the stress in one of the gears. The volume of the speed reducer, μ_1, which translates to its weight, is given as follows.

$$\mu_1(x) = 0.7854x_1x_2^2(10x_3^2/3 + 14.933x_3 - 43.0934)$$

$$- 1.508x_1(x_6^2 + x_7^2) + 7.477(x_6^3 + x_7^3) + 0.7854(x_4x_6^2 + x_5x_7^2) \quad (6.49)$$

The stress in one of the gears, μ_2, is given as follows.

$$\mu_2(x) = \frac{\sqrt{(745x_4/x_2x_3)^2 + 1.69 \times 10^7}}{0.1x_6^3} \quad (6.50)$$

Ten inequality constraints are imposed by considering gear and shaft design practices. An upper limit on the bending stress of the gear tooth is imposed (g_1). The contact stress of the gear tooth is constrained using g_2. Constraints g_3 and g_4 are upper limits on the transverse deflections of the shafts. Dimensional constraints due to space limitations are imposed using g_5 through g_7. The constraints g_8 and g_9 are design requirements on the shaft based on past experience. The stress in the gear shaft is constrained by g_{10}. The constraints are given as follows:

$$g_1 \equiv \frac{27}{x_1 x_2^2 x_3} \leq 1 \tag{6.51}$$

$$g_2 \equiv \frac{397.5}{x_1 x_2^2 x_3^2} \leq 1 \tag{6.52}$$

$$g_3 \equiv \frac{1.93 x_4^3}{x_2 x_3 x_6^4} \leq 1 \tag{6.53}$$

$$g_4 \equiv \frac{1.93 x_5^3}{x_2 x_3 x_7^4} \leq 1 \tag{6.54}$$

$$g_5 \equiv \frac{x_2 x_3}{40} \leq 1 \tag{6.55}$$

$$g_6 \equiv \frac{x_1}{12 x_2} \leq 1 \tag{6.56}$$

$$g_7 \equiv \frac{5 x_2}{x_1} \leq 1 \tag{6.57}$$

$$g_8 \equiv \frac{1.5 x_6 + 1.9}{x_4} \leq 1 \tag{6.58}$$

$$g_9 \equiv \frac{1.1 x_7 + 1.9}{x_5} \leq 1 \tag{6.59}$$

$$g_{10} \equiv \frac{\sqrt{(745 x_5/x_2 x_3)^2 + 1.575 \times 10^8}}{0.1 \times 1,100 x_7^3} \leq 1 \tag{6.60}$$

All the seven design variables have upper and lower bounds (b_1 through b_7), as given below.

$$b_1 \equiv 2.6 \leq x_1 \leq 3.6 \tag{6.61}$$
$$b_2 \equiv 0.7 \leq x_2 \leq 0.8 \tag{6.62}$$
$$b_3 \equiv 17 \leq x_3 \leq 28 \tag{6.63}$$
$$b_4 \equiv 7.3 \leq x_4 \leq 8.3 \tag{6.64}$$
$$b_5 \equiv 7.3 \leq x_5 \leq 8.3 \tag{6.65}$$
$$b_6 \equiv 2.9 \leq x_6 \leq 3.9 \tag{6.66}$$
$$b_7 \equiv 5.0 \leq x_7 \leq 5.5 \tag{6.67}$$

Based on the above information, answer the following questions.

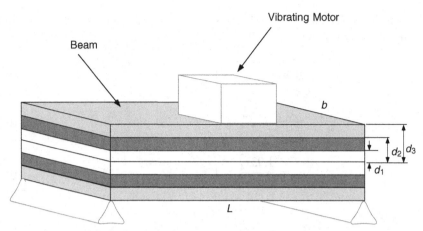

Figure 6.13. Sandwich Beam Designed with Vibrating Motor

(a) Find the design that yields the minimum weight for the speed reducer, subject to the constraints discussed above.

(b) Find the design that yields the minimum stress for the gear, subject to the constraints discussed above.

(c) Now, try to minimize the weight and the stress simultaneously. Use the weighted sum method to form the AOF. Minimize the AOF for a set of 10 evenly spaced weights between zero and one. Can a complete representation of the Pareto frontier be obtained?

(d) How can a better representation of the Pareto frontier be obtained? (Hint: Consider a larger set of evenly spaced weights between zero and one when compared to Part (c). Increase the number of points until you can obtain a complete representation of the Pareto frontier). Discuss the advantages/disadvantages of using the technique discussed above, especially in large scale problems.

6.10 You have been recently hired as an optimization expert for the company CoBeams. This company specializes in the design of sandwich beams. These sandwich beams are composites made up of different material layers. The engineering properties of the beam, due to its composite nature, are well suited for certain design applications. A particular order has been requested by one of the company's major customers. The beam under consideration is idealized as a pinned-pinned sandwich beam that supports a motor as shown in Fig. 6.13.

A vibratory disturbance (at v Hz) is imparted from the motor onto the beam. The beam is of length L and width b. The variables, d_1 and d_2, respectively, locate the contact of Materials one and two, and two and three. The variable, d_3, locates the top of the beam. The mass density, Young's modulus, and cost per unit volume for Materials one, two, and three, are respectively denoted by the triplets (ρ_1, E_1, c_1), (ρ_2, E_2, c_2), and (ρ_3, E_3, c_3).

The overall objective is to design the preceding sandwich beam in such a way as to passively minimize the vibration of the beam that results from the disturbance ($v = 10\text{Hz}$). Minimizing the vibration will require maximizing the fundamental frequency, f_0, of the beam. The optimal solution should be such that the fundamental frequency is maximized economically (*i.e.*, at minimum cost, c).

In the design of the plant, the quantities and objectives of interest are as follows:

Fundamental frequency maximization:

$$\max_x f_0 = (\pi/2L^2)(EI/\mu)^{\frac{1}{2}} \tag{6.68}$$

where

$$EI = (2b/3)[E_1 d_1^3 + E_2(d_2^3 - d_1^3) + E_3(d_3^3 - d_2^3)] \tag{6.69}$$

$$\mu = 2b[\rho_1 d_1 + \rho_2(d_2 - d_1) + \rho_3(d_3 - d_2)] \tag{6.70}$$

The minimization of the cost is given as

$$\min_x c = 2bL[c_1 d_1 + c_2(d_2 - d_1) + c_3(d_3 - d_2)] \tag{6.71}$$

The mass is given as

$$M = \mu L \tag{6.72}$$

The width of Layer 2 is given as

$$d_{21} = d_2 - d_1 \tag{6.73}$$

The width of Layer 3 is given as

$$d_{32} = d_3 - d_2 \tag{6.74}$$

The design parameters are

$$x = [d_1 \quad d_2 \quad d_3 \quad b \quad L] \tag{6.75}$$

1. Develop a function that allows you to enter the design variables, d_1, d_2, d_3, b, and L, for the corresponding values of f_0, c, M, d_{21}, and d_{32}. Save this function as vbeam.m.

Table 6.3. *Constants and Design Variable Constraints*

Constants	Constraints	
	Inequality	Equality
$\rho_1 = 100\text{kg/m}^3$	$0.01 \leq d_1 \leq 0.3\text{m}$	$M = 1{,}845\text{kg}$
$\rho_2 = 2{,}770\text{kg/m}^3$	$0.01 \leq d_2 \leq 0.35\text{m}$	$d_3 = 0.345\text{m}$
$\rho_3 = 7{,}780\text{kg/m}^3$		
$E_1 = 1.6 \times 10^9\text{Pa}$	$0.3 \leq b \leq 0.7\text{m}$	
$E_2 = 70 \times 10^9\text{Pa}$	$3 \leq L \leq 6\text{m}$	
$E_3 = 200 \times 10^9\text{Pa}$		
$c_1 = 500\ \$/\text{m}^3$	$d_{21} \geq 0.001$	
$c_2 = 1{,}500\ \$/\text{m}^3$	$d_{32} \geq 0.001$	
$c_3 = 800\ \$/\text{m}^3$		

2. The customer's specified conditions (*e.g.*, material properties and dimensions) are expressed in Table 6.3.

3. To ensure your function is working correctly, enter the following values for the design variables: $d_1 = 0.3$, $d_2 = 0.35$, $d_3 = 0.4$, $b = 0.4$, and $L = 5$. Your output should give $f_0 = 112.684{,}9$, $c = 1{,}060$, $M = 2{,}230$, $d_{21} = 0.05$, and $d_{32} = 0.05$

4. The customer has also given you a budget of \$500. Use the weighted sum method to maximize $\mu_1 = f_0$, minimize $\mu_2 = c$, and determine the weights to obtain an optimal design cost of $\approx$ \$500, if possible. Use the values of the design variables given in Part 2 above as initial values. (Hint: Form the aggregate objective function, $f(x) = w_1\mu_1 + w_2\mu_2$, where $w_2 = 1 - w_1$. Vary the value of w_1, ranging between 0.85 and 1, with incremental changes of 0.01. Determine which set of weights in this range gives the closest value of $c = 250$.)

5. Plot a Pareto frontier that contains 100 distinct solutions. (Note: Due to numerical issues, the generation of, say, 100 Pareto solutions might produce significantly less than 100 distinct solutions.)

6. From the plot generated in 4, form a table that provides the different values of the design objectives for the corresponding design variables.

7. Another customer interested in a similar beam design wants to know, as soon as possible, what fundamental frequency can be achieved with a design budget of \$600. He also wishes to know how the design variables will be affected. By referring to your table from 5, provide the requested information for the customer.

Advanced Problems

6.11 **Optimization of a Simple Two Bar Truss:** Minimize the square of the deflection at node P and minimize the total structural volume (see Fig. 6.14); subject to constraints that ensure the normal stresses and beam cross sectional areas are within acceptable levels. Plot the Pareto frontier for this problem.

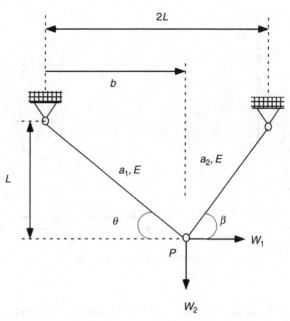

Figure 6.14. Schematic of Simple Two-Bar Truss

Design Parameters: a_1 = cross sectional area of bar 1 (m²); a_2 = cross sectional area of bar 2 (m²); and b = horizontal distance from the leftmost part of the truss to node P (m).

Constant Numerical Values: L = 18.288 m (60 ft); W_1 = 4.45 × 10⁵ N (100 kips); W_2 = 4.45 × 10⁶ N (1,000 kips); E = 1.99 × 10¹¹ Pa (29 × 10³ kpsi); S_{max} = 3.79 × 10⁹ Pa (550 kpsi) (Maximum allowable stress); $a_{i_{min}}$ = 5.16 × 10⁻⁴ m², i = 1, 2; $a_{i_{max}}$ = 1.94 × 10⁻³ m², i = 1, 2; b_{min} = 9.144 m (30 ft); and b_{max} = 27.432 m (90 ft).

Design Objectives: μ_1 = the square of the displacement of node P (m²); μ_2 = the total structural volume (m³).

Determining Nodal Displacements in Bars: In order to formulate the problem, you first need to calculate the stiffness matrix of the system. You have only one node, so you will have a 2 × 2 stiffness matrix. The relation between the nodal displacement (u) and the axial deflections in the bars (δ) is as follows:

$$\delta = Au, \text{ where } A = \begin{bmatrix} \cos\theta & \sin\theta \\ -\cos\beta & \sin\beta \end{bmatrix}$$

$$\delta = \begin{bmatrix} \delta_1 \\ \delta_2 \end{bmatrix}, u = \begin{bmatrix} u_1 \\ u_2 \end{bmatrix}$$

The force balance at the node yields:

$$A^T f = W, \text{ where } W = \begin{bmatrix} W_1 \\ W_2 \end{bmatrix}, f = \begin{bmatrix} f_1 \\ f_2 \end{bmatrix}$$

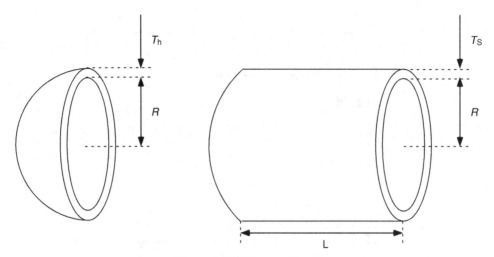

Figure 6.15. Pressure Vessel

where W is the applied load, and f is the axial force in each bar.

You also have, from elasticity equations: $f = C\delta$

where $C = \frac{E}{L} \begin{bmatrix} a_1 \sin\theta & 0 \\ 0 & a_2 \sin\beta \end{bmatrix}$

Substituting the expression for f into the force balance equation:

$A^T C\delta = W, (A^T CA)u = W$

Consequently: $u = (A^T CA)^{-1}W$. Use the above information to solve the problem.

6.12 Design a cylindrical pressure vessel that can store 750 ft^3 of compressed air at a pressure of 30 psi. The design variables are: (i) the radius, R; (ii) the length of the cylindrical shell, L; (iii) the shell thickness, T_s; and (iv) the head thickness T_h. The basic pressure vessel configuration is shown in Fig. 6.15 below.

The main cylindrical shell of the pressure vessel is fabricated in two halves of rolled steel plate. The two halves are joined together by a longitudinal weld. Manufacturing limitations restrict the length of the shell to be at a maximum of 20 ft. The pressure vessel has two end caps, which are hemispherical, forged, and welded to the shell. All of the welds are single-welded butt joints with a backing strip. The material is carbon steel.

The radius can range from 25 to 150 inches, and the length of the cylindrical shell from 25 to 240 inches. The thickness of the shell must be between 1 and $1\frac{3}{8}$ inches, and the thickness of the head must be between $\frac{5}{8}$ and 1 inch. According to the ASME Boiler and Pressure Vessel Codes, the thickness of the shell must be at least 0.0193 times the radius, and the thickness of the head must be at least 0.00954 times the radius.

1. Minimize the total cost of the pressure vessel, which is a combination of the welding, material, and forming costs. The total cost of the system is given as

$$C = 0.622,4T_s RL + 1.778,1T_h R^2 + 3.166,1T_s^2 L + 19.84T_s^2 R \quad (6.76)$$

Determine the optimal cost, as well as the optimal values, for the design variables.

2. What other design objectives do you think could be considered in this problem? Develop an equation for the hoop stress in the cylindrical section.

3. Minimize this hoop stress in the pressure vessel. Provide the minimum hoop stress value, as well as the optimal values, for the design variables.

4. You are now asked to minimize both objectives (cost and hoop stress) simultaneously. Plot the Pareto frontier for this bi-objective problem. Determine whether and when the solutions you obtained in No. 1 and No. 3 belong to the Pareto frontier of the bi-objective problem.

5. Modify your code for the bi-objective problem to ensure that the hoop stress does not exceed the maximum stress for carbon steel ($\sigma_{max} = 35,000$ psi). Do you expect the Pareto solutions to be different from those of No. 4 with the inclusion of this additional constraint? To be certain, repeat No. 4 with your modified code. Is the new Pareto frontier different from the one you obtained in No. 4? Explain why or why not.

6.13 The company Modern Energy specializes in new energy efficient systems for commercial and domestic buildings. You have been recently hired as an optimization expert to be a part of their research and development department. The head of your design team spawned an ambitious idea that uses windows to remove heat from a building. He calls this design the Active Window [9]. He claims that if an active cooling system is placed around the window, then it might be possible for that window to remove heat from the building in addition to removing the heat that was allowed to enter in the first place. He suggests using thermoelectric (TE) units aligned around the perimeter of the window as shown in Fig. 6.16 (see Refs. [2, 10, 11, 12, 11]).

TE units are solid state devices that use thermocouples to convert electrical energy into thermal energy or vice versa. For a given input current, the TE unit absorbs heat at its cold side and dissipates it at its hot side. The amount of heat that can be absorbed by the TE unit, Q_{te}, for a input current, I_{te}, is given as

$$Q_{te} = 2N \left[\alpha I_{te} T_c - \frac{I_{te}^2 \rho}{2G} - \kappa \Delta T G \right] \quad (6.77)$$

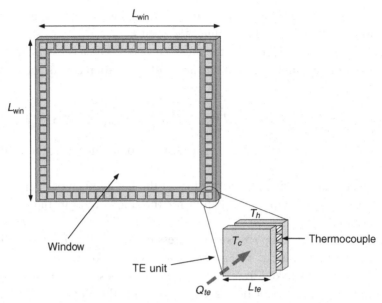

Figure 6.16. Proposed Schematic of Active Window

and the supplied voltage, V_{te}, as

$$V_{te} = 2N \left[\frac{I_{te}\rho}{G} + \alpha \Delta T \right] \tag{6.78}$$

where N is the number of thermocouples, α is the Seebeck coefficient, T_c is the cold side temperature of the TE unit, ρ is the resistivity, ΔT is the temperature difference between the hot and cold side of the TE unit, κ is the thermal conductivity, and G is the geometry factor, which is the area to thickness ratio of the TE unit.

For practical and innovative purposes, it is necessary that this design provides an efficiency that is equivalent to or greater than existing cooling systems. Cooling systems are usually rated by their coefficient of performance (COP). The COP is defined as the amount of heat that can be absorbed by a cooling device per unit of input power. Good cooling systems have a COP of approximately 3.

The head of your design team has assigned you the task of doing an initial study on the Active Window. He is interested in finding out the maximum amount of heat the Active Window can remove from a building under energy efficient conditions. The Active window will be considered feasible only if it can, at the very least, remove the heat it allowed to enter the building for a COP $\geq$ 3. Experts on window heat transfer are available to supply you with information regarding the heat allowed by a particular window.

The head of your design team suggests carrying out your study based on three TE units for which pertinent information is provided in Table 6.4. Information in Table 6.4 corresponds to $T_c = 297$ K.

Table 6.4. *Geometrical Properties for the Different Thermoelectric Units*

TE No.	N	G	L_{te}	I_{max}	V_{max}	T_h	α	ρ	κ
1	125	0.00184	0.0244	8.75	14.1	310.4	$2.02e-4$	$1.01e-5$	1.51
2	125	0.00282	0.0244	12.18	13.4	317.3	$2.02e-4$	$1.01e-5$	1.51
3	31	0.00473	0.01565	24.13	3.6	305.4	$2.02e-4$	$1.01e-5$	1.51

In Table 6.4, T_h (K) is the hot side temperature of the TE unit. Before you begin your study, he would like you to note the following:

1. Despite the fact that the TE units are aligned physically in series around the perimeter of the window, they can be connected electrically in parallel and series. N_{tep} and N_{tes} correspond to the number of TE units connected in parallel and series, respectively.
2. The total length occupied by the TE units is the product of the length of one TE unit, L_{te} (m), and the total number of TE units, $N_{tep}N_{tes}$.
3. The total length occupied by the TE units should not exceed the perimeter of the window.
4. I_{te} and V_{te} should not exceed the TE maximum current, I_{max}, and TE maximum voltage, V_{max}, respectively. These values are specified by the manufacturer, and are given in Table 6.4
5. The total heat absorbed by the Active Window, Q (W), is given as

$$Q = N_{tep}N_{tes}Q_{te} \tag{6.79}$$

$$Q \geq 1 \tag{6.80}$$

6. The Power input to the Active Window, P (W), is given as

$$P = N_{tep}N_{tes}I_{te}V_{te} \tag{6.81}$$

$$P \geq 1 \tag{6.82}$$

7. The COP for the Active Window should be above 3 *when possible*, or otherwise as close to 3 as possible. Specifically, we have

$$COP = \frac{Q}{P} \tag{6.83}$$

Use I_{te}, N_{tep}, and N_{tes} as design variables with the following bounds:

$$I_{te} \geq 0 \tag{6.84}$$

$$N_{tep}, N_{tes} \geq 1 \tag{6.85}$$

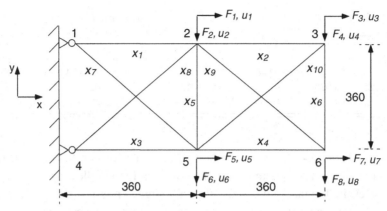

Figure 6.17. Ten-Bar Truss

The head of your design team wants a report that contains all the necessary information. You can complete your report by doing the following:

1. Maximize Q for a window with $L_{win} = 1$ m, satisfying all the conditions previously stated for each TE unit. Can all the TE units satisfy the conditions imposed? Develop a table that provides the optimum values for the design variables, V_{te}, Q, P, and the COP for the TE units that satisfy the constraints.

2. If the experts claim that a window with $L_{win} = 1$ m allows 581 W to enter the building, what conclusions can you draw based on the feasibility of the Active Window with respect to the TE units?

3. Remove the constraint for COP and maximize Q for each TE unit. Develop a table similar to that in No. 1 above.

4. With the constraint for COP removed, plot a Pareto frontier that maximizes Q and minimizes P for each TE unit. Identify the point on each plot that corresponds to the maximum COP achievable by that TE unit.

5. Compare the value of Q for TE No. 3 corresponding to the maximum COP with that from No. 1. What observation can be made?

Graduate Level Problems

6.14   (Note: This problem could also be used as an *Advanced* problem). Figure 6.17 provides the schematic representation of a ten-bar truss. The ten members are connected to each other at six nodes, represented by the numbers 1 through 6. You can assume that the truss lies and deflects in one plane, and that every node has only two degrees of freedom (the displacements along the horizontal and vertical axes). The left-hand-side of the truss is fixed to the wall, yielding zero displacements for Nodes 1 and 4. The displacements along the horizontal and vertical axes for the remaining nodes are represented by u_1 through u_8. F_1 through F_8 represent loads applied to these nodes, respectively. The cross sectional areas of the truss members are given by x_1 through x_{10}. The stresses induced in the ten bars are given by

$$\{\sigma\} = [S]\{u\} \qquad (6.86)$$

where, $\{\sigma\} = \{\sigma_1, \sigma_2, ...\sigma_{10}\}^T$ is the stress vector, and $\{u\} = \{u_1, u_2, ...u_8\}^T$ is the displacement vector. Matrix $[S]$ is given by Eq. 6.96 at end of this problem. The displacements vector u is evaluated using the finite element formulation

$$\{F\} = [K]\{u\} \qquad (6.87)$$

where, the force vector is $\{F\} = \{F_1, F_2, ...F_8\}^T$. The stiffness matrix $[K]$ is given by Eq. 6.14 at the end of this problem. The maximum allowable stresses are $(\sigma_{max}^t, \sigma_{max}^c) = (150\,MPa, -150\,MPa)$ (tension/compression), and the maximum allowable deflections are $(u_{max}^t, u_{max}^c) = (15\,mm, -15\,mm)$ (tension/compression). The cross sectional areas should be between 5 mm^2 and 200 mm^2.

1. Minimize the total material volume of the truss when $F_i = 2KN$ (i=1 to 8) and $E = 2 \times 10^5\,N/mm^2$. The optimization problem statement is given by

$$\min_{\{x\}} \quad V \qquad (6.88)$$

   subject to

$$\{\sigma_{max}^c\} \le \{\sigma\} \le \{\sigma_{max}^t\} \qquad (6.89)$$

$$\{u_{max}^c\} \le \{u\} \le \{u_{max}^t\} \qquad (6.90)$$

$$\{x_{min}\} \le \{x\} \le \{x_{max}\} \qquad (6.91)$$

2. Minimize the maximum deflection of the truss. The optimization problem statement is given as follows.

$$\min_{\{x\}} \quad max(\{u\}) \qquad (6.92)$$

   subject to

$$\{\sigma_{max}^c\} \le \{\sigma\} \le \{\sigma_{max}^t\} \qquad (6.93)$$

$$\{u_{max}^c\} \le \{u\} \le \{u_{max}^t\} \qquad (6.94)$$

$$\{x_{min}\} \le \{x\} \le \{x_{max}\} \qquad (6.95)$$

3. Generate a two-objective Pareto frontier for this problem with at least 100 design points. Plot it, discuss the tradeoffs that it presents to you, and do so in quantitative terms.

Matrices: For the above problem, the [S] matrix is given as

$$[S] = \frac{E}{360} \begin{bmatrix} 1 & 0 & 0 & 0 & 0 & 0 & 0 & 0 \\ -1 & 0 & 1 & 0 & 0 & 0 & 0 & 0 \\ 0 & 0 & 0 & 0 & 0 & 0 & 1 & 0 \\ 0 & 0 & 0 & 0 & 1 & 0 & -1 & 0 \\ 0 & -1 & 0 & 0 & 0 & 0 & 0 & 1 \\ 0 & 0 & 0 & -1 & 0 & 1 & 0 & 0 \\ 0 & 0 & 0 & 0 & 0 & 0 & 0.5 & 0.5 \\ 0.5 & -0.5 & 0 & 0 & 0 & 0 & 0 & 0 \\ -0.5 & -0.5 & 0 & 0 & 0.5 & 0.5 & 0 & 0 \\ 0 & 0 & 0.5 & -0.5 & 0 & 0 & -0.5 & 0 \end{bmatrix} \qquad (6.96)$$

For the above problem, the [K] matrix is given as follows (with $z = 2\sqrt{2}$):

$$[K] = \frac{E}{360}[K_a | K_b] \qquad (6.97)$$

where

$$[K_a] = \begin{bmatrix} x_1 + x_2 + \frac{x_8 + x_9}{z} & \frac{-x_8 + x_9}{z} & -x_2 & 0 \\ \frac{-x_8 + x_9}{z} & x_5 + \frac{x_8 + x_9}{z} & 0 & 0 \\ -x_2 & 0 & x_2 + \frac{x_{10}}{z} & -\frac{x_{10}}{z} \\ 0 & 0 & -\frac{x_{10}}{z} & x_6 + \frac{x_{10}}{z} \\ -\frac{x_9}{z} & -\frac{x_9}{z} & 0 & 0 \\ -\frac{x_9}{z} & -\frac{x_9}{z} & 0 & -x_6 \\ 0 & 0 & -\frac{x_{10}}{z} & \frac{x_{10}}{z} \\ 0 & -x_5 & \frac{x_{10}}{z} & -\frac{x_{10}}{z} \end{bmatrix} \qquad (6.98)$$

$$[K_b] = \begin{bmatrix} -\frac{x_9}{z} & -\frac{x_9}{z} & 0 & 0 \\ -\frac{x_9}{z} & -\frac{x_9}{z} & 0 & -x_5 \\ 0 & 0 & -\frac{x_{10}}{z} & \frac{x_{10}}{z} \\ 0 & -x_6 & \frac{x_{10}}{z} & -\frac{x_{10}}{z} \\ x_4 + \frac{x_9}{z} & \frac{x_9}{z} & -x_4 & 0 \\ \frac{x_9}{z} & x_6 + \frac{x_9}{z} & 0 & 0 \\ -x_4 & 0 & x_3 + x_4 + \frac{x_7 + x_{10}}{z} & \frac{x_7 - x_{10}}{z} \\ 0 & 0 & \frac{x_7 - x_{10}}{z} & x_5 + \frac{x_7 + x_{10}}{z} \end{bmatrix} \qquad (6.99)$$

BIBLIOGRAPHY OF CHAPTER 6

[1] C. A. Coello Coello, G. B. Lamont, and D. A. Van Veldhuizen. *Evolutionary Algorithms for Solving Multi-objective Problems*. Springer, 2nd edition, 2007.

[2] K. Miettinen. *Nonlinear Multiobjective Optimization*. Springer, 1998.

[3] Y. Collette and P. Siarry. *Multiobjective Optimization: Principles and Case Studies*. Springer, 2011.

[4] G. P. Rangaiah and A. Bonilla-Petriciolet. *Multi-objective Optimization in Chemical Engineering: Developments and Applications*. John Wiley and Sons, 2013.

[5] A. Messac. From the dubious construction of objective functions to the application of Physical Programming. *AIAA Journal*, 38(1):155–163, January 2000.

[6] C. A. Mattson, A. A. Mullur, and A. Messac. Smart Pareto filter: Obtaining a minimal representation of multiobjective design space. *Engineering Optimization*, 36(6):721–740, 2004.

[7] K. E. Lewis, W. Chen, and L. C. Schmidt. *Decision Making in Engineering Design*. ASME Press, 2006.

[8] J. Golinski. Optimal synthesis problems solved by means of nonlinear programming and random methods. *Journal of Mechanisms*, 5(3):287–309, 1970.

[9] T. Harren-Lewis, S. Rangavajhala, A. Messac, and Junqiang Zhang. Optimization-based feasibility study of an active thermal insulator. *Building and Environment*, 53:7–15, 2012.

[10] R. Khire. *Selection-Integrated Optimization (SIO) Methodology for Adaptive Systems and Product Family Optimization*. PhD thesis, Rensselaer Polytechnic Institute, Troy, New York, September 2006.

[11] A. Messac, R. S. Birthright, T. Harren-Lewis, and S. Rangavajhala. Optimizing thermoelectric cascades to increase the efficiency of thermoelectric windows. In *4th AIAA Multidisciplinary design optimization specialist conference. Schaumburg, IL*, 2008.

[12] T. Harren-Lewis, S. Rangavajhala, A. Messac, and Junqiang Zhang. Optimization-based feasibility study of an active thermal insulator. *Building and Environment*, 53:7–15, 2012.

7

Numerical Essentials

7.1 Overview

In this chapter, you will be exposed to perhaps some of the most important issues in optimization. Interestingly, most optimization books and courses leave it up to the students to learn many of these issues simply by chance, leaving a broad set of the important practical topics unaddressed. We are referring here to *numerical conditioning issues* which strongly control the success of computational optimization.

We will not engage in highly mathematical and theoretical issues of numerical optimization. Instead, we will learn about the tangible and practical issues that directly affect our success in formulating and solving optimization problems. Fortunately, for most of this chapter, all we need is the knowledge of first-year college mathematics.

Keeping these issues in mind will often make the difference between success and failure. That is: (i) easily applying optimization *successfully*, or (ii) experiencing great frustration in trying to obtain an adequate solution *unsuccessfully*. In addition, we will learn how to control the resulting accuracy of our results, a critical issue in practice. The topics explored here include: numerical conditioning, scaling, finite differences, automatic differentiation, termination criteria, and sensitivities of optimal solutions, as well as examples that illustrate how to handle these issues in practice (Refs. [1, 2, 3].

7.2 Numerical Conditioning—Algorithms, Matrices and Optimization Problems

We begin by asking the following questions: What is *numerical conditioning*? And *how does it relate to optimization*? As we have learned, optimization depends on the numerical evaluation of the performance of the system being optimized. This performance evaluation typically involves coding, simulation, or software-based mathematical analysis, often also involving matrix manipulations (*e.g.*, involving Hessians). For example, linear programming involves extensive matrix manipulations. Therefore, understanding the numerical properties of the matrices and of the algorithms used in optimization codes is important. From a practical point of view, we can think of a numerically *well-conditioned* problem or matrix as one that lends itself to easy numerical computation. Conversely, we can think of a numerically *ill-conditioned* problem or matrix as one that lends itself to difficult numerical computation. A well-conditioned problem is also said to be *well-posed*. Numerical conditioning can also refer to the property of a particular *algorithm*, as you will see shortly. A well-conditioned algorithm is likely to converge relatively easily, while an ill-conditioned algorithm may converge after many more iterations or may not converge at all. In other words, numerical conditioning may refer to the facility with which an algorithm converges. Numerical conditioning may also have practical consequences on the quality or accuracy of the resulting solution.

The immediate questions that come to mind at this point are: How do we know if an algorithm, a matrix, or an optimization problem is well-conditioned or not? How do we quantify numerical conditioning? Well, there is some good news:

(i) *Algorithms*: Since this book is primarily concerned with the practical aspects of formulating and applying optimization, we will not be learning about how to make algorithms well-conditioned. This is an advanced topic of numerical computation. Fortunately, if we use reputed optimization codes, the algorithmic numerical conditioning properties are usually adequate.

(ii) *Matrices*: For our purpose, the numerical conditioning of a matrix is quantified by the *condition number*. For symmetric matrices (*e.g.*, Hessians (Eq. 2.42)), the condition number is the square of the ratio of the highest to the lowest eigenvalues. The case of non-symmetric matrices involves *singular values*, which is an advanced topic that does not directly concern us. A condition number with an order-of-magnitude of one is desirable, while a much higher condition number is a concern. The numerical properties of matrices play an important role in optimization algorithms. Sometimes an optimization run that does not converge will report that the "Hessian is ill-conditioned" (see Sec. 2.5.4 for the definition the Hessian). The strong relevance of a function's Hessian becomes fully evident in our study of the more advanced aspects of optimization presented in Part IV of this book; specifically, Chapters 12 and 13.

(iii) *Optimization Problems*: Posing our problems well is critically important. Two theoretically equivalent problems can have radically different numerical properties, as described later. Fortunately, we can generally deal with problem conditioning through proper *scaling*, which we will study in Sec. 7.3.

7.2.1 Reasons Why the Optimization Process Sometimes Fails

There are many reasons why optimization runs sometimes fail to converge to an adequate solution. The following are the prevailing ones that you should keep in mind:

1. The problem has a coding bug – a software/programming error;
2. The problem is ill conditioned – poorly scaled;
3. The problem is incorrectly formulated (*e.g.*, missing a constraint);

4. The problem posed does not reflect a physical design that is realistic (you can't fool mother nature!);
5. The algorithm used is not appropriate, or not sufficiently robust, for the problem at hand.

In the event of non-convergence, or indicate convergence to a solution that is not to one's liking, the above items should be explored – roughly in the order presented. Next, we briefly comment on each of the above items. **(1)** Regarding coding errors, one simply needs to employ the debugging strategy of personal choice. This issue concerns computer coding in general, and is not exclusive to optimization. **(2)** The problem of scaling is addressed in detail later in this chapter. **(3)** As far as the formulation of the optimization problem is concerned, the material in this book is of direct help, in particular, the previous chapter on multiobjective optimization. Proper formulation is also an issue of common sense. Failing to include a constraint, for example, could yield a design that is not desirable. **(4)** The problem of seeking an unrealistic design is one that should be carefully examined. Often, relaxing the constraints will allow the search process to explore physically feasible designs. Another possible cause for unknowingly seeking an unrealistic design is the inappropriateness of the objective function (*e.g.*, wrong weights in the weighted sum approach). As a final example, we could be trying to design a small table to support an elephant in a way that is impossible. All the modeling equations and optimization formulations issues might be seemingly fine, but we are simply asking for the impossible. **(5)** The final item presented concerns the appropriateness of the algorithm. For example: (i) the algorithm might not be sufficiently robust for problems of poor numerical conditioning; (ii) the algorithm might not be appropriate for problems of large dimensions; (iii) the algorithm might be limited to solving specific types of problems: continuous, discrete, or integer variables. (iv) the algorithm might not work well for *noisy* (*i.e.*, non-smooth) objective functions or constraints.

Before we present the approaches to address numerical conditioning issues in optimization, it is important that we first learn about certain numerical problems that can occur independently of the optimization process. Specifically, we find that the matrices themselves can be problematic, and so can the way that they are used in a given algorithm. These two issues are addressed next.

7.2.2 Exposing Numerical Conditioning Issues—Algorithms and Matrices

Through simple examples, we illustrate how we must concern ourselves with matrices and algorithm issues, in addition to those directly related to optimization. We provide an example of how numerical conditioning issues can affect us in a seemingly simple case. This telling example will sensitize us to the critical nature of numerical issues. For the sake of simplicity of presentation, we only use a 3×3 matrix. We also use a numerical computation that is simple and readily understood. In practice, matrices are much larger and computations are much more complex. In spite of the simplicity of the present case, the numerical difficulties presented are quite serious.

Consider Matrix A, which depends on α, given by

$$A = \begin{bmatrix} 1/\alpha & 0 & 0 \\ 0 & (\alpha+1)/2 & (\alpha-1)/2 \\ 0 & (\alpha-1)/2 & (\alpha+1)/2 \end{bmatrix} \tag{7.1}$$

The three eigenvalues of Matrix A can be evaluated as: $1/\alpha$, 1, and α. As a result, the condition number of A (the square of the ratio of the highest to the lowest eigenvalue since A is symmetric) is $1/\alpha^4$ when $\alpha \le 1$ and α^4 when $\alpha \ge 1$. Therefore, we should expect that for a very low or a very high value of α, we may experience numerical difficulties, particularly when the algorithm within which it is being used is not well conditioned.

To explore the numerical properties of Matrix A, let n be any positive integer, and consider the expressions

$$A_1 = (A \cdot A^{-1})^n - I = [0] \tag{7.2}$$

$$A_2 = A^n \cdot (A^{-1})^n - I = [0] \tag{7.3}$$

where I is the identity matrix. Using elementary linear algebra, we can indeed verify that both A_1 and A_2 are identically equal to 3×3 zero matrices. We can further write the scalar equations

$$\| A_1 \| = \| (A \cdot A^{-1})^n - I \| = 0 \tag{7.4}$$

$$\| A_2 \| = \| A^n \cdot (A^{-1})^n - I \| = 0 \tag{7.5}$$

where we use the maximum norm defined as $\| M \| = \max\{| m_{ij} |\}$, with m_{ij} denoting the ij-th entry of Matrix M.

We make the important observation that the zero answers in Eqs. 7.4 and 7.5 are exact only from a theoretical standpoint. When we compute $\| A_1 \|$ and $\| A_2 \|$ using a computer, we immediately observe that the numerical results depart markedly from the theoretical answers. To illustrate this important point, we present Table 7.1, where the incorrect answers are in bold face. In this table, we vary the parameter α in Matrix A, as well as the power n in the A_1 and A_2 expressions.

We further make three specific observations: **(1)** Even for high values of the condition number C_n, the quantity $\| A_1 \|$ is evaluated accurately. In fact, $\| A_1 \|$ is evaluated accurately for all the cases presented in Table 7.1. **(2)** Even for values of the condition number that are less than 100 ($\alpha = 0.4$), the quantity $\| A_2 \|$ is unacceptably inaccurate, for $n = 50$. **(3)** Even though $\| A_1 \|$ and $\| A_2 \|$ are theoretically both equal to zero, the computation of $\| A_1 \|$ is more numerically robust than that of $\| A_2 \|$. Finally, **(4)** in general, the stability of the algorithm and the condition numbers of the matrices involved can greatly impact the accuracy of the solutions obtained. For example, *in optimization*, how we pose a constraint (numerically) can impact the success of the optimization. Next, we expose the critical need for scaling in the following example.

Table 7.1. *Ill-Conditioned Matrices and Algorithms* $(\| M \| = \max_{ij}\{| m_{ij} |\})$

α	C_n	$\|A_1\| = \|(A \cdot A^{-1})^n - I\|$			$\|A_2\| = \|A^n \cdot (A^{-1})^n - I\|$		
		n			n		
		10	20	50	10	20	50
0	∞	—	—	—	—	—	—
10^{-6}	10^{24}	$3e^{-10}$	$6e^{-10}$	$1e^{-9}$	**1**	**1**	**1**
10^{-3}	10^{12}	0	0	0	**1**	**1**	**1**
0.4	39	$1e^{-15}$	$2e^{-15}$	$6e^{-15}$	$5e^{-13}$	$4e^{-9}$	**1**
0.8	2.4	0	0	0	$1e^{-15}$	$2e^{-15}$	$4e^{-12}$
0.95	1.2	$3e^{-17}$	$7e^{-17}$	$2e^{-16}$	$7e^{-16}$	$9e^{-16}$	$3e^{-15}$
1	1	0	0	0	0	0	0
2	16	0	0	0	0	0	0
10	10^4	$4e^{-15}$	$9e^{-15}$	$2e^{-14}$	$1e^{-15}$	**1**	**1**
10^2	10^8	0	0	0	**1**	**1**	**1**
10^5	10^{20}	$4e^{-11}$	$7e^{-11}$	$2e^{-10}$	**$5e^{33}$**	**$10e^{83}$**	**$4e^{233}$**

7.2.3 Exposing Numerical Conditioning Issues—Optimization Problems

We provide a simple example of an optimization problem for which proper scaling is essential. We begin by stating the general optimization problem formulation as follows.

PROB-7.2-GOPF: General Optimization Problem Formulation

$$\min_{x} \; f(x) \tag{7.6}$$

subject to

$$g(x) \leq 0 \tag{7.7}$$

$$h(x) = 0 \tag{7.8}$$

$$x_l \leq x \leq x_u \tag{7.9}$$

Next, we consider the following seemingly trivial optimization problem given by

$$\min_{x_1, x_2} \; f(x_1, x_2) = (x_1 - 3.67 \times 10^{-6})^2 + (x_2 - 3.67 \times 10^{-7})^2 \tag{7.10}$$

subject to

$$x_1 - 2x_2 = 0 \tag{7.11}$$

$$0 \leq x_i \leq 10^{-4} \quad (i = 1, 2) \tag{7.12}$$

Using the techniques later presented in this chapter on scaling, we find that the solution to this problem is $x = 10^{-6} \times \{3.083, 1.541\}$. The important message here is that, without proper scaling, the solution produced by MATLAB could be deemed incorrect. Specifically, MATLAB converged to the solution $x = 10^{-6} \times \{4.948, 2.474\}$, which has strongly inaccurate values of x. (We note that different MATLAB settings may lead to different equally erroneous answers). The danger here is that there is no indication that we are dealing with an incorrect solution. This simple example points to the importance of using various strategies to increase our confidence in the solutions obtained by optimization codes. Indeed, one of the more important ways to increase confidence is to implement proper scaling as mentioned earlier in this section, and presented in the next section.

7.3 Scaling and Tolerances for Design Variables, Constraints and Objective Functions

This section deals directly and explicitly with the all important subject of scaling in optimization. We all understand that it is important to learn how to formulate an optimization problem in particular cases, which may range from finance to engineering; and to learn useful theoretical aspects of optimization. However, unless we also pay special attention to the numerical issues of scaling, we may have serious trouble in practice. Fortunately, the information that is presented in this section provides the essence of what we need to know in practice.

Specifically, in practice, we need to know about scaling for (i) design variables, (ii) objective functions, and (iii) constraints. Depending on the problem, one, two, or all three of the above items may be critical. What do we mean by "critical"? We mean that with scaling, we may obtain faster convergence to the correct minimum. Without scaling, we may experience non-convergence or premature termination at a poor design, and not even realize that this is the case.

For convenience, we discuss scaling in the context of the MATLAB optimization routine fmincon, for which the optimization problem is posed as follows.

PROB-7.2-MATLAB: fmincon Optimization Model

$$\min_{x} \quad f(x) \tag{7.13}$$

subject to

$$c(x) \leq 0 \tag{7.14}$$

$$ceq(x) = 0 \tag{7.15}$$

$$A x \leq b \tag{7.16}$$

$$Aeq\ x = beq \tag{7.17}$$

$$LB \leq x \leq UB \tag{7.18}$$

where (i) A and Aeq are *matrices* that define the left-hand-sides of the linear constraints, (ii) b and beq are *vectors* that define the right-hand-sides of the linear constraints, (iii) c(x) and ceq(x) are vectors of nonlinear constraints, and (iv) LB and UB denote bounds on the design variables. Equation 7.18 defines what is called

Main file with scaling

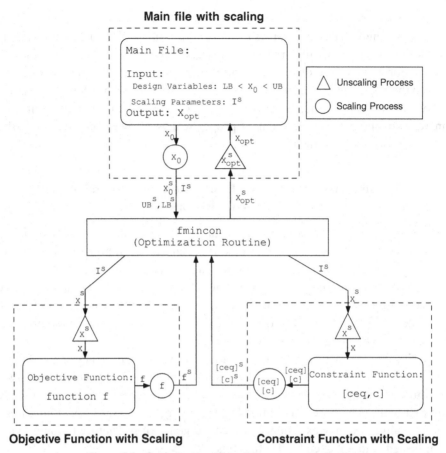

Figure 7.1. Optimization Flow Diagram—with Scaling

the *side constraints*. It is useful to note the similarities between **PROB-7.2-GOPF** and **PROB-7.2-MATLAB**. The latter provides the flexibility of differentiating between linear and nonlinear constraints. Doing so can provide significant numerical efficiencies. That is, it is numerically efficient to put the linear constraints in Eqs. 7.16 and 7.17, even though linear constraints can also be put in Eqs. 7.14 and 7.15; however, the reverse is not true. This MATLAB notation is the one used in Figs. 5.4 and 7.1. The former has no scaling, while the latter does.

Figure 7.1 illustrates the optimization process using fmincon (or any code that conforms with the **PROB-7.2-MATLAB** problem definition). The general scaling process is discussed with the aid of Fig. 7.1. Scaling is discussed in terms of its three generic components: (i) design variables, (ii) objective functions, and (iii) constraints. Upon examining Fig. 7.1, we make the following important observations: (i) The variables that immediately *enter* and *leave* the optimization routine (fmincon) are scaled variables, and (ii) these variables are immediately unscaled if/when they are used in the main file, the objective function, or the constraint function modules. Stated differently, the variables are scaled as they enter fmincon, and they are unscaled as they leave fmincon for use anywhere else. The variables are scaled as they enter `fmincon` because the optimization routine needs to use variables in the desirable numerical ranges to operate properly. However, the variables must regain their unscaled values when they are used in the actual model models. Figure 7.1 depicts both the unscaling processes (triangle) and the scaling processes (circle). *Note that in this chapter, ( )^s denotes the scaled version of the variable ( ), except in the cases of* α^s, β^s, and γ^s, which are prescribed scaling constants.

In Fig. 7.1, we describe *where* scaling takes place for the design variables. We note that, in a similar fashion, scaling also takes place for the quantities `c(x)`, `ceq(x)` and `f(x)`, as shown in Fig. 7.1. The quantity I^s in Fig. 7.1 represents the set of scaling parameters that are used for the scaling and unscaling of various quantities. This set includes α^s, β^s, and γ^s, which are all defined in this section. Next, we discuss the important issue of *numerical accuracy representation*, which will be followed by the development of *scaling approaches*.

7.3.1 Understanding the Accuracy of the Reported Results

An important aspect of our work is to understand (i) the accuracy of the results we obtain, and (ii) how to report these results. These issues entail two related components. The first involves understanding *and controlling* the accuracy of the results produced by the optimization code using scaling. The second involves understanding the inherent accuracy of the physics-based models used in the optimization. We discuss each next.

Accuracy of Results of Optimization Code:

As we will learn in this section, we have the ability to control the accuracy of the results that the optimization code produces. We can do so with *scaling*, in conjunction with prescribing certain *parameter settings* in the optimization code. As we examine the numbers that the optimization code produces, we keep in mind that the computer has in its memory many more digits than we report for that number. It is up to us to decide/understand how many digits of the number are *significant* (*i.e.*, meaningful).

As we *report the numbers*, we may choose to either put zero digit in the place of the meaningless digits or use exponent notation. For example, we may use 12,345,000, or 1.2345×10^7, for five significant digits (see Table 7.2), even though the first 8 digits in the computer memory might be 12,345,678. Note that as we decided to report five significant digits of the computer results, we did so simply because we determined that the *optimization* converged to within five significant digits. This convergence does not tell us that these five significant digits are *physically* meaningful. Next, we discuss this physical aspect in greater detail.

Accuracy Results with Physics-Based Model:

In addition to the accuracy of the optimization code results discussed above, we must also concern ourselves with the accuracy of the physics-based models being used. For example, if we obtain the optimal value of the deflection of a beam as $1.234,56 \times 10^{-6}$, perhaps only 3 or 4 of these digits might be physically meaningful, even though all 6 might be significant in the optimization code convergence.

Table 7.2. *Numerical Accuracy Definitions*

Generic Number	DA	NSD
0.012,3	10^{-4}	3
100.012,345,6	10^{-7}	10
123,000	10^3	3

DA: Decimal Accuracy
NSD: No. of Significant Digits of Accuracy

If we wish to know how many digits are physically meaningful, we need to ask the structural engineer who developed the model. He or she might tell us that the deflection model that we are using only provides an accuracy of approximately four significant digits. In this case, we would only report to the outside world the number 1.234×10^{-6}. We conclude by noting that if (i) we report n_r significant digits of accuracy, (ii) the physics-based model produces n_{pm} significant digits of accuracy, and (iii) the optimization code converges to within n_{op} significant digits of accuracy, then it is advisable to use n_r such that

$$n_r \leq \min\{n_{pm}, n_{op}\} \qquad (7.19)$$

where n_r need not be the same for each resulting number (*e.g.*, design variable). That is, the accuracy for *deflection* might be different from the accuracy for *stress* in a given model. Next, we present the scaling approach (the *how*) for design variables, objective functions, and behavioral constraints.

7.3.2 Design Variable Scaling—Order of Magnitude (DV-1)

Design variables scaling is often desirable anytime the order of magnitude of the design variables is much higher or much lower than one. Similarly, when the actual/unscaled values of the design variables are in the desirable order of magnitude, no scaling is typically necessary. As illustrated in Fig. 7.1, the scaling of the design variables takes place in three places: (i) the main file, (ii) the objective function evaluation, and (iii) the constraints evaluations. Although there are sophisticated approaches for scaling, in most practical cases, we can simply multiply each design variable by a constant that will (i) bring it close to 1, and/or (ii) address some tolerance or accuracy issues to be discussed later.

To bring the magnitude of the design variable close to "one", we use the following trivial form of design variable scaling

$$x_i^s = \alpha_i^s x_i \qquad (i = 1, ..., n_x) \qquad (7.20)$$

where n_x is the number of design variables, and α_i^s is a constant that is chosen to make the quantity $\alpha_i^s x_i$ approximately equal to one (or on the order of one). Therefore,

Eq. 7.20 represents the scaling process that takes place in the circles of Fig. 7.1 that pertain to the design variables. Conversely, the unscaling process takes the form $x_i = x_i^s / \alpha_i^s$ in the triangles. The upper and lower bounds of the design variables (in the *side constraints*) can be scaled as

$$\{\mathsf{LB}_i^s, \mathsf{UB}_i^s\} = \alpha_i^s \{\mathsf{LB}_i, \mathsf{UB}_i\} \qquad (i = 1, ..., n_x) \tag{7.21}$$

Note that in the case where the initial design variable is several orders of magnitude different from the final or optimal value, it may be necessary to perform more than one optimization run, each with an updated scaling (*i.e.*, each time using the final design variable value of one run as the initial value of the next run).

Let's further clarify. Say we have $x_{3-initial} = 123{,}45.67$, but $x_{3-optimal} = 0.012{,}345$ (much lower magnitude). In this case, to perform the optimization, we determine the scaling factor using $x_{3-initial}$, which has a higher magnitude. This scaling factor might not yield an accurate answer for $x_{3-optimal}$, which has a significantly lower magnitude than $x_{3-initial}$. To increase confidence in our results, we determine a new scaling factor using the possibly inaccurate results obtained for x_3 from the first optimization run. We then perform a second optimization run using the updated scaling factor. This second optimization run will generally yield more accurate results.

7.3.3 Design Variable Scaling—Tolerance Definition (DV-2)

Before we begin, let us clarify our terminology pertaining to accuracy. Specifically, we recall the definitions of Decimal Accuracy (DA) and Number of Significant Digits (NSD). We observe the definitions of these terms by inspecting Table 7.2, where three generic numbers are examined.

Next, we proceed with the development approach to scaling design variables. As previously discussed, the scaling of the design variables only needs to take into account their orders of magnitude. The proper order of magnitude of the design variables will promote improved convergence of the algorithms. However, having the proper order of magnitude alone will not address the all important issue of the *desired accuracy of each variable*. To explain this issue, let us consider two scenarios:

Scenario 1

If we are optimizing the thickness of a sheet of paper with an order of magnitude of 10^{-3} (say, 0.001,234,567), we might need the *decimal accuracy* (or tolerance) of the pertinent design variable to be 10^{-6}, yielding the number 0.001,234, which has 4 *significant digits*.

Scenario 2

Similarly, if we are optimizing the annual revenue of a company with an order of magnitude of 10^9 (say, \$1,234,567,891.234), we might need the decimal accuracy (or tolerance) of the pertinent design variable to be 10^3, yielding the number \$1,234,567,000, which has 7 significant digits.

7.3.4 Design Variable Scaling—Optimization Code Decimal Accuracy Setting (DV-3)

As we see, the *number of significant digits* of a number depends on its *decimal accuracy*, as generated by the code. All codes, in fact, provide us with the means to specify this decimal accuracy. In the case of MATLAB, the optimization parameter setting called TolX governs the *decimal accuracy* of the entries of the design variable vector x. The default value of TolX is 10^{-6}. Accordingly, in the case of Scenario 1 above, the nominal result would be $0.001,234$; however, in the case of Scenario 2 above, the nominal result would be $\$1,234,567,891.234,000$. At this point, we make two important observations. In Scenario 1, the code needs to produce only four significant digits, which poses no particular difficulty. In Scenario 2, however, the code needs to produce 16 significant digits of accuracy, which is problematic. In the latter case, the code will either not converge to this many significant digits, or it may take an unduly long time to do so.

7.3.5 Design Variable Scaling—Combining Order of Magnitude and Desired Tolerance (DV-4)

We discuss the order of magnitude and the desired tolerance of design variables in terms of the resulting number of significant digits of accuracy. We again discuss these concerns with the aid of Scenarios 1 and 2.

In Scenario 1, if we are satisfied with four correct digits as presented above, all is well. If, instead, we would prefer five significant digits of accuracy, we can perform scaling and let $\alpha_i^s = 10$ (Eq. 7.20) for the pertinent variable. In this case, the code now works with $x_i^s=0.012,345,67$; and since TolX$=10^{-6}$, the code will yield the result $x_i^s=0.012,345$, which has five significant digits of accuracy, as desired. Recall that we need to unscale using the relation $x_i = x_i^s/\alpha_i^s$, yielding the unscaled/correct value $x_i=0.001,234,5$, which has the same number of significant digits of accuracy (*i.e.*, five). We note that an alternative approach for obtaining five significant digits (alternative to scaling) is to let TolX$=10^{-7}$, which would again yield $x_i=0.001,234,5$. However, this approach is not generally advised since changing TolX in MATLAB affects all the design variables, and not just x_i. *In addition, significantly altering the default tolerance values of any code may result in unpredictable consequences on the operation of the code.*

In Scenario 2, if we would like to have 7, rather than 16, significant digits of accuracy, we can simply scale and let $\alpha_i^s = 10^{-9}$, leading to $x_i^s=1.234,567,891,234$, yielding the final value $x_i^s=1.234,567$, which has 6 decimal digits of accuracy (TolX$=10^{-6}$) and 7 significant digits of accuracy, as desired. Again, recall that we need to unscale using the relation $x_i = x_i^s/\alpha_i^s$, yielding the unscaled/correct value $x_i=\$1,234,567,000$, which has the same number of significant digits (*i.e.*, seven).

7.3.6 Design Variable Scaling—Setting Scaling Parameters (DV-5)

As noted above, the magnitude of the design variable x_i has a direct impact: (i) on the success of the optimization (*i.e.*, convergence), (ii) on the computation time, and (iii) on the accuracy of the final result (*e.g.*, tolerance). We note that the magnitude that is most important is that of the design variable *at the end* of the optimization, where the desired tolerance is meaningful. To bring the final/optimal design variable to an order of magnitude of one, we need to divide the design variable by a number that is approximately equal to its optimal value. As such, the user could specify his or her best guess as to what the typical optimal value of x_i might be. When we do so, the optimal value will have approximately six decimal digits of accuracy (TolX=10^{-6}), and seven significant digits of accuracy, since the optimal value is approximately on the order of one. Specifically, we let

$$\alpha_i^s = 10^{n_{dvi}}/x_{i-typical} \qquad (i = 1, ..., n_x) \qquad (7.22)$$

in Eq. 7.20 (with TolX=10^{-6}), where n_{dvi} denotes an exponent parameter setting for the *i*-th design variable. Using Eq. 7.22 is expected to yield approximately 7 + n_{dvi} significant digits of accuracy. If, instead, we let $\alpha_i^s = 10^{n_{dvi}}$, we will simply have *increased* the accuracy of the *i*-th design variable by n_{dvi} digits compared to its accuracy without scaling. When n_{dvi} is negative, we will have *decreased* the accuracy by n_{dvi} digits. Finally, we note that the absolute value of n_{dvi} should generally be no greater than three, to avoid requesting excessive accuracy.

In the case of MATLAB, the designer is allowed to specify the *typical* values of the design variables of the vector x by prescribing the vector TypicalX. Using the vector TypicalX and the scalar TolX, one is expected to obtain the desired number of significant digits of accuracy. Therefore, one has the option of setting the above two MATLAB parameters, or to simply set a scaling parameter for each design variable. The latter approach offers more flexibility.

As we examine the development presented above, a word of caution is in order. When we predict a given number of significant digits, our prediction is only approximate. We also note that different codes behave differently, and different versions of the same code will often behave differently. However, the general ideas presented in this section embody the important message that we need to know.

We conclude with two important notes. In the case where a given initial or final design variable is nearly zero, it may be unnecessary to perform scaling for that design variable. Moreover, in the case where the design variables are nearly of the same order of magnitude (and different from one), with similar tolerances, it may be acceptable to use a single value of the constant α for all design variables.

7.3.7 Objective Function Scaling

Most optimization codes are designed to perform well when the objective function value is on the order of one. While many codes offer significantly more flexibility, it is nevertheless safer to let the objective functions take on the desired order of magnitude through appropriate scaling. For values that lie significantly outside of this range, various aspects of the optimization algorithm become ill-conditioned. The finite difference derivatives, for example, take on undesirably large or small magnitudes.

For the purpose of optimization, the scaling of the objective function only needs to be performed after its evaluation in the objective function module, before it is passed to `fmincon`. This scaling is illustrated in the circular functional block after the objective function evaluation in Fig. 7.1. The mathematical expression for this scaling can take the form

$$f^s(x) = \beta^s f(x) \tag{7.23}$$

where β^s is a constant chosen so as to bring $f(x)$ within the desirable range. At the conclusion of the optimization run, `fmincon` returns the optimal value of the objective function. We recall that this is the scaled value. Therefore, before we use it, we must unscale it in the main file (the unscaling is not shown in Fig. 7.1) by simply using the equation $f = f^s/\beta^s$.

Much of the discussion regarding accuracy in the earlier subsections for the design variables also applies in the case of the objective function, and will not be repeated here in detail. The number of significant digits of accuracy of the objective function is obtained with the aid of scaling, in conjunction with the MATLAB setting called `TolFun`. The default value of `TolFun` is 10^{-6}. `TolFun` represents the *decimal accuracy* of the objective function $f(x)$.

In a manner similar to the discussion above, we obtain the desired accuracy of the objective function by scaling with

$$\beta^s = 10^{n_{obj}}/f_{typical} \tag{7.24}$$

(`TolFun`$=10^{-6}$), where n_{obj} denotes an exponent parameter setting, which is expected to yield approximately $7 + n_{obj}$ significant digits of accuracy. The constant $f_{typical}$ represents a number that is of the order of magnitude of the optimal value of the objective function. If, instead, we let $\beta^s = 10^{n_{obj}}$, we will simply have *increased* the accuracy of the objective function by n_{obj} digits when compared to its accuracy without scaling. When n_{obj} is negative, we will have *decreased* the accuracy by n_{obj} digits. Again, we note that the absolute value of n_{obj} should generally be no more than three to avoid requesting excessive accuracy. In the case where the optimal value of the objective function is near zero, scaling may be unnecessary.

We conclude this discussion on objective function scaling by noting that some codes set the accuracy of the objective function by prescribing the *number of significant digits* (say, 6), rather than the *decimal accuracy* (say, 10^{-6}) at convergence. For example, if the optimal value of the objective function is f_{opt}=1,234.123,456,789, the former case will yield f_{opt}=1,234.12, while the latter case will be f_{opt}=1,234.123,456 (without scaling). On the other hand, if we have f_{opt}=0.000,001,234,567, the former case will yield f_{opt}=0.000,001,234,56, while the latter case will be f_{opt}=0.000,001 (without scaling). We can see the relative advantages of each approach. However, with scaling, both approaches can be made to yield the same answer.

7.3.8 Behavioral Constraints Scaling

Behavioral constraints, in some sense, describe aspects of the system behavior. They do so in the form of imposed constraints that can be equalities or inequalities. In *PROB-7.2-MATLAB*, the behavioral constraints are expressed by Equations 7.14 to 7.17. As stated earlier, the first two constraints are nonlinear, while the last two constraints are linear. Equation 7.18 concerns the design variables (*i.e.*, *side constraints*), and is addressed earlier under design variable scaling. In keeping with the scope of this book, we will only explicitly discuss the scaling of nonlinear behavioral constraints. The essence of the following discussion on nonlinear behavioral constraints also applies to linear behavioral constraints. However, linear constraints also entail other complex numerical conditioning issues that are often automatically addressed in the linear programming algorithm, and do not require our attention here. Next, we address nonlinear equality constraints. The same techniques apply to inequality constraints. Consider the vector of constraints

$$\texttt{ceq(x)} = 0 \qquad (7.25)$$

In MATLAB, as in most optimization codes, the above constraint is satisfied to within a given *decimal accuracy*. In MATLAB, the pertinent tolerance parameter is called $\texttt{TolCon}$, and its default value is $\texttt{TolCon} = 10^{-6}$. Specifically, in satisfying the constraints in Eq. 7.25, the optimization algorithm will stop after the relation

$$|\texttt{ceq(x)}_j| \leq \texttt{TolCon} \qquad (j = 1, ..., n_{ec}) \qquad (7.26)$$

is satisfied, where n_{ec} denotes the number of nonlinear equality constraints. Consequently, for example, $\texttt{ceq(x)} = 10^{-7}$ will be considered *numerically* acceptable. However, this situation might be *physically* unacceptable. Consider the following two scenarios.

Scenario 1
Consider the constraint $2x_1^2 + x_2 = 0$. Assume that the normal/good values of the design variables are on the order of 10^{-7}, say, $x_1 = x_2 = 2.5 \times 10^{-7}$. By inspection, we see that these values of x do not satisfy the constraint (x_2 is not equal to $-2x_1^2$.) In this case, we have $2x_1^2 + x_2 = 2.500,001,25 \times 10^{-7}$ and Eq. 7.26 yields $2.500,001,25 \times 10^{-7} \leq$

10^{-6}, which *numerically* satisfies the constraint. This is in disagreement with our view that the constraint is not satisfied.

To correct this situation, we multiply the left- and right-hand-sides of the constraint by 10^6, which yields $(2x_1^2 + x_2) \times 10^6 = 0$. We let $\text{ceq}(\text{x})_1 = (2x_1^2 + x_2) \times 10^6$. Substituting the values of x in $\text{ceq}(\text{x})_1$, Eq. 7.26 yields $2.500,001,25 \times 10^{-1} \leq 10^{-6}$, which is clearly not satisfied – in agreement with our view. The important message here is that, by multiplying the constraint by a constant, we can enforce satisfaction of the constraint within the accuracy that we desire.

Scenario 2

Consider the constraint $x_1 = 2x_2$, which yields $\text{ceq}_1 = x_1 - 2x_2$. Assume that the design variables are on the order of 10^6; and the optimization yields $x_1 = 2,222,222.33$, and $x_2 = 1,111,111.11$. For practical purposes, we might consider these values of x to satisfy the constraint. However, Eq. 7.26 yields $0.11 \leq 10^{-6}$, which is clearly not satisfied – in disagreement with our view.

We could satisfy the constraint by increasing TolCon to be greater than 0.11, but such a high value of TolCon will likely compromise the operation of the optimization code and adversely affect the other constraints. Alternatively, could try to force the code to generate a very high number of digits of accuracy, say $x = \{2,222,222.222,222,2; 1,111,111.111,111,1\}$, but this approach would require too many iterations and often will not converge.

To correct this situation, we follow a path similar to the previous approach. We multiply the left- and right-hand-sides of the constraint by 10^{-6}, yielding $(x_1 - 2x_2) \times 10^{-6} = 0$. We let $\text{ceq}(\text{x})_1 = (x_1 - 2x_2) \times 10^{-6}$. Substituting the values $x = \{2,222,222.33; 1,111,111.11\}$ in $\text{ceq}(\text{x})_1$, Eq. 7.26 yields $0.11 \times 10^{-6} \leq 1.0 \times 10^{-6}$, which is clearly satisfied – in agreement with our view. The message here is the same as above; namely, that by multiplying the constraint by a constant, we can enforce the satisfaction of the constraint within the accuracy that we desire. In doing so, we save unnecessary computation and we increase the likelihood of successful convergence.

Constraint Scaling Approach: The above discussion leads us to a simple approach to scaling equality constraints. We simply let

$$\text{ceq}(\text{x})_i^s = \gamma_i^s \text{ceq}(\text{x})_i \qquad (i = 1, ..., n_{ec}) \tag{7.27}$$

where $\text{ceq}(\text{x})_i$ denotes the *i*th entry of $\text{ceq}(\text{x})$, $\text{ceq}(\text{x})_i^s$ is the scaled value of $\text{ceq}(\text{x})_i$, γ_i^s is a constant that is used to increase or decrease the degree of constraint satisfaction, and n_{ec} denotes the number of equality constraints. At this point, an important question arises: What is a good value for γ_i^s? As we can see from the above discussion, the answer depends on (i) the accuracy to which we need the constraint (Eq. 7.25) to be satisfied, and (ii) the value of the *constraints* setting of the optimization code that we are using (TolCon in the case of MATLAB). We can address this question in various ways.

First, the most direct answer to this question is to simply let $\gamma_i^s = 10^n$ or 10^{-n} if we wish to *increase* or *decrease* the accuracy of the constraint satisfaction, respectively. The greater the integer n, the greater the increased or decreased constraint satisfaction.

The next approach involves a pseudo-normalization scheme. We let

$$\gamma_i^s = 10^{n_{eqi}} / \text{ceq}_{i-nominal} \tag{7.28}$$

where $\text{ceq}_{i-nominal}$ is a *nominal* value of ceq_i and n_{eqi} denotes an exponent parameter setting for the i-th constraint. We note the similarities between Eqs. 7.22, 7.24 and 7.28. However, while values for $x_{typical}$ and $f_{typical}$ can generally be estimated, the *typical/nominal* value of ceq_i is known and is zero. Therefore, we cannot divide by it in a fashion similar to Eqs. 7.22 and 7.24. In the present case, we used what we call $\text{ceq}_{nominal}$. To determine good values of $\text{ceq}_{nominal}$, we are guided by the constraint Scenarios 1 and 2 above, and present the following three representative cases:

$$\text{ceq}(x)_i \equiv r(x) - 1.23 \times 10^{-8} = 0; \text{ceq}_{i-nominal} = 1.23 \times 10^{-8} \tag{7.29}$$

$$\text{ceq}(x)_i \equiv r(x) - 1.23 \times 10^{8} = 0; \text{ceq}_{i-nominal} = 1.23 \times 10^{8} \tag{7.30}$$

$$\text{ceq}(x)_i \equiv 2x_1 + x_2 = 0; \text{ceq}_{i-nominal} = x_{2-typical} \tag{7.31}$$

where $x_{2-typical}$ is a typical value of x_2, used to make one term in $\text{ceq}(x)_i$ be close to "one." If we let $n_{eqi} = 0$ in Eq. 7.28, in each of the three cases above, the value chosen for $\text{ceq}_{i-nominal}$ will be such that one term in ceq_i^s will be on the order of one. Since "one" is much larger than TolCon, the remaining terms in $\text{ceq}(x)_i$ will attempt to cancel the "one" term as much as possible, thereby promoting a meaningful satisfaction of the constraint. In the cases where we wish to further increase or decrease constraint satisfaction, we can simply choose a positive or negative integer for n_{eqi} in Eq. 7.28, as previously discussed.

We again note that all scaling functions above are presented as simple multiplicative constants – a proportional function. These comments conclude the presentation of the scaling approach. Next, we provide useful information regarding practical MATLAB implementation, followed by some insightful examples.

7.3.9 Setting MATLAB Optimization Options and Scaling Parameters: Syntax

In our approach to scaling, we discussed the need to sometimes change certain optimization settings. Here, we provice the syntax for changing them in MATLAB. We can change them in the following ways:

Several at once:

```
options = optimset('TolFun', 1e-8, 'TolX', 1e-7, 'TolCon', 1e-6)
```

In this case, several settings are used simultaneously and assigned to the set called "options," which will be used in the optimization function call "fmincon." *One at a time*:

```
options1 = optimset
options2 = optimset(options1, 'TolFun', 1e-8)
options3 = optimset(options2, 'TolX', 1e-7)
options = optimset(options3, 'TolCon', 1e-6)
```

To obtain more thorough information regarding the MATLAB settings parameters; we can type, on the MATLAB prompt, "help optimset." The options used are: *Display, TolX, TolFun, TolCon, DerivativeCheck, Diagnostics, FunValCheck, GradObj, GradConstr, Hessian, MaxFunEvals, MaxIter, DiffMinChange and Diff-MaxChange, LargeScale, MaxPCGIter, PrecondBandWidth, TolPCG, TypicalX, Hessian, HessMult, HessPattern, PlotFcns,* and *OutputFcn.*

Fortunately, we rarely have to deal with most of these parameters. To obtain general information regarding the optimization settings; we can type, at the MATLAB prompt, "help optimoptions."

The left-hand-sides of any of the above optimset calls define a set of "options" that can be used in the fmincon call, which will be used during optimization as follows.

```
[xopt,fopt] = fmincon('fun',x0,A,B,Aeq,Aeq,LB,UB,'nonlcon',
                    options)
```

In addition to setting the "options" parameters as performed above, we also need to define the scaling parameters and pass them to the various routines. In Fig. 7.1, these scaling parameters are defined by I^s. Specifically, the scaling parameters α^s (alphas), β^s (betas), and γ^s (gammas) can be defined in the main calling function and passed to fmincon, the constraint function nonlcon.m, and the objective function objfun.m. The commands are:

```
[xopts,fopts] = fmincon('objfun',x0s,A,B,Aeq,Aeq,LB,UB,
                'nonlcon',... options, alphas, betas, gammas);

[Cs,Ceqs]  = nonlcon(xs, alphas, betas, gammas);

fs  = objfun(xs, alphas, betas, gammas);
```

This presentation of the MATLAB syntax concludes our discussion of scaling approaches. Next, we provide some simple examples. A larger example is presented in Sec. 7.7.

7.3.10 Simple Scaling Examples

Here we perform the scaling for some simple examples.

Example 1

Consider the optimization problem of Eqs. 7.10 to 7.12. As previously stated, the MATLAB answer with scaling is $x = 10^{-6} \times \{3.083, 1.541\}$ and $f = 1.724 \times 10^{-12}$, while the MATLAB answer without scaling is $x = 10^{-6} \times \{4.948, 2.474\}$ and $f = 6.0714 \times 10^{-12}$. This is a case where scaling is indeed required or, possibly, the tolerance parameters must be changed. We previously discussed why scaling is generally the preferable approach.

Let us scale the design variables, the objective function, and the constraint:

(i) From Eqs. 7.20 and 7.22, and observing the orders of magnitude of the design variables, we let $n_{dvi} = 0$ and $x_{i-typical} = 10^{-7}$, with $(i = 1, 2)$.

(ii) From Eqs. 7.23 and 7.24, and observing the orders of magnitude, we let $n_{obj} = 0$ and $f_{typical} = 10^{-12}$.

(iii) Similarly, according to Eqs. 7.27 and 7.28, we let $n_{eqi} = 0$ and $ceq_{nominal} = 10^{-6}$, using $x_{2-typical}$, according to Eq. 7.31.

Using these scaling parameters, MATLAB indeed yields the accurate answers above.

We note that, in some cases, we may need to perform some preliminary investigation in order to obtain the pertinent orders of magnitude of the various quantities needed to implement the scaling. Finally, we note that the answers we obtain should not be sensitive to the scaling parameters we used. That is, if we moderately increase the scaling, we should obtain similarly more accurate answers. If our answers change significantly, then our scaling might be inadequate.

Example 2

Consider a slightly modified version of the optimization problem of Eq. 7.10 used in Example 1, where the objective function includes a constant term. From optimization theory, the design variables of the modified problem should not change from their original values; only the objective function should change. Let us examine the resulting pitfalls. The modified version reads

$$\min_{x_1, x_2} \quad f(x_1, x_2) = 1.111 \times 10^{-8} + (x_1 - 3.67 \times 10^{-6})^2$$

$$+ (x_2 - 3.67 \times 10^{-7})^2 \qquad (7.32)$$

subject to

$$x_1 - 2x_2 = 0 \qquad (7.33)$$

$$0 \leq x_i \leq 10^{-4} \quad (i = 1, 2) \qquad (7.34)$$

Without scaling, the MATLAB answers are $x = 10^{-6} \times \{4.948, 2.474\}$ and $f = 1.112 \times 10^{-8}$, which can be deemed incorrect. Note the low orders of magnitude of the various quantities. We also note that this problem poses numerical issues, in part, because the objective function is fairly *flat* near the minimum. We implement scaling in accordance with our previous discussion:

(i) For the design variables (Eqs. 7.20 and 7.22), we let $n_{dvi} = 0$ and $x_{i-typical} = 10^{-7}; (i = 1, 2)$.

(ii) For the objective function (Eqs. 7.23 and 7.24), we let $n_{obj} = 0$ and $f_{typical} = 10^{-8}$.

(iii) For the equality constraint (Eqs. 7.27 and 7.28), we let $n_{eqi} = 0$ and $ceq_{nominal} = 10^{-6}$.

This scaling yields a MATLAB answer of $x = 10^{-6} \times \{1.200, 0.600\}$ and $f = 1.1116 \times 10^{-8}$, which is deemed incorrect. We know that it is incorrect because, in this particular case, we know the correct answer from above. It is important to note that we could have easily been *fooled* into thinking that, since we implemented scaling, our answer is correct. The way we deal with this is to make sure we develop confidence in any answer we obtain through optimization, a topic we discuss in Sec. 7.6.3. In the case of scaling, what we have done thus far is *applied guidelines*. We used $n_{obj} = 0$ in accordance with the guidelines provided. We need to explore whether changing n_{obj} (or changing any of the other scaling parameters) is not necessary. Indeed, if we let $n_{obj} = 2$, MATLAB yields $x = 10^{-6} \times \{3.0828, 1.541\}$ and $f = 1.111 \times 10^{-8}$, which is correct.

As a final comment, we wish to emphasize that the material of this section is a comprehensive presentation of information that we would learn over perhaps years of trial and error in optimization, which, to our knowledge, is not provided in any other textbook in this form. What we have provided is a set of general guidelines and insights into the numerical processes that will be very helpful to you. We have not given strict predictions of computational outcomes as it might well be impossible to do so. In fact, we should sometimes expect the unexpected. Final numerical results depend on too many unknown parameters for us to make predictions with any kind of certainty. However, we have provided you with a set of useful tools and understandings that should be of critical assistance in practice. **An example of a larger scaling problem is presented in Section 7.7.**

7.4 Finite Difference

This section presents finite difference and the important role it plays in computational optimization. The first subsection introduces the fundamentals of finite difference and the second presents its pertinent accuracy issues.

7.4.1 Fundamentals of Finite Difference

A particular class of optimization algorithms is known as *gradient-based algorithms*. *This class of algorithms uses the gradient of the objective function and of the constraints* in the search for the optimal solution. As discussed in Sec. 2.5.4, the gradient of a scalar-valued function is a vector of the partial derivatives of a function with respect to each of the design variables. Specifically, the gradient of a scalar function $f(x)$, where x is an n-dimensional column vector, is given by

$$\frac{\partial f}{\partial x} = \left\{ \frac{\partial f}{\partial x_1}, \cdots, \frac{\partial f}{\partial x_n} \right\}^T \tag{7.35}$$

The gradient vector is used not only to govern the search, but also to help decide when the optimum is reached and the search terminated. When the optimum is reached, we say that the *optimality condition* is satisfied.

Generally, the value of the objective function is evaluated using a complex computer code. As such, we do not have an explicit analytical expression for the objective function that can be used to evaluate its gradient. As a result, an adequate approximation is used instead. This approximation is referred to as a *finite difference* derivative, as opposed to an *analytical* derivative [4]. The finite difference derivative of $f(x)$ at a point x_0 is given by

$$\frac{\partial f}{\partial x}(x_0) \approx \frac{\Delta f}{\Delta x}(x_0) = \left\{ \frac{\Delta f_0}{\Delta x_1}, \cdots, \frac{\Delta f_0}{\Delta x_n} \right\}^T \tag{7.36}$$

with

$$x_0 = \{x_{10}, x_{20}, \cdots, x_{n_x 0}\}^T \tag{7.37}$$

where Δx_i is a *small* deviation of x_i about x_{i0}, and Δf_0 is the corresponding variation in $f(x)$. The quantity $\frac{\Delta f_0}{\Delta x_i}$ can be evaluated using three typical approaches: (i) forward difference, (ii) backward difference, or (iii) central difference. We express each as follows:

Forward Difference

$$\frac{\Delta f_0}{\Delta x_i} = \frac{f(x_0 + \Delta x_i) - f(x_0)}{\Delta x_i} \tag{7.38}$$

Backward Difference

$$\frac{\Delta f_0}{\Delta x_i} = \frac{f(x_0) - f(x_0 - \Delta x_i)}{\Delta x_i} \tag{7.39}$$

Central Difference

$$\frac{\Delta f_0}{\Delta x_i} = \frac{f(x_0 + \frac{\Delta x_i}{2}) - f(x_0 - \frac{\Delta x_i}{2})}{\Delta x_i} \tag{7.40}$$

Further, we note that the finite difference approximation entails an error that can be partially explained by the equation

$$\frac{\partial f}{\partial x_i}(x_0) = \frac{\Delta f_0}{\Delta x_i} + \propto (\Delta x_i)^2 + HOT \tag{7.41}$$

where $\propto (\Delta x_i)^2$ is a term proportional to $(\Delta x_i)^2$ that is ignored by the finite difference approximation above, and *HOT* represents additional *Higher Order Terms*

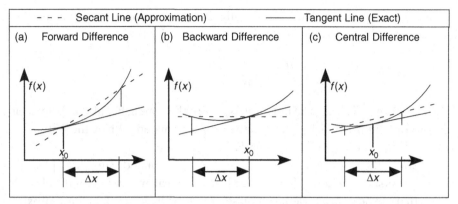

Figure 7.2. Graphical Representation of Finite Difference Approximation

(proportional to $(\Delta x_i)^{n_h}$; $n_h > 2$) that are also ignored. The smaller the magnitude of Δx_i, the more negligible the ignored terms become and the more accurate the finite difference, at least in theory. However, as we will see later in this section, there is a limit to the acceptable smallness of Δx_i in practice.

It is interesting to think of the three finite difference evaluation options, both in mathematical terms and in geometrical/graphical terms. Equations 7.38, 7.39, and 7.40 provide the expressions for the mathematical evaluations of the forward, backward, or central difference, respectively. Similarly, Figs. 7.2 (a), (b), and (c) provide the respective graphical interpretations of these finite differences, in the case where x is a scalar.

We make the following observations:

(i) The solid line represents the *tangent line* at the point x_0. The slope of the tangent line is exactly the derivative of $f(x)$ at the point x_0.

(ii) The dashed line represents the so-called *secant line*. The slope of the secant line is equal to the finite difference value.

(iii) As Δx tends to zero, the secant line converges to the tangent line; and the finite difference (Eqs. 7.38, 7.39, or 7.40) converges to the gradient (Eq. 7.35). However, as we will see shortly, excessively small values of Δx pose some numerical difficulties.

(iv) *Number of Function Calls – Objective Function*: When x is a scalar, the gradient is also a scalar (or a one-dimensional vector). In this case, the forward, backward, and central difference evaluations require two function calls each (see Eqs. 7.38, 7.39, and 7.40). In the case where the vector x has dimension n_x, then the finite difference approximation of the gradient requires $n_x + 1$ function evaluations for forward and backward difference (*i.e.*, one evaluation at x_0, and n_x evaluations obtained after deviating each of the n_x entries of the vector x). Note that the case of central difference requires $2n_x$ function evaluations. In this latter case, there is no evaluation at x_0; we instead deviate each variable forward and backward (see Eq. 7.40).

(v) *Number of Function Calls – Constraints*: When we have constraints, and we also use a finite difference approximation for the gradient in the constraints, it may lead to a large number of constraint functions evaluation. For example, if we have n_{eq} constraints and we use forward difference, the finite difference evaluations will require $n_{eq}(n_x + 1)$ constraint function evaluations.

(vi) The central difference option generally yields more accurate answers (see Fig. 7.2), but also requires more function evaluations as discussed above.

This discussion leads us to the all important topic of the accuracy of the finite difference approximation.

7.4.2 Accuracy of Finite Difference Approximation

The success of all gradient-based optimization algorithms, as their names suggest, strongly depends on the accuracy of the evaluated gradient vector. An important question at this point is: How accurate is this finite difference approximation? This is a critical question, since gradient-based optimization is one of the most popular approaches in practice.

Assume that the function $f(x)$ has n_{sda} significant digits of accuracy. As a rule of thumb, the number of digits of accuracy of derivatives drops by half ($n_{sda}/2$). For the second derivatives, it drops by another half ($n_{sda}/4$). Please keep in mind that this is indeed a *rule of thumb*. In practical cases, the situation could be much worse or much better. The resulting finite difference values may become useless in the optimization algorithm, and result in serious convergence difficulties.

Optimizing with Experimental Data

An important practical situation of interest occurs when experimental data is used for the objective function or the constraint functions. This situation may have low accuracy (say six digits.) In this case, the first derivative may only have three digits of accuracy, and the second derivative might be practically unusable in an optimization algorithm.

In these cases, it may be useful to first develop so-called *response surfaces* of the resulting data. Once obtained, it might be more reliable to optimize using these surfaces, which are essentially a best-fit of the data available. At a basic level, the situation is straightforward; that is, (i) form a best-fit function of the data (make the best fit as smooth as possible) and (ii) optimize using this best fit function. This approach works quite well. The details of this topic are beyond the scope of this book. References [5] and [6] offer representative works in the area. The first is a fundamental book on response surface methodology. The second provides response surface information from the perspective of design of experiments, within one concise chapter.

How to Impact Finite Difference Accuracy

We can impact finite difference accuracy in three basic ways. Fortunately, most optimization codes perform well in promoting the maximum accuracy of the obtained results. However, there is much that we can also do. The three basic approaches are as follows.

(1) *Adequate Scaling.* Adequate scaling (as previously discussed) will address: (ii) the magnitude of the objective functions, (i) the magnitude of the design variables, (iii) the magnitude of the constraints, and (iv) the pertinent setting parameters of the optimization codes. When these issues are addressed, finite difference will tend to perform more effectively.

(2) *Forward, Backward, or Central Difference*: As previously discussed, forward and backward differences provide similar accuracies, while central difference provides greater accuracy. However, the central difference is more computationally intensive. Depending on the computer labor involved in evaluating the objective functions and constraints, we may decide to choose one option vs. another.

(3) *The Magnitude of* Δx: This last consideration is the most critical. Too small or too large a magnitude will result in excessive inaccuracies. Let us consider each scenario.

i Too large a magnitude of Δx makes the secant line too distinct from the tangent line. Their corresponding slopes become too dissimilar (*i.e.*, the finite difference and gradient become too different). This situation is readily seen in Fig. 7.2.

ii Too small a magnitude of Δx also makes the corresponding Δf too inaccurate. Let us explain. Assume that the function $f(x)$ possesses eight significant digits of accuracy. (This means that all digits beyond the 8th are useless.) Let $f(x_0) =$ **1.234,567,8**12,34, where the first eight (bold) digits are accurate/significant and the last four digits are useless/incorrect. Let us consider two cases.

   CASE 1: Assume that Δx is so small that $f(x_0 + \Delta x) =$ **1.234,567,8**25,67. In the case of forward difference, $\Delta f_0 = f(x_0 + \Delta x) - f(x_0)$. The key observation here is that, using the above numbers, we obtain $\Delta f_0 =$ **0.000,000,0**13,01. *However, only the bold digits are meaningful, leaving no accuracy at all.* This evaluation of the finite difference will have no digit of accuracy. This loss of accuracy is referred to as a ***cancelation error***. Cancelation errors may occur when we subtract two numbers that have similar magnitudes and an insufficient number of digits of accuracy.

   CASE 2: Assume that Δx is now sufficiently large so that $f(x_0 + \Delta x) =$ **1.234,688,9**25,57, where the change in $f(x)$ takes place in the last four significant digits, leaving the first four unchanged. In this case, $\Delta f_0 =$ **0.000,121,1**13,23. We have four significant digits of accuracy in the finite difference computation. *This is half the number of digits of accuracy in $f(x)$, which reflects a good compromise between too large or too small a value of Δx.*

Selection of Finite Difference Method.

When using the MATLAB Optimization Toolbox, one can select how to evaluate the finite difference derivatives. The pertinent option is set up using the command `optimset`. For the function `fmincon`, two options are available: 'forward'

Table 7.3. *Example of Finite Difference Accuracy* $(f(x) = 4x^4 + 2x^3 + 1/x)$

x	Δx	$\frac{\partial f}{\partial x}$	$\frac{\Delta f}{\Delta x}$ (Forward Diff.)		$\frac{\Delta f}{\Delta x}$ (Central Diff.)	
			Value	DoA	Value	DoA
	10^{-2}	1579.4506172	1584.588	2	1579.452	6
	10^{-4}	"	1579.502	4	1579.4506175	10
4.5	10^{-6}	"	1579.451	6	1579.450616	9
	10^{-8}	"	1579.45056	7	1579.4508	7
	10^{-10}	"	1579.451	6	1579.451	6
	10^{-12}	"	1579.792	4	1579.338	4
	10^{-14}	"	1546.141	2	1682.565	1
	10^{-2}	−49382.715926	−15325.670	0	210526.316	0
	10^{-4}	"	−48309.179	1	−49388.813	4
	10^{-6}	"	−49371.744	3	−49382.717	7
$4.5e^{-3}$	10^{-8}	"	−49382.606	5	−49382.715923	10
	10^{-10}	"	−49382.7147	7	−49382.71587	8
	10^{-12}	"	−49382.720	6	−49382.749	6
	10^{-14}	"	−49379.878	3	−49385.562	4
	10^{-16}	"	−49169.557	2	−49737.992	2
	10^{-18}	"	−28421.709	0	−85265.128	0

DoA: Number of Digits of Accuracy

(the default) and 'central' (about the center). The command to select the central difference method is shown as follows.

```
options = optimset('FinDiffType','central')
```

Example of Finite Difference Accuracy

Table 7.3 provides a specific example of the number of significant digits of accuracy available in a given finite difference case. We perform forward and central finite difference. The results are self explanatory. We note that MATLAB was used, and that it provides 16 significant digits of accuracy by default. The finite difference is evaluated for two values of x. We make the following observations:

(1) The 3rd column reports 11 (of the available 16) significant digits of accuracy for the gradient – full MATLAB accuracy. These results are used as a benchmark to determine finite difference accuracy.

(2) For very small or very large values of Δx, the number of Digits of Accuracy (DoA) is expectedly lower. In fact, for two cases, we have no accuracy at all (Number of DoA = 0).

(3) For mid-range values of Δx, the accuracy is generally the highest.

(4) Central difference is predictably more accurate than forward difference.

(5) Generally, the finite difference yields approximately six to eight digits of accuracy. This is somewhat less than one-half of 16 – the accuracy of $f(x)$ in MATLAB.

(6) These trends are shown graphically in Fig. 7.3, where the horizontal axis depicts $-\log \Delta x$ (smaller values of Δx on the right and larger values on the left).

$$f(x) = 4x^4 + 2x^3 + \frac{1}{x}$$

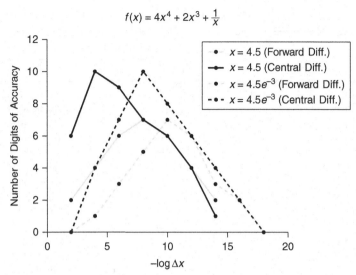

Figure 7.3. Finite Difference—Number of Digits of Accuracy

We conclude with the following important message: (i) When $\triangle x$ is too small, we have large errors because $f(x)$ has limited accuracy (16 DoA in MATLAB). (ii) When $\triangle x$ is too large, we have large errors because the secant line departs from the tangent line. (iii) The greatest accuracy we can generally obtain for the finite difference is approximately one-half the accuracy of the function $f(x)$. With this understanding of the practical limitations of finite difference, we now turn to another option that is largely immune to these numerical issues: *Automatic Differentiation*.

7.5 Automatic Differentiation

As discussed in Sec. 7.4 above, optimization codes that are gradient-based need the differentiation of the objective function and of the constraints with respect to the design variables to allow them to march toward the optimum. In most cases, we simply let the code evaluate this differentiation using finite difference. In simple cases, we could manually write a code that evaluates these differentiations analytically, as opposed to letting the code use finite difference. Let us refer to this option as *Analytical Differentiation*. We could then provide the analytical differentiation code to MATLAB, or to any other optimization software that we are using (if it has the capability to use that code). In more complicated cases, it is generally much easier and more practical to let the optimization code use finite difference to obtain the needed differentiations. However, as we see from Sec. 7.4, finite difference can be computationally intensive, as well as computationally ill-conditioned, leading to potential complications.

We note important tradeoffs between analytical differentiation and finite difference:

Analytical Differentiation

(i) It is highly efficient, computationally.
(ii) It promotes faster and more certain convergence, and is numerically stable.

(iii) It is more difficult and time consuming to implement, in practice, for problems of realistic complexity.

(iv) It is generally not implemented in optimization codes and must be explicitly implemented by the user.

Finite Difference

 (i) It is computationally intensive.

(ii) It generally converges more slowly than with the finite difference approach, and is more likely to lead to numerical instabilities.

(iii) It is much easier to implement in practice.

(iv) It is generally an integral component of the gradient-based optimization codes, requiring no user coding.

Automatic Differentiation

Interestingly, there is yet another approach that offers us the best of both worlds. It is the *Automatic Differentiation* approach. Specifically, it offers the benefits of analytical differentiation and finite difference without their respective complications. It essentially employs analytical differentiation. However, it does not require the user to develop the analytical expressions for the derivatives manually. These derivatives are *automatically* developed by another software – the *Automatic Differentiation* software. We explain in detail with the help of Fig. 7.4.

In Fig. 7.4, we explain and contrast Automatic Differentiation and Finite Difference. For the sake of presentational simplicity, we only discuss the differentiation of the objective function. We note that the very same discussion applies to the differentiation of the constraints. Figure 7.4 is divided into four blocks: (A), (B), (C), and (D). Blocks (A) and (C) describe the finite difference option, and Blocks (B) and (D) describe the automatic differentiation option. In addition, Blocks (A) and (B) describe the work/development that must be done by the engineer before optimization can proceed, while Blocks (C) and (D) depict the software modules that are required for the optimization to proceed. Let us describe each option.

Finite Difference

Block **(A)** describes quite a familiar situation. Namely, that we must have a software routine available to evaluate the objective function that we will provide to the optimization code for the optimization to proceed.

Block **(C)** describes a situation where (i) the objective function evaluation code is used by the optimization code, and (ii) the finite difference operation is performed *within* the optimization code. The remainder is self explanatory.

Automatic Differentiation

Block **(B)** explains the preparatory work that is required before optimization can proceed. An Automatic Differentiation (AD) Software is used as depicted in the figure. The function evaluation routine is available to us. This code becomes the input to the AD software. The AD software then uses a series of chain rule applications throughout the objective function code, and generates another code as its output. This output code is specifically designed to *generate the gradient vec-*

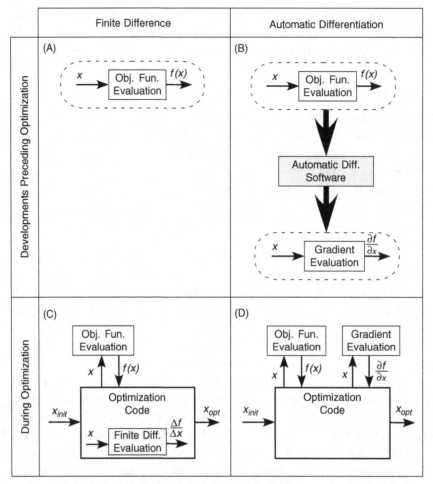

Figure 7.4. Automatic Differentiation vs. Finite Difference Approximation

tor, given a current value of x. In summary, at the end of the work of Block (B), we have available to us two codes: one that evaluates the objective function, given x, and one that generates the gradient of $f(x)$, given x. At this point, it is critically important to keep in mind that this gradient is evaluated analytically and not through finite difference. We also note that sometimes the generated gradient code provides two outputs: (i) the gradient and (ii) the objective function. In this case, the originally used objective function evaluation function is no longer needed. (In Fig. 7.4, these objective functions and gradients are generated by two codes).

Block **(D)** describes the optimization process when no finite difference takes place. Instead, the gradient is evaluated during the optimization process using the function generated by the Automatic Differentiation Software in Block (B). The well-known automatic differentiation software called ADIFOR may be used, which we describe next.

ADIFOR (**A**utomatic **DI**fferentiation in **FOR**tran) was developed by the Mathematics and Computer Science Division at Argonne National Laboratory and the Center for Research on Parallel Computation at Rice University. ADIFOR is a software tool that performs automatic differentiation for Fortran 77 programs. Given a Fortran 77 source code and a user's specification of the dependent and independent variables, ADIFOR will generate a code that computes the partial derivatives of all of the specified dependent variables with respect to all of the specified independent variables. ADIFOR also outputs the objective function value. Details on the background and implementation of ADIFOR are provided in Refs. [7, 8].

As the C programming language increased in popularity, an analog of ADIFOR was developed. ADIC is a tool that implements automatic differentiation on a ANSI C code. Reference [9] describes the architecture of ADIC and how to use it, with the help of an example. In Ref. [10], an enhancement for ADIFOR and ADIC that computes Hessians is discussed.

7.6 Other Important Numerical and Computational Issues

In this section, we discuss several useful numerical issues that do not readily fall under the previously discussed topics. These are:

1. Sensitivity of optimum solution [11],
2. Termination of the optimization run – criteria and causes,
3. Level of confidence in the optimal results obtained,
4. Computational burden and problem dimension,
5. Additional numerical pitfalls.

7.6.1 Sensitivity of Optimal Solutions in Nonlinear Programming

Sensitivity to System Parameters

Understanding how sensitive our optimum solution is to small changes in various system parameters (that are not design variables) is generally important. This information provides us with the means to potentially further improve our design by reconsidering the values of these parameters. The simplest way to assess this sensitivity is to run the optimization code with different values of the potentially sensitive parameters, and to observe the resulting optimal solution. Alternatively, an indication of the sensitivity can be obtained by simply evaluating the objective function and the constraints for different values of these parameters that are near their nominal values. It is important to note that undue sensitivity to a system parameter may potentially indicate pitfalls in the physical design, which should be carefully examined. This aspect of sensitivity is one of continuing research in the community, and will not be further addressed here.

In the case of linear programming, there are also important practical sensitivity issues that should be considered. These issues are discussed in Sec. 11.7.4.

Sensitivity to Design Variables

The more urgent aspect of the sensitivity of the optimal solution is that with respect to the design variables. The generated optimum solution can be highly sensitive to the design variables in a way that indicates the presence of numerical issues, and that the solution obtained might not be adequate or mathematically optimal. Extreme sensitivity to design variables may also actually indicate that the problem is not well posed, or may also indicate the presence of pitfalls in the physical design – as in the case of system parameters discussed above. Stated differently, the problem might be a purely computational issue that has no bearing on the actual design, or might indicate pitfalls in the actual design. In the case where the problem is of a numerical nature, we can simply employ the techniques presented in Sec. 7.3. If the computational issues persist after proper scaling, then a consideration of the physical design may be advisable.

Sensitivity to Weights in the Objective Function

Undue sensitivity to the weights in the objective function may also be a sign of complications that should be explored. Careful scaling of the objectives and of the design variables will generally be helpful. In the case where we are dealing with the weights in the weighted sum approach, our physical insight into the meaning of the weights can be helpful, and we may reconsider our stated preferences in forming the objective function. However, the mere knowledge of the sensitivity to these weights may give us insight into the nature of our optimal solution with respect to our preferences. In the case where the weights are part of more complex nonlinear objective functions, the issues are more involved.

7.6.2 Optimization Termination Criteria and Optimization Termination Causes

Optimization codes terminate the convergence process for numerous reasons. Some are desirable and generally indicate a successful outcome. Others indicate non-convergence or convergence to a non-optimal solution. Yet other termination causes call for further investigation. The information on page 159 (regarding possible reasons for the optimization process to fail) is directly relevant to our present discussion. Let us consider various termination scenarios.

Scaling

As previously discussed, scaling is arguably the most important step we can take to maximize the likelihood of convergence to the optimum solution. Therefore, the wealth of pertinent information provided in Sec. 7.3 should be exploited to the fullest.

A poorly scaled problem will usually force the optimization code to terminate at a non-optimal solution, while the designer might not even be aware of this serious situation.

Optimization Settings

All good optimization codes provide the user with the ability to influence the conditions under which the optimization process will terminate. To influence the code, the designer has at his/her disposal various settings. Different codes have different versions of these settings. Pertinent MATLAB settings include:

1. The max number of function evaluations (MaxFunEvals). Its default value is 100*(Number of Variables).
2. Maximum number of iterations (MaxItr). The default is 400.

Even though you may not have a complete understanding of these settings, increasing them might be helpful when MATLAB informs us that the code terminated because these maximum values have been reached. Please see the end of Sec. 7.3 to determine how to change these settings. Please note that the default values of these settings may change from one version of MATLAB to the other.

7.6.3 Developing Confidence in Optimization Results

Upon obtaining a final solution from an optimization run, it is important to take steps to develop appropriate confidence in the validity of the answer. These steps may include:

1. Being satisfied that the previous discussion in Sec. 7.2.1 on page 159 regarding why the optimization process sometimes fails have been addressed.
2. Implementing multiple starts with very different initial points to avoid or identify local minima, particularly in the case of gradient-based algorithms. It may be required to update the scaling for each different start.
3. Critically examine the physical validity/feasibility of the optimum solution. Common sense alone will often point to potential issues.

7.6.4 Problem Dimension and Computational Burden

The problem size or dimension, which can be represented by the number of design variables, is directly related to its computational intensity (see Sec. 15.3). The fidelity of the models employed is also a determining factor for the computation resources needed. In addition, the more computationally intensive the problem is, the more likely it is to involve numerical issues. Therefore, whenever possible, we should explore strategies to reduce the problem dimension [12]. Design variable linking is one venue to this end (see Sec. 15.3.1). The proper fidelity of the models that should be used is an important continuing research activity in the community, and should be considered in any serious system optimization effort.

7.6.5 Additional Numerical Pitfalls

Additional potential numerical pitfalls are briefly mentioned here:

1. Some constraints can be violated during the optimization, and become satisfied during the optimization process. This is perfectly fine. However, some others may cause the optimization algorithm to misbehave or fail. These include constraints that lead to invalid results in the given context. Examples are:
   - The violation of a constraint that the outer diameter of a pipe must be larger than its inner diameter, resulting in a non-physical design.
   - The violation of a constraint that the argument of a square root must be positive, potentially resulting in a non-physical design.
2. Division by a quantity that vanishes.
3. Taking the square root of a negative quantity.
4. Having the diameter of a bar become negative.
5. Having a mass take on a negative value.
6. Numerical cancelation, as discussed in the context of finite difference (see Sec. 7.4).
7. Matrices becoming *singular to working precision.*
8. Starting the optimization too close to its optimum, which can cause convergence problems.

7.7 Larger Scaling Example: Universal Motor Problem
7.7.1 Universal Motor Problem Definition

Let us consider the design of a universal motor [13, 14]. A universal motor is simply an electric motor that functions on both AC and DC supplies to produce a torque. The ratio of the power input to the power output of the motor gives a measure of the motor efficiency, and it is known that increasing the windings in the field and armature of the motor will increase the efficiency. However, this increase in windings is associated with an increase in the mass of the motor. Since these motors are widely used in kitchen appliances, a motor having a large mass is undesirable. The universal motor is to be optimally designed in an effort to maximize its efficiency and minimize its mass.

Our universal motor problem is governed by the following design variables:

1. Number of turns of wire on the armature, N_a,
2. Number of turns of wire on the field, N_f,
3. Cross sectional areas of the wires on the field, A_{wf},
4. Cross sectional area of the wires on the armature, A_{wa},
5. Electric current, I,

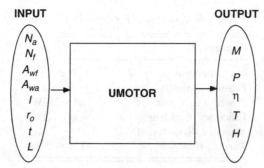

Figure 7.5. Input and Output for the Program `umotor.m`

6. Outer radius of the stator, r_o,
7. Thickness of the stator, t, and
8. Stack length, L

Finding the values of mass, M, power, P, efficiency, η, torque, T, and magnetizing intensity, H, of the motor requires the design variables as inputs. No equations are given to you to compute the values of mass, power, efficiency, torque, and magnetizing intensity of the motor. However, you are given a MATLAB function `umotor.m`, in the book website (`www.cambridge.org/Messac`), that allows you to enter all the design variables in a row vector (lets call it x) to get the resulting mass, M, power, P, efficiency, η, torque, T, and magnetizing intensity, H, of the motor as shown in Fig. 7.5. Thus

$$x = [N_a \quad N_f \quad A_{wf} \quad A_{wa} \quad I \quad r_o \quad t \quad L]$$

We will use the outputs of `umotor.m` (see the book website `www.cambridge.org/Messac`) in our objective function and constraints. Anytime you need the values of mass, M, power, P, efficiency, η, torque, T, and magnetizing intensity, H, of the motor, you should call the function `umotor.m`.

The side limits on the design variables are as follows:

1. $100 \leq N_a \leq 1{,}500$ turns
2. $1 \leq N_f \leq 500$ turns
3. $0.01 \times 10^{-6} \leq A_{wf} \leq 1.0 \times 10^{-6} \mathrm{m}^2$
4. $0.01 \times 10^{-6} \leq A_{wa} \leq 1.0 \times 10^{-6} \mathrm{m}^2$
5. $0.1 \leq I \leq 6$ Amps
6. $0.01 \leq r_o \leq 0.10$m
7. $0.0005 \leq t \leq 0.10$m
8. $0.000566 \leq L \leq 0.10$m

In addition to the design variables bounds, we have additional constraints presented in Table 7.4.

Your task is to obtain the Pareto frontier for the two objectives, mass and efficiency. Use a set of 100 evenly spaced weights, and use the initial guess of $X0 = (LB + UB)/2$ for all Pareto points. Prepare a MATLAB code that generates

Table 7.4. *Constraints for Universal Motor*

Constraints	Value
Magnetizing intensity, H	$H < 5{,}000$
Feasible geometry	$r_o > t$
Power of each motor, P	$P = 300\text{W}$
Efficiency of each motor, η	$\eta \geq 0.15$
Mass of each motor, M	$M \leq 2.0\text{kg}$
Torque, T	$T = 0.25\text{Nm}$

the Pareto frontier for this problem. You will realize that `fmincon` is unable to converge for some Pareto points. If you use the given assumptions regarding the initial guess and weights, you will obtain several solutions that are not Pareto, as seen in Fig. 7.6(a).

This difficulty occurs because the design variables are of different orders of magnitude, thereby causing numerical difficulties for `fmincon`. *In this exercise, you learn a method to overcome the numerical issues caused by design variable scaling and obtain a good representation of the Pareto frontier.*

7.7.2 Design Variable Scaling

One effective yet simple approach to overcome the above discussed numerical issues is to **scale** each design variable in `fmincon` so that it is on the order of one. The scaled design variables are used only for numerical purposes, so that `fmincon` will converge. We need to make sure that the original (or unscaled) design variables are used to compute the objective function and constraint values.

The scaled design variables used by the `fmincon` optimizer will need to be unscaled to their original magnitudes when executing the objective function and constraint functions. Note that the function `umotor.m` needs the original design variables to yield meaningful results for mass, M, power, P, efficiency, η, torque, T, and magnetizing intensity, H, of the motor. Figure 7.6(b) provides the correct Pareto frontier that results from scaling. Figure 7.6 illustrates side-by-side Pareto frontier plots with and without scaling, which demonstrates the dramatic impact that scaling can have. The procedure for implementing the scaling of the design variables in the MATLAB code is shown in Fig. 7.7.

In addition to scaling and unscaling in the objective and constraint files, we need to do the following. In the main file, reset the upper and lower bounds to correspond with our scaled design variables. You may also wish to specify an initial point that lies within the lower and upper bounds of the scaled design variables.

7.8 Summary

This chapter provided an analytical, intuitive, and computational discussion of the major numerical issues encountered in practical implementation computational optimization. Effective methods that address these issues were also presented. These

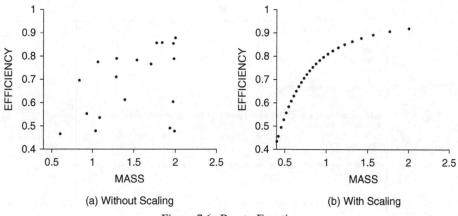

Figure 7.6. Pareto Frontier

methods included numerical conditioning of algorithm, matrices, and the optimization problem; followed by methods used to scale and to set the tolerances for different optimization parameters (*e.g.*, variables, objective functions and constraint functions). The example provided at the end illustrates the application of such scaling techniques. This chapter also provides a description of the primary methods used to estimate function gradients (*e.g.*, finite differences). The chapter concluded with a discussion of some of the other important numerical considerations toward successful application of optimization, such as termination criteria and solution sensitivity. Importantly, these issues can collectively *make or break* optimization in practice!

7.9 Problems

Warm-up Problems

7.1 Give five reasons why we need to pay attention to numerical issues in our engineering activities, which include design, analysis, and optimization. Not all five should be from the book.

7.2 What does it mean for a problem to be (i) well-posed?, (ii) ill-conditioned? Explain each in approximately 200 words.

7.3 Provide three important reasons why optimization codes sometimes do not work well. Explain each in detail in approximately 300 words.

7.4 Evaluate A_1 in Eq. 7.2, while A is given by Eq. 7.1. Prove that Eq. 7.2 indeed holds true for any positive integer n.

7.5 Let $\alpha = 0.5$ in Eq. 7.1. Evaluate A_1 in Eq. 7.2 with $n = 2$. Perform this calculation without a computer, and show all your intermediate steps. This problem will also help you recall elementary matrix algebra. Please do a quick review if some related concepts have become a bit rusty in your mind.

Main file

```
% Upper and lower bounds without scaling
%LB=[100,   1,    0.01e-6,  0.01e-6,  0.1,  0.01,  0.0005,...
0.000566];
%UB=[1500,  500,  1.00e-6,  1.00e-6,  6,    0.1,   0.1,   ...
0.1];

% Initial guess without scaling
%x0 =(LB+UB)/2;

% Upper and lower bounds with scaling (note the difference between scaled and
unscaled LB and UB)
LBS=[0.1,  0.01,  0.01,  0.01,  0.01,  1.0,  0.05,  0.0566];
UBS=[1.5,  5.00,  1.00,  1.00,  0.60,  10,   10,    10.000];

% Initial guess with scaling (note the difference between scaled and unscaled X0)
X0 = (LBS+UBS)/2;

%.......
%... Pareto frontier generation code....
%......

% The solutions you get from fmincon are scaled design variables. To compute the
% objective functions, you need to call umotor.m, which needs design variables
% with actual magnitudes

Na  = x(1)*1000;
Nf  = x(2)*100;
Awf = x(3)*1e-6;
Awa = x(4)*1e-6;
I   = x(5)*10;
r0  = x(6)*1e-2;
t   = x(7)*1e-2;
L   = x(8)*1e-2;

y = [Na Nf Awf Awa I r0 t L];

[M,P,n,T,H] = u_motor(y);

%....record your objective function values, use these to
%plot the Pareto frontier
```

Constraint file after Scaling

```
function [C,eq] = con_motor1(x,w)

% "x" is scaled design variable
% Unscale the design variables to their actual magnitudes to
% compute constraints

Na  = x(1)*1000;     % Wire turns on armature
Nf  = x(2)*100;      % Wire turns on the field
Awf = x(3)*1e-6;     % Field wire c/s section
Awa = x(4)*1e-6;     % Armature wire c/s section
I   = x(5)*10;       % Current
r0  = x(6)*1e-2;     % Stator outer radius
t   = x(7)*1e-2;     % Stator thickness
L   = x(8)*1e-2;     % Stack length

y = [Na Nf Awf Awa I r0 t L];

% Call umotor.m with the actual magnitudes
[M,P,n,T,H] = u_motor(y);

%...define the constraints using [M,P,n,T,H]
```

Objective function file after Scaling

```
function [f] = obj_motor1(x,w)

% "x" is scaled design variable
% Unscale the design variables to their actual magnitudes
% to compute objective values
Na  = x(1)*1000;     % Wire turns on armature
Nf  = x(2)*100;      % Wire turns on the field
Awf = x(3)*1e-6;     % Field wire c/s section
Awa = x(4)*1e-6;     % Armature wire c/s section
I   = x(5)*10;       % Current
r0  = x(6)*1e-2;     % Stator outer radius
t   = x(7)*1e-2;     % Stator thickness
L   = x(8)*1e-2;     % Stack length

y = [Na Nf Awf Awa I r0 t L];

% Call umotor.m with the actual magnitudes
[M,P,n,T,H] = u_motor(y);

%...define the objective functions
%using [M,P,n,T,H]
```

Figure 7.7. Scaling of Design Variables

7.6 How can you evaluate the condition number of a symmetric matrix; and how does this number relate to its numerical conditioning properties?

7.7 Use MATLAB to solve the optimization problem defined by Eqs. 7.10, 7.11, and 7.12. Do so without implementing any scaling. Compare the solution you

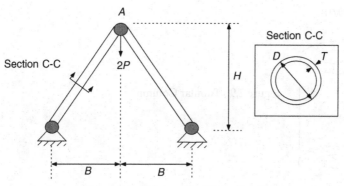

Figure 7.8. Schematic of Two-Bar Truss

obtained with the solution $x = 10^{-6} \times \{3.083, 1.541\}$. Which solution yields a lower value of the objective function? Explain.

7.8   Consider the general optimization problem statement defined by Eqs. 7.13 through 7.18. (a) Provide the mathematical equation for a constraint that can be interchangeably used either in Eq. 7.15 or Eq. 7.17. Explain. Is it computationally more desirable to use this constraint in Eq. 7.15 or Eq. 7.17? Explain. (b) Provide the mathematical equation for a constraint that can be used in Eq. 7.15, but not in Eq. 7.17. Explain.

Intermediate Problems

7.9   For Matrix A given in Eq. 7.1, generate the Table 7.1 – while adding a column for $n = 100$. Please keep in mind that you may not obtain numbers that are exactly equal to those in the table, but you should be able to observe that the conclusions will be the same. Submit your clearly organized MATLAB codes.

7.10   For Matrix A given by

$$A = \begin{bmatrix} (1+\alpha)/\alpha & (1-\alpha)/\alpha \\ (1-\alpha)/\alpha & (1+\alpha)/\alpha \end{bmatrix} \qquad (7.42)$$

generate the Table 7.1 while adding a column for $n = 100$. Discuss any interesting observations that you can make from the results. Submit your clearly organized MATLAB codes.

7.11   The two bar truss shown in Fig. 7.8 is to be designed so that: (1) it is as light as possible, and (2) the vertical deflection at node A is as low as possible. The dimensions H and D are the design variables. The following parameters are given:

$$B = 30 \; inch$$

$$T = 0.1 \; inch$$

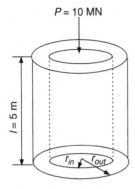

P = 10 MN

l = 5 m

r_{in} r_{out}

Figure 7.9. Tubular Column

$$P = 33 \times 10^3 \; lb$$

$$\text{Young's modulus, } E = 30 \times 10^6 \; psi$$

$$\text{Material density, } \rho = 0.3 \; lb/in^3$$

$$\text{Maximum allowable stress, } \sigma_A = 100 \times 10^3 \; psi$$

The bounds on H and D are given as:

$$0.5 \leq D \leq 5 \; inch$$

$$0.5 \leq H \leq 50 \; inch$$

The vertical deflection at node A is given as follows:

$$u = \frac{P(B^2 + H^2)^{\frac{3}{2}}}{\pi DTEH^2}$$

Assume that the cross sectional area of the beam is πDT, and use this expression for your analysis. Answer the following questions:

1. What are the most important failure modes in this problem?
2. Clearly write down the mathematical expressions for your constraints.
3. Formulate the bi-objective optimization problem, and obtain the Pareto frontier.

7.12 Consider the tubular column shown in Fig. 7.9.

1. Design the column to support the load P at minimum mass, m, and minimum cost, C, by choosing optimal values for the outer radius, r_{out}, and the inner radius, r_{in}. Ensure that your column avoids buckling; P should be less than the critical buckling load, P_{cr}. In addition, ensure that the load induced stress, σ, does not exceed the maximum stress, σ_{max}, of the material. Use the expressions given below to aid in your design optimization.

    Material Properties:

$$E = 207 \text{ GPa (Young's Modulus)}$$
$$\rho = 783 \text{ kg/m}^3 \text{ (Density)}$$

$$\sigma_{max} = 248 \text{ MPa}$$

Design Metrics:

$$m = \pi \rho l (r_{out}^2 - r_{in}^2)$$
$$C = \frac{100 r_{out}}{r_{out}^2 - r_{in}^2}$$

Design Parameters:

$$0.01 \leq r_{in} \leq 0.08 \text{ m}$$
$$0.04 \leq r_{out} \leq 0.2 \text{ m}$$

Other Expressions:

$$\sigma = \frac{P}{\pi (r_{out}^2 - r_{in}^2)}$$
$$P_{cr} = \frac{\pi^3 E (r_{out}^4 - r_{in}^4)}{16 l^2}$$

2. Plot the Pareto frontier for this problem. Ensure you obtain a complete Pareto frontier.

7.13 MATLAB code for universal motor with and without scaling:
1. Write a MATLAB code to generate the Pareto frontier of the universal motor problem discussed in this chapter. Do so *without* design variable scaling. In other words, reproduce the results shown in Fig. 7.6(a). Use the initial guess of $X0 = (LB + UB)/2$ for all Pareto points, where LB and UB are the original design variable bounds. Note that you may not get the exact identical point, but the conclusion will be clear.
2. Now, scale the design variables as described previously (refer to Fig. 7.7). Use the initial point specified in Fig. 7.7 for all Pareto points. Generate the Pareto frontier using scaled design variables. In other words, reproduce the results shown in Fig. 7.6(b).

7.14 Consider the following nonlinear programming (NLP) problem:

$$\min_{x} f(x) = x_1^2 + x_1 x_2 + 2x_2^2 - 6x_1 - 2x_2 - 12x_3$$

subject to

$$g_1(x) = 2x_1^2 + x_2^2 \leq 15$$
$$g_2(x) = x_1 - 2x_2 - x_3 \geq -3$$

1. Use the default settings for the function `fmincon` in the MATLAB Optimization Toolbox to obtain an optimal solution, x^*, to this problem. Was your solution process able to converge?
2. Make the following changes to your code and the various search options in the MATLAB Optimization Toolbox to try and obtain the optimal solution.
   a) Make sure that the medium scale SQP algorithm, with Quasi-Newton update and line-search, is the selected solver in `fmincon`. Although you may not yet know these algorithms, you can proceed with this problem. The details of these algorithms are presented later in the book.

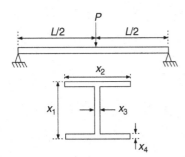

Figure 7.10. Schematic of Simply Supported I-Beam

    b) Set the solution point tolerance, function tolerance, and constraint tolerance in `fmincon` to 10^{-7}.

    c) Specify the initial guess as $x_0 = [1 \quad 1 \quad 1]^T$.

    d) Make sure that the inequality constraint g_2 is treated as a linear constraint by `fmincon`.

    e) Solve the NLP by using a finite difference approximation of the gradients of the objective function and constraints first. Then, resolve the problem by providing explicit expressions for the gradients of the objective function and constraints (ask `fmincon` to check for the gradients before running the optimization in this latter case).

3. Record the optimal solution for the design variables and the objective function. Ensure that your problem converges.

7.15 Optimize the simple I-beam shown in Fig. 7.10. The symmetrical I-beam cross section is also shown.

Find the optimal cross section dimensions that minimize the structural mass and minimize the mid span static deflection. As shown in Fig. 7.10, the I-beam is simply supported at both ends and is subjected to an applied concentrated load, P. The final design must be able to sustain a maximum bending stress of 160 MPa, and the beam dimensions must be within the acceptable limits given below.

Acceptable Beam Dimensions: $0.10 \text{ m} \leq x_1 \leq 0.80 \text{ m}$; $0.10 \text{ m} \leq x_2 \leq 0.50 \text{ m}$; $0.009 \text{ m} \leq x_3 \leq 0.05 \text{ m}$; $0.009 \text{ m} \leq x_4 \leq 0.05 \text{ m}$.

Constant Numerical Values: $P = 600 \text{ kN}$; $L = 2 \text{ m}$; $E = 200 \text{ GPa}$; $\rho = 7,800 \text{ kg/m}^3$.

1. Provide a clear optimization problem statement.

2. Use the weighted sum method to complete the following table. W1 denotes the weight for the mass objective, and W2 denotes the weight for the deflection objective.(Hint: Your code should cater to any scaling problems.

3. Plot the Pareto frontier for this problem.

5. Show the points obtained in No. 2 on your plot.

7.16 Problem 6.10 of the Multiobjective chapter presents numerical challenges that can be addressed with proper scaling of the design variables, objective functions, and/or constraints. First solve Parts 1 through 6 of that problem

Table 7.5. *Optimum Design Results for Different Objective Weights*

W1	W2	Mass (M)	Deflection (δ)	x1	x2	x3	x4
0	1						
0.25	0.75						
0.5	0.5						
0.75	0.25						
1	0						

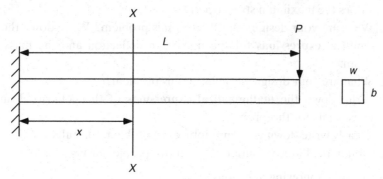

Figure 7.11. Schematic of Cantilever Beam

using appropriate scaling. Comment on the behavior of the Pareto frontier with and without scaling.

Advanced Problems

7.17 Consider the task of designing a cantilever beam with a rectangular cross section subjected to the load P, as shown in Fig. 7.11. We are interested in finding the cross sectional dimensions, b and w, that can **safely support the load** P. At the same time, we would like the beam to: (1) be as light as possible and (2) have the least possible deflection (see Fig. 7.11).

The following information is given to you:

1. $P = 600$ kN,
2. $L = 2$ m,
3. Young's Modulus E of the beam material $= 200$ GPa,
4. Density of the beam material $= 7800$ kg/m^3,
5. Maximum allowable stress of the beam material $= 160$ MPa,
6. Maximum and minimum acceptable values for b and w are: $b_{min} = 0.1$ m, $b_{max} = 0.8$ m, $w_{min} = 0.1$ m, and $w_{max} = 0.5$ m.
7. The expression for the deflection of a cantilever beam at any Section $X - X$

$$\delta = \frac{Px^3}{3EI}; \quad I = \frac{wb^3}{12}$$

where x is the length of the beam measured from the fixed end (see Fig. 7.11),

8. The expression for the bending stress at the Section $X - X$ is given as

$$S = \frac{P(L-x)b}{2I}$$

Answer the following questions:

1. At what value of x does the maximum deflection occur? At what value of x does the maximum stress occur?
2. What are your design objectives in this problem? Write down the **mathematical expressions** for the maximum deflection and the mass of the beam.
3. What are your design variables in this problem?
4. Write down the **mathematical expressions** of the behavioral and side constraints for this problem.
5. Clearly write down your multiobjective problem formulation.
6. Obtain the Pareto frontier of the above problem using `fmincon`.

7.18 Answer the following questions:

   (i) Discuss two reasons why the MATLAB tolerance parameter `typicalX` can be helpful to use in an optimization problem.
   (ii) Discuss two reasons why the MATLAB tolerance parameter `tolx` can be helpful to use in an optimization problem.
   (iii) Develop an optimization problem for which you do not obtain the optimum solution without scaling the design variables.
   (iv) For the problem you just developed in Part (iii), use the scaling techniques presented in this chapter to obtain the optimum.
   (v) Now, instead of scaling the design variables, obtain the optimum solution simply by using the tolerance parameters `typicalX` and `tolx`. The size of the problem is up to you. Note that much of the learning will take place in the process of generating the problem.

BIBLIOGRAPHY OF CHAPTER 7

[1] E. Cheney and D. Kincaid. *Numerical Mathematics and Computing*. Cengage Learning, 2012.
[2] T. Sauer. *Numerical Analysis*. Pearson Education, 2012.
[3] M. Grasselli and D. Pelinovsky. *Numerical Mathematics*. Jones and Bartlett Publishers, 2008.
[4] M. Nuruzzaman. *Finite Difference Fundamentals in MATLAB*. Createspace Independent Pub., 2013.
[5] R. H. Myers, D. C. Montgomery, and A. C. M. Christine. *Response Surface Methodology: Process and Product Optimization Using Designed Experiments*. John Wiley and Sons, 3 edition, 2009.
[6] D. C. Montgomery. *Design and Analysis of Experiments*. Wiley, 8th edition, 2012.

[7] C. H. Bischof, A. Carle, G. F. Corliss, A. Griewank, and P. Hovland. ADIFOR: Generating derivative code from Fortran programs. *Scientific Programming*, 1:11–29, 1992.

[8] C. H. Bischof, A. Carle, P. Khademi, and A. Mauer. ADIFOR 2.0: Automatic differentiation of Fortran 77 programs. *IEEE Computational Science and Engineering*, 3:18–32, 1996.

[9] C. H. Bischof, L. Roh, and A. J. Mauer-Oats. ADIC: An extensible automatic differentiation tool for ANSI-C. *Software: Practice and Experience*, 27(12):1427–1456, 1997.

[10] J. Abate, C. H. Bischof, L. Roh, and A. Carle. Algorithms and design for a second-order automatic differentiation module. In *International Symposium on Symbolic and Algebraic Computation*, pages 149–155, 1997.

[11] J. R. R. A. Martins, P. Sturdza, and J. J. Alonso. The complex-step derivative approximation. *ACM Transactions on Mathematical Software (TOMS)*, 29(3):245–262, 2003.

[12] P. M. Zadeh, V. V. Toropov, and A. S. Wood. Metamodel-based collaborative optimization framework. *Structural and Multidisciplinary Optimization*, 38(2):103–115, 2009.

[13] T. W. Simpson, J. R. A. Maier, and F. Mistree. Product platform design: Method and application. *Research in Engineering Design*, 13(1):2–22, 2001.

[14] S. J. Chapman. *Electric Machinery and Power System Fundamentals*. McGraw-Hill, 2002.

8

Global Optimization Basics

8.1 Overview

The objective of global optimization is to find the best local solution among the known or discoverable set of local optima. The best local optimum is called the global optimum. Formally, global optimization generally seeks a global solution to an unconstrained or constrained optimization problem. Global search techniques are often essential for many applications (*e.g.*, advanced engineering design, data analysis, financial planning, process control, risk management, and scientific modeling). Highly nonlinear engineering models often entail multiple local optima.

A global optimization problem can be defined as

$$\min_{x} \quad f(x) \tag{8.1}$$

subject to

$$g(x) \leq 0 \tag{8.2}$$

$$h(x) = 0 \tag{8.3}$$

$$x_l \leq x \leq x_u \tag{8.4}$$

The function $f(x)$ represents the objective function. The constraints include the inequality constraint function $g(x)$, the equality constraint function $h(x)$, and the side constraints.

The unique challenges presented by global optimization are discussed in Sec. 8.2. Three typical methods employed to solve global optimization problems are presented in Sections 8.3, 8.4, and 8.5. Section 8.6 illustrates how to solve global optimization problems using the MATLAB Global Optimization Toolbox. Helpful pertinent references for a more advanced treatment include (Refs. [1, 2]).

8.2 Practical Issues in Global Optimization

As explained in Chapter 2, objective functions can have global minima and local minima. When this occurs, the problem is classified as a multimodal optimization problem. The objective of global optimization is to find the global optima.

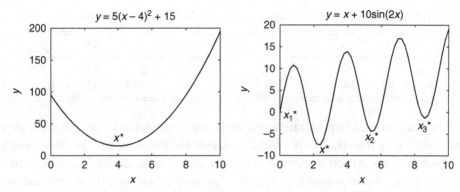

$$y = 5(x-4)^2 + 15 \qquad\qquad y = x + 10\sin(2x)$$

Figure 8.1. Unimodal and Multimodal Objective Functions

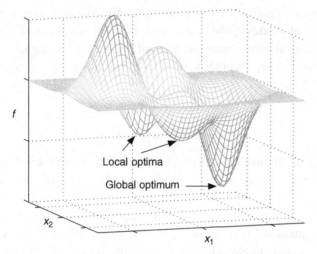

Figure 8.2. The 3D Surface Plot of a Multimodal Function

Figure 8.1 presents two different types of one-dimensional objective functions. The variable, x, is in the interval between 0 and 10. The function illustrated in Fig. 8.1(a) is unimodal. It only has one local minimum, which is its global minimum. The function shown in Fig. 8.1(b) is multimodal. The four points, x_1^*, x_2^*, x_3^*, and x^*, are all local minima. The function value of x^* is the lowest of all the local minima. It is the global minimum of the multimodal objective function.

The global minimum of the unimodal objective function in Fig. 8.1(a) can be found using the gradient-based optimization methods presented in Chapters 12 and 13. In the case of the multimodal objective function presented in Fig. 8.1(b), it is difficult to readily find the global minimum. Gradient-based optimization methods will simply yield one of the local minima, depending on the starting point used. Finding the global optimum is not guaranteed when using gradient-based algorithms.

Figure 8.2 provides the 3D surface mesh of a two-dimensional multimodal function. Note that this function has several local minima and one global minimum.

Table 8.1. *Design Variable Combinations and Their Objective Function Values*

x_1	1	1	1	3	3	3	5.5	5.5	5.5
x_2	2	3.5	5.5	2	3.5	5.5	2	3.5	5.5
$f(x)$	45	54	94	5	14	54	11.25	20.25	60.25

In practical global optimization problems, the situation can be more complicated and challenging than the above examples. Objective functions may be non-smooth or discontinuous. At some points or in certain intervals, the derivatives of objective functions may not be available or may be expensive to compute. For some real life engineering design problems, it is often challenging to determine whether a highly nonlinear function is unimodal or multimodal before starting the optimization. More importantly, even if a problem is known to be multimodal, is is generally not possible to know how many modes/optima there are. Gradient-based optimization methods, in their conventional forms, may not be appropriate for these global optimization problems. Derivative-free methods, such as evolutionary algorithms, may be more appropriate in these cases.

8.3 Exhaustive Search

Exhaustive search algorithms enumerate all of the possible candidates for optimization problems, and the best solution is the global optimum. Exhaustive search may be applicable to optimization problems that are comprised of a manageable number of variable combinations. The time required for conducting an exhaustive search may dramatically increase as the number of candidate solutions increases. The following example implements a global optimization problem exhaustive search process.

Example: Use exhaustive search to find the global optimum of the following optimization problem.

$$\min f(x) = 5(x_1 - 4)^2 + 4(x_2 - 2)^2 \tag{8.5}$$

subject to

$$x_1 \in \{1, 3, 5.5\} \tag{8.6}$$

$$x_2 \in \{2, 3.5, 5.5\} \tag{8.7}$$

Each of the design variables has three possible values. In selecting x_1 and x_2 from the feasible sets, there are nine possible combinations. The nine combinations and their objective function values are listed in Table 8.1.

Comparing the nine objective function values in Table 8.1, we find that the global minimum of $f(x)$ is 5. The optimal solution is $x_1 = 3$ and $x_2 = 2$.

8.4 Multiple Start

The multiple start approach uses gradient-based methods to find local and global minima. It generates multiple starting points, and stores local and global solutions found during the search process.

Several variations of the multiple start method are available in the literature. The basic multiple start method consists of the following steps:

1. Generate multiple starting points. The points can be either uniformly distributed within predefined bounds or generated using a sampling algorithm.

2. Filter out the infeasible starting points. This step is optional. If applied, this step may help reduce the total computational expense required to find the corresponding feasible optima.

3. For each starting point, use a gradient-based optimization method to search for a local minimum.

4. Save the multiple local minima returned by Step 3.

5. Compare all the local minima. The local minimum with the lowest objective function value is considered the global minimum. However, since there is no guarantee that we have obtained the complete set of local minima, we cannot be certain that we have obtained the global minimum.

The function shown in Fig. 8.1(b) is a multimodal objective function. The multiple start method can be used to find its global optimum.

Example: The optimization problem shown in Fig. 8.1(b) is stated as follows. Use the multiple start method to find its global optimum.

$$\min f(x) = x + 10\sin(2x) \tag{8.8}$$

subject to

$$0 \le x \le 10 \tag{8.9}$$

Choose $0, 1, 2, 3, 4, 5, 6, 7, 8, 9$, and 10 as starting points. These eleven points are uniformly distributed within the specified bounds. Note that the practical implementation of the multiple start method can demand a significantly higher number of starting points. In this example, fmincon is used to solve the optimization problem corresponding to each starting point. The MATLAB codes, including the main file, the objective function file, and the constraint function file, are given below.

1. **Main file**

```
clear
clc

% Define Linear constraints
```

Table 8.2. *Local Optima for 11 Starting Points*

x_0	0	1	2	3	4	5	6	7	8	9	10
x^*	0	8.61	2.33	0	5.47	5.47	0	0	8.61	0	2.33
$f^*(x)$	0	-1.37	-7.66	0	-4.52	-4.53	0	0	-1.37	0	-7.66

```
A=[ ]; B=[ ]; Aeq=[]; Beq=[];

% Define bounds constraints
LB=[0]; UB=[10];

% Define optimization options
options=optimset('largescale','off','MaxFunEvals', ...
200000,'display','off','MaxIter',1e6);

% Define starting points
xsta = [0 1 2 3 4 5 6 7 8 9 10];
k=11;

% Optimization from 11 starting points
for i=1:1:k
    x0=xsta(i)
    [xopt,fopt(i)]=fmincon('MS_Func', x0, A, B, ...
    Aeq, Beq, LB, UB, 'MS_cons', options)
end
```

2. **Objective function file**

```
function [f]=MS_Func(x)


f = x+10*sin(2*x);
```

3. **Constraint function file**

```
function [C Ceq]=MS_cons(x)


C = [];
Ceq = [];
```

The optimization yields 11 local optima, some of which are redundant. They are listed in Table 8.2. In the table, x_0 represents the starting points, x^* represents the corresponding local optimal values of the variable, and $f^*(x)$ represents the corresponding local optima of $f(x)$.

By comparing the local optima given by the multiple start method (as reported in Table 8.2), the global minimum is found to be -7.66. The corresponding optimal value of x is 2.33. It is important to note that, in a multiple start approach, since the individual optimization runs do not depend on each other, the runs can be executed in parallel. This strategy will allow you to significantly reduce the net computing lapse time by taking advantage of the parallel and distributed computing capabilities widely available today. MATLAB itself provides pertinent capabilities in the form of its "Parallel Computing Toolbox."

8.5 Role of Genetic Algorithms in Global Optimization

Evolutionary algorithms are population-based optimization algorithms inspired by the principles of natural evolution. When compared with other optimization algorithms, the advantage of evolutionary algorithms is that they do not make limiting assumptions about the underlying objective functions. The objective functions are treated as black-box functions. Furthermore, the definition of objective functions does not require significant insight into the structure of the design space.

Darwin's theory of evolution identified the principles of natural selection and survival of the fittest as driving forces behind biological evolution. His theory can be summarized as follows [3].

Variation
There is variation between individuals in a population.
Competition
Resources are limited. In such an environment, there will be a struggle for survival among individuals.
Offspring
Species have great fertility. They produce more offsprings than can grow to adulthood.
Genetics
Organisms pass genetic traits to their offspring.
Natural selection
Those individuals with the most beneficial traits are more likely to survive and reproduce.

Evolutionary algorithms are based on the principles of biological evolution described in Darwin's theory. Genetic Algorithms (GAs) are a class of evolutionary algorithms, also known as population-based metaheuristic optimization algorithms. GA was first conceived by J.H. Holland in [4]. GA uses a population of solutions, whose individuals are represented in the form of chromosomes. The individuals in the population go through a process of simulated evolution to obtain the global optimum.

The GA repeatedly modifies a set of solutions or individuals in the course of its entire run. At each iteration, the genetic algorithm selects individuals from the current population to serve as parents based on certain criteria. The parents are then used to create the next generation of individuals, called children. Over successive generations, the population evolves toward an optimal solution or Pareto frontier, depending on the type of problems and the type of GA being used. The procedure for a genetic algorithm is illustrated in Fig. 8.3.

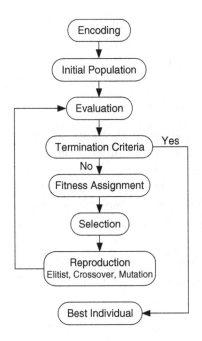

Figure 8.3. Procedure for a Genetic Algorithm

The terms used in the procedure in Fig. 8.3 are explained below:

Encoding
Encoding is a way to represent individual solutions in evolutionary algorithms. Typically, individual solutions are coded as a finite fixed length string. Binary numbers are usually used as codes. This string is also known in the literature as a chromosome. For example, a binary number, 10001, represents the decimal number 17. The conversion from the binary number to the decimal number is given by
$1 \times 2^4 + 0 \times 2^3 + 0 \times 2^2 + 0 \times 2^1 + 1 \times 2^0 = 16 + 1 = 17$.

Initial population
The algorithm begins by generating a population of individuals in the design space. Prior to population initialization, the designer must choose the number of individuals in each population and the number of bits in the encoding process. Both of these decisions are extremely important in promoting the success of the GA-based optimization. For example, too large a population can lead to undesirable computational expense, while too small a population can lead to premature convergence to a local optimum or suboptimal point. Further description of these features and issues of GAs can be found in the foundational book on GA by Goldberg [5].

Evaluation
Computation of the objective values for the individual solutions.

Optimization criteria
The stopping criteria of the algorithm. Examples of stopping criteria include the number of generations, the computation *time* limit, and the function tolerance.

Fitness assignment

There are several choices of fitness assignment. In a rank-based fitness assignment, the individuals are sorted according to their objective values. It creates an order among the individuals.

Selection

A selection criterion filters out the candidate solutions with poor fitness and retains those with acceptable fitness to enter the reproduction process with a higher probability.

Reproduction

A new generation in the genetic algorithm is created through reproduction from the previous generation. Three mechanisms (elitist, crossover, and mutation) are primarily used to create a new generation.

Elitist

The individuals with the best fitness values in the current generation are guaranteed to survive in the next generation.

Crossover

In this technique, a part of the encoded string of one individual is exchanged with the corresponding string part of another individual. There are many approaches to performing the crossover operation. Suppose there are two individuals, 10101 and 11001. Exchange the first two bits of the two individuals. The offspring of the two individuals are 11101 and 10001.

Mutation

Mutated child solution is generated from a single parent by randomly reversing some bits from 0 to 1, or vice versa. For example, through mutation, the 2nd bit and the 5th bit of 10101 are reversed. The new offspring is 11100.

Best individual

The global optimum that satisfies the termination criteria.

The steps of GA are illustrated in more detail in Chapter 19. The multimodal function shown in Fig. 8.1(b) is solved using MATLAB GA Solver as follows.

Example: The optimization problem presented in Fig. 8.1(b) is stated as follows.

$$\min f(x) = x + 10sin(2x) \tag{8.10}$$

subject to

$$0 \leq x \leq 10 \tag{8.11}$$

To set up this problem in MATLAB, using the ga command, three M-files are generated: a main file, an objective function file, and a constraint function file. Note that this file structure is similar to the one used with $fmincon$.

The main file contains the initializations, bounds, options, and the ga command. The objective function file contains the objective or the fitness function definition. The constraint file contains the nonlinear inequality and equality constraints. The files are reported below.

1. **Main file**

```
clear
clc

% Define Linear constraints
A=[ ]; B=[ ]; Aeq=[]; Beq=[];

% Define bounds constraints
LB=[0]; UB=[10];

% Number of design variables
nvars = 1;

% Optimization function ga
[x,fval] = ga(@GA_Func,nvars,A,B,Aeq,Beq,LB,UB,@GA_cons);

display(x)
display(fval)
```

2. **Objective function file**

```
function [f]=GA_Func(x)

f = x+10*sin(2*x);
```

3. **Constraint function file**

```
function [C Ceq]=GA_cons(x)

% Define inequality constraints
C = [];

% Define equality constraints
Ceq = [];
```

The global optimum of the objective function obtained by MATLAB is −7.6563. The optimum value of the variable is 2.3312. The result is the same as that obtained using the multiple start method.

This problem can also be solved using the ga command from the graphical user interface of the GA Solver. Please note that the folder that contains the fitness function and the nonlinear constraint function should be selected as the MATLAB **Current Directory**. Figure 8.4 shows how to set up the GA solver to solve the problem. This screen can be opened by typing optimtool('ga') in the Command Window. Alternatively, you can type optimtool in the Command Window; then choose the ga solver option from the top left dropdown menu. Set @GA_Func as the fitness function. Set @GA_cons as the nonlinear constraint function. Since there is no nonlinear constraint for this problem, we can leave it

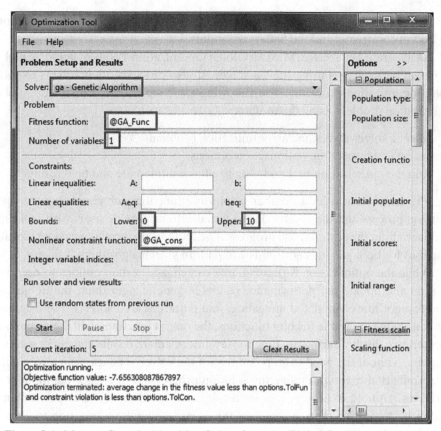

Figure 8.4. MATLAB Genetic Algorithm Solver from the Global Optimization Toolbox

blank. Set the number of variables as 1. Set the lower bound as 0 and the upper bound as 10.

After 5 iterations, the optimization is terminated. The global optimum value of the objective function is −7.6563.

8.6 MATLAB Global Optimization Toolbox

The MATLAB Global Optimization Toolbox provides methods to search for global solutions to problems that contain multiple maxima or minima. The optimization solvers in the toolbox include global search, multiple start, pattern search, genetic algorithms, and simulated annealing solvers. Using the toolbox, one can select a solver and define an optimization problem. Genetic algorithms and pattern search solvers can also be customized. For genetic algorithms, initial population and fitness scaling options can be modified and parent selection, crossover, and mutation functions can be defined by the users. For pattern search, polling, searching, and other functions can be defined.

The MATLAB Multiobjective Genetic Algorithm Solver can be used to solve multiobjective optimization problems to generate Pareto frontiers. This solver can be used to solve either smooth or nonsmooth optimization problems, with or without the bound constraints and the linear constraints.

As illustrated in Sec. 8.5, MATLAB Global Optimization Toolbox can solve global optimization problems using the ga solver. It also provides other solvers to solve global optimization problems. The following steps are required to solve a global optimization problem using the toolbox.

1. Select a solver and define an optimization problem.
2. Set up and inspect the optimization options.
3. Run the optimization tool and visualize the intermediate and final results.

The toolbox includes a number of plotting functions for visualizing the optimization process and the results. These visualizations provide users with real-time feedback about the optimization progress. The toolbox also provides custom plotting functions for both genetic algorithms and pattern search algorithms.

While the optimization is running, one can change certain options to refine its solution and update the performance results in genetic algorithms, multiobjective genetic algorithms, simulated annealing, and pattern search solvers. For example, one can enable or disable the plot functions, the output functions, and the command-line iterative display during run time to view intermediate results and query solution progress, without the need to stop and restart the solver. The user can also modify the termination criteria to refine the solution progression or reduce the number of iterations required to achieve a desired tolerance based on run time performance feedback. An introductory webinar of the MATLAB global optimization toolbox is provided in Ref. [6].

If the Parallel Computing Toolbox is available, it can be used in conjunction with the Global Optimization Toolbox to reduce computation time using parallel processing. Built-in support for parallel computing accelerates the objective and constraint function evaluation in genetic algorithms, multiobjective genetic algorithms, and pattern search solvers.

This chapter provided an introductory presentation of global optimization, with sufficient information to tackle elementary problems. This book also presents more advanced approaches that can be used in more practical contexts. The pertinent chapters are: Discrete Optimization Basics (Chapter 9), Discrete Optimization (Chapter 14), and Evolutionary Algorithms (Chapter 19).

8.7 Summary

Most engineering design problems are non-linear in nature, and the resulting nonlinear problems often involve multiple local optima. Global optimization is the process of identifying the best of these local optima, otherwise known as the global optimum, in such nonlinear problems. This chapter introduced the major global optimization approaches, starting with basic approaches such as exhaustive search to advanced approaches such as genetic algorithms. The chapter concluded with an overview of the MATLAB Global Optimization Toolbox.

8.8 Problems

Warm-up Problems

8.1 Use the exhaustive search technique to find the global optimum for the following optimization problem.

$$\min f(x) = (x_1 - 6)^2 + 2(x_2 - 3)^2 \tag{8.12}$$

subject to

$$x_1 \in \{1, 3, 5.5, 7\} \tag{8.13}$$
$$x_2 \in \{2, 3.5, 5\} \tag{8.14}$$

8.2 Use the multiple start method to find the global optimum for the following problem.

$$\min f(x) = 2x + 8\cos(3x) \tag{8.15}$$

subject to

$$0 \le x \le 10 \tag{8.16}$$

8.3 Use GA to find the global optimum for the following problem.

$$\min f(x) = x + 7\cos(2x) \tag{8.17}$$

subject to

$$0 \le x \le 10 \tag{8.18}$$

Intermediate Problems

8.4 Consider the following optimization problem. Solve it using the GA Solver in MATLAB, using both the m-files and the graphical user interface.

$$\min f(x) = 5x_1^2 - 23x_1x_2 - 9x_2^2 + 9x_3^2 + 15 \tag{8.19}$$

subject to

$$x_1^2 + x_2^2 = 10 \tag{8.20}$$
$$x_1x_3 \le 100 \tag{8.21}$$
$$3x_1 + 4x_3 = 100 \tag{8.22}$$
$$-100 \le x_1, x_2, x_3 \le 100 \tag{8.23}$$

8.5 Learn the `peaks` command from the MATLAB tutorial available on www. mathworks.com. In this optimization problem, the lower bounds of the two variables are −3 and their upper bounds are 3. Now use GA to determine the global maximum of the `peaks` function within the defined region. Generate a 3D plot of the `peaks` command using `mesh`, and label the global maximum on the plot. Turn in your M-file and the plots.

BIBLIOGRAPHY OF CHAPTER 8

[1] M. Locatelli and F. Schoen. *Global Optimization: Theory, Algorithms, and Applications.* SIAM, 2013.

[2] C. A. Floudas and P. M. Pardalos. *Recent Advances in Global Optimization.* Princeton University Press, 2014.

[3] S. Sumathi, T. Hamsapriya, and P. Surekha. *Evolutionary Intelligence: An Introduction to Theory and Applications with MATLAB.* Springer, 2008.

[4] J. H. Holland. *Adaptation in Natural and Artificial Systems: An Introductory Analysis with Applications to Biology, Control, and Artificial Intelligence.* MIT Press, 1992.

[5] D. E. Goldberg. *Genetic Algorithms in Search, Optimization, and Machine Learning.* Addison-Wesley Publishing Company, Inc, 1989.

[6] The MathWorks, Inc. Global optimization with MATLAB products. `http://www.mathworks.com/videos/global-optimization-with-matlab-products-81716.html`, 2011.

9

Discrete Optimization Basics

9.1 Overview

Previous chapters focused on methods to solve optimization problems that involve continuous design variables. Real life engineering design problems often involve discrete choices. For example, the number of components in a product has to be a positive integer. In this case, corresponding design variables must be restricted to a set of given discrete numbers. This type of optimization problem is called a discrete optimization problem.

Discrete optimization is broadly defined in Sec. 9.2. Section 9.3 describes how to use the exhaustive search approach to solve discrete optimization problems. Section 9.4 illustrates the relaxation approach. Section 9.5 introduces some advanced approaches to address discrete problems, which include genetic algorithms, simulated annealing, and the branch and bound method. The chapter concludes with a summary in Sec. 9.6. More advanced approaches for discrete optimization are presented in Chapter 14, entitled Discrete Optimization.

9.2 Defining Discrete Optimization

Discrete optimization can address many real-world problems and has been applied in a wide range of fields, such as industrial operations, transportation, and finance. A generic discrete optimization problem can be defined as

$$\min_{x,y,z} \ f(x,y,z) \tag{9.1}$$

subject to

$$g(x,y,z) \leq 0 \tag{9.2}$$

$$h(x,y,z) = 0 \tag{9.3}$$

$$x \in Z^m \tag{9.4}$$

$$y \in R^n \tag{9.5}$$

$$z \in A \tag{9.6}$$

where x, y, and z are the design variable vectors; $f(x)$ is the objective function; $g(x)$ and $h(x)$ are inequality constraints and equality constraints, respectively; Z^m is a set of given feasible integers; R^n is a set of real numbers; and A stands for a combinatorial set resulting from given feasible discrete values. Depending on the existence of x, y, and z, discrete optimization problems can be classified into the following five categories.

1. **Pure integer programming problems:** Only x exists. The design variables only take on integer values (see Refs. [1, 2]).
2. **Mixed-integer programming problems:** Both x and y exist. Some design variables take on integer values, while others are allowed to take on continuous values.
3. **Discrete non-integer optimization problems:** Only z exists. The design variables are allowed to be selected from a given set of discrete values.
4. **Binary programming problems:** Only x exists, and it can only take on a value of either 0 or 1. These problems are also called zero-one programming problems.
5. **Combinatorial optimization problems:** Only z exists. The possible feasible solutions are defined by a combinatorial set resulting from the given feasible discrete values.

For a purely finite discrete optimization problem (no continuous variables), all the feasible design variable combinations are known before hand. Theoretically, using exhaustive search, all the possible designs can be evaluated to determine the optimum in the purely finite-discrete optimization problem. In practice, this approach (exhaustive search in Sect. 9.3) only works for very small problems. For medium-size problems and industrial scale problems, this approach becomes computationally prohibitive. Consider a binary programming problem that has 100 design variables. The number of its possible solutions is $2^{100} = 1.27 \times 10^{30}$. In Sec. 9.5, evolutionary algorithms and the branch and bound method are introduced. These methods are particularly effective in solving practical discrete optimization problems.

9.3 Exhaustive Search

Exhaustive search is a straightforward approach that can be leveraged to solve small scale discrete problems. It enumerates all of the feasible candidates. The best solution among them is the optimum.

Exhaustive search is viable when solving optimization problems with a manageable number of combinations. The following example illustrates how to solve a combinatorial optimization problem using exhaustive search. Chapters 8 and 14 provide further pertinent information.

Example: Use exhaustive search to find the optimum of the following combinatorial optimization problem.

$$\min_{x,y,z} \ f(x) = 5(x_1 - 4)^2 + 4x_2x_3 \tag{9.7}$$

Table 9.1. *Design Variable Combinations and Their Objective Function Values*

x_1	x_2	x_3	$f(x)$	x_1	x_2	x_3	$f(x)$
1	2	4	77	1	2	7	101
1	3.5	4	101	1	3.5	7	143
1	5.5	4	133	1	5.5	7	199
3	2	4	37	3	2	7	61
3	3.5	4	61	3	3.5	7	103
3	5.5	4	93	3	5.5	7	159
5.5	2	4	43.25	5.5	2	7	67.25
5.5	3.5	4	67.25	5.5	3.5	7	109.25
5.5	5.5	4	99.25	5.5	5.5	7	165.25

subject to

$$x_1 \in \{1, 3, 5.5\} \tag{9.8}$$

$$x_2 \in \{2, 3.5, 5.5\} \tag{9.9}$$

$$x_3 \in \{4, 7\} \tag{9.10}$$

The number of feasible combinations of the three design variables is $3 \times 3 \times 2 = 18$. These 18 combinations and their objective function values are listed in Table 9.1.

When comparing the 18 objective function values in Table 9.1, the minimum of $f(x)$ is estimated to be 37. The optimal solution is: $x_1 = 3$, $x_2 = 2$, and $x_3 = 4$.

9.4 Relaxation Approach

The relaxation approach assumes that all the design variables in a problem are continuous, and solves the problem using continuous optimization techniques. After the optimal solution to the relaxed continuous optimization problem is obtained, the optimum variable values are rounded off to the nearest feasible discrete solution.

Please note that the solution obtained by this approach can often be sub-optimal. Furthermore, this method does not guarantee that the solution is feasible.

Example: Consider the following linear discrete optimization problem.

$$\min_{x} \quad -2x_1 + x_2 \tag{9.11}$$

subject to

$$6x_1 - 4x_2 \leq 15 \tag{9.12}$$

$$x_1 + x_2 \leq 5 \tag{9.13}$$

$$x_1, x_2 \geq 0 \tag{9.14}$$

$$x_1, x_2 \in Z \tag{9.15}$$

By ignoring the constraint in Eq. 9.15, the problem can be solved as a continuous optimization problem. In this example, MATLAB function linprog is used to solve the relaxed linear programming problem. The MATLAB code is given below.

```
clear
clc

% objective function
f = [-2; 1];

% Define Linear constraints
A=[6 -4; 1 1]; B=[15; 5];
Aeq=[]; Beq=[];

% Define bounds constraints
LB=[0 0]; UB=[];

[x, fopt] = linprog(f, A, B, Aeq, Beq, LB, UB)
```

The final solution obtained is: $x_1 = 3.5$ and $x_2 = 1.5$. Rounding off to the nearest integer yields an integer optimal solution of $x_1 = 4, x_2 = 2$. However, they do not satisfy the inequality constraint given as Eq. 9.13. However, if we round off the continuous solution to $x_1 = 3, x_2 = 1$, we obtain the correct solution.

The following example illustrates how the relaxation approach could yield suboptimal or infeasible solutions.

Example: Use the relaxation approach to solve the following discrete linear programming problem. Examine whether the solution is correct.

$$\min_{x} \quad -5x_1 - x_2 \tag{9.16}$$

subject to

$$10x_1 + x_2 \leq 20 \tag{9.17}$$

$$x_2 \leq 2 \tag{9.18}$$

$$x_1, x_2 \geq 0 \tag{9.19}$$

$$x_1, x_2 \in Z \tag{9.20}$$

Using the relaxation approach, the constraint in Eq. 9.20 is ignored. The relaxed optimization problem is solved using MATLAB function linprog. The MATLAB code is given below.

```
clear
clc
```

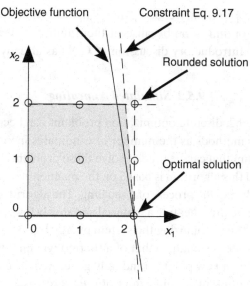

Figure 9.1. The Actual Optimum Solution

```
% objective function
f = [-5; -1];

% Define Linear constraints
A=[10 1; 0 1]; B=[20; 2];
Aeq=[]; Beq=[];

% Define bounds constraints
LB=[0 0]; UB=[];

[x, fopt] = linprog(f, A, B, Aeq, Beq, LB, UB)
```

The optimal solution of the relaxed optimization problem is $x_1 = 1.8$ and $x_2 = 2$. Rounding off to the nearest integer yields an integer optimal solution of $x_1 = 2$ and $x_2 = 2$.

The optimization problem can also be solved graphically as seen in Fig. 9.1. From this figure, note that the rounded solution lies in the infeasible region of the design space. The optimal solution is $x_1 = 2$ and $x_2 = 0$. This example demonstrates that the relaxation approach can be misleading.

9.5 Advanced Options: Genetic Algorithms, Simulated Annealing, and Branch and Bound

9.5.1 Genetic Algorithms

Genetic Algorithms (GAs) are a family of computational algorithms inspired by the principles of evolution described in Darwin's theory. GAs were first conceived by

J.H. Holland in 1975 [3]. Binary coded genetic algorithms are particularly popular choices for discrete optimization because of their ability to deal directly with discrete search spaces. Introductory discussion of GA has already been presented in Chapter 8.

9.5.2 Simulated Annealing

As discussed in Sec. 9.2, discrete optimization problems can become unmanageable using combinatorial methods as the number of candidates increases exponentially. An effective algorithm that can be used to solve these problems is simulated annealing. The idea behind this algorithm is based on the manner in which liquids freeze or metals recrystalize during the process of annealing. The algorithm mimics the metallurgical process of annealing: heating a material and slowly lowering the temperature to decrease defects, thus minimizing the system energy [4]. At the beginning of this method, the initial state is similar to that of a thermodynamic system. At each iteration of the algorithm, a new point is randomly generated. The distance of the new point from the current point, or the extent of the search, is based on a probability distribution. The scale of the distribution is proportional to the temperature. The algorithm accepts all new points that lower the energy; but with a certain probability, also accepts points that raise the energy. By accepting points that raise the energy, the algorithm avoids being trapped in local minima, and is capable of global exploration for better solutions. An annealing schedule is selected to systematically decrease the temperature as the algorithm proceeds. As the temperature decreases, the algorithm reduces the extent of its search to progressively converge to a minimum.

Simulated annealing has been used in various combinatorial optimization problems and has been particularly successful in circuit design problems. For more information on simulated annealing, refer to Chapter 19. More details and examples are also provided in [5].

9.5.3 Branch and Bound

The branch and bound method is a basic technique used to solve discrete programming problems. This method is based on the observation that the enumeration of integer solutions has an inherent tree structure. It enumerates candidate solutions systematically for a discrete optimization problem. Refer to Chapter 14 for more information on this method.

The procedure for the branch and bound method used to solve a *linear integer programming problem* is illustrated in Fig. 9.2 (see Refs. [1, 2]).

The technical terms used in Fig. 9.2 are explained below.

Relaxed continuous LP: A relaxed continuous linear programming (LP) problem is formulated by ignoring the integer constraints. The resulting optimal solution may have some non-integer variable values. If the resulting LP solution has only integer values, the obtained solution is the integer optimal solution.

Ceil: The notation of $\lceil x \rceil$ is defined as the ceiling function. This function returns the smallest integer value that is greater than or equal to x. For example, $\lceil 5.14 \rceil = 6$, $\lceil 10 \rceil = 10$, and $\lceil -8.6 \rceil = -8$.

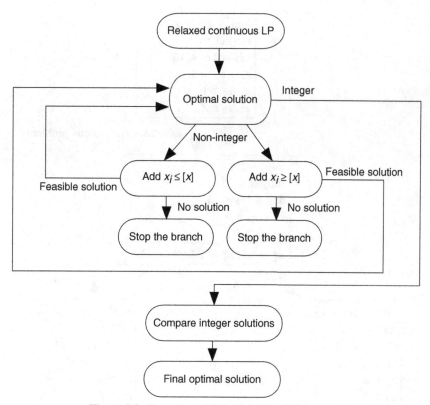

Figure 9.2. Branch and Bound Method Flowchart

Floor: The floor function is denoted $\lfloor x \rfloor$, which returns the largest integer that is less than or equal to x. For example, $\lfloor 5.14 \rfloor = 5$, $\lfloor 10 \rfloor = 10$, and $\lfloor -8.6 \rfloor = -9$.

Add a ceil or a floor: For those design variables with decimal parts in the optimal result, two subproblems are created by imposing a *ceil* or a *floor* on the design variable values, respectively. The following constraint is added to the first subproblem: $x_i \leq \lfloor x \rfloor$. The second subproblem is formulated by adding the constraint $x_i \geq \lceil x \rceil$. The two subproblems are then solved as continuous problems. The solutions of the two subproblems are then examined for fractional parts, and the process is repeated.

Branching: The above process that adds a ceil or a floor is called branching. For a given variable, the branching process is repeated until the relaxed continuous problem with the additional constraints yields either an integer solution or an infeasible solution. The branching process is repeated for all the variables that have fractional solutions.

Bounds: The added ceil and floor constraints are called bounds on the variable values.

A basic implementation of the branch and bound method for linear integer programming problems consists of the following steps (see Refs. [1, 2]). Note that the given procedure only applies to linear integer programming problems.

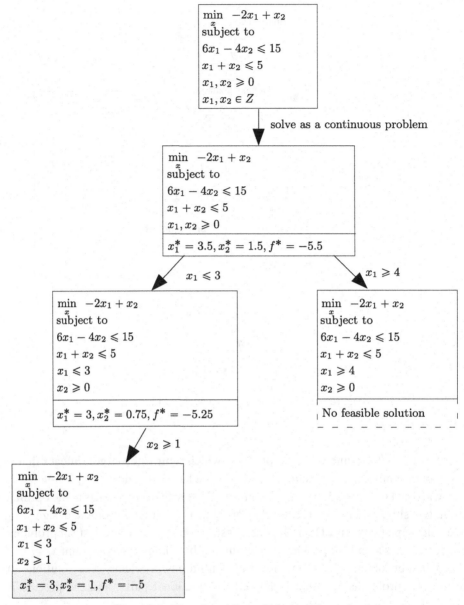

Figure 9.3. Branch and Bound Method Example

1. Formulate a relaxed continuous linear programming problem by ignoring the integer constraints. The relaxed problem is comprised of continuous variables.

2. If the solution to the above relaxed continuous (linear optimization) problem only involves integers, it is the optimal solution. If the solution has non-integer variables, go to the next step

3. Select one non-integer variable and generate two subproblems (two branches). For one of them, add a ceil constraint to the selected non-integer variable and, for the other branch, add a floor constraint to it.

4. If a branch has no feasible solutions, stop this branch. If the solution only has integers, it becomes a candidate for the final optimal solution. If the solution has non-integer values, go back to Step 3.

5. Once the branching process is completed for all the variables, compare all the integer solutions obtained by the different branches. The best solution is considered the final optimal solution.

Example: Solve the following linear discrete optimization problem using the branch and bound method.

$$\min_x \ -2x_1 + x_2 \tag{9.21}$$

subject to

$$6x_1 - 4x_2 \le 15 \tag{9.22}$$

$$x_1 + x_2 \le 5 \tag{9.23}$$

$$x_1, x_2 \ge 0 \tag{9.24}$$

$$x_1, x_2 \in Z \tag{9.25}$$

Figure 9.3 presents the subproblems and their solutions when the branching process begins with the variable x_1. The gray-shaded boxes are the integer solutions obtained during this branching. Note the branching process after the optimal solution $x_1^* = 3, x_2^* = 0.75$, and $f^* = -5.25$. The two further possible branches are $x_2 \le 0$ and $x_2 \ge 1$. The first branch, $x_2 \le 0$, is not feasible, and is not solved further. The second branch, $x_2 \ge 1$ yields the optimal integer solution, $x_1^* = 3, x_2^* = 1$, and $f^* = -5$.

9.6 Summary

In this chapter, a new class of optimization problems was introduced – the discrete optimization problem. Several simple approaches were illustrated with examples. These approaches can be readily applied to solve discrete optimization problems.

9.7 Problems

9.1 Formulate a discrete optimization problem based on your real-world engineering design experience. Solve it using the appropriate optimization method.

9.2 Use exhaustive search to find the optimum for the following discrete optimization problem.

$$\min f(x) = x_1^2 + 3(x_2 - 1)^2 \tag{9.26}$$

subject to

$$x_1 \in \{1, 2.5, 6\} \tag{9.27}$$

$$x_2 \in \{-2, 3\} \tag{9.28}$$

9.3 Consider the two discrete optimization problems given in Sec. 9.4. Reproduce the results of the two problems using the MATLAB function `linprog`. Solve the second problem graphically, and show the actual optimum on your figure.

9.4 Reproduce the results for the example given in Sec. 9.5.3. Turn in your M-file and results.

9.5 Consider the linear discrete optimization problem given in Sec. 9.5.3. Apply the branch and bound method starting with x_2.

BIBLIOGRAPHY OF CHAPTER 9

[1] J. K. Karlof. *Integer Programming: Theory and Practice*. CRC Press, 2005.
[2] D. S. Chen, R. G. Batson, and Y. Dang. *Applied Integer Programming: Modeling and Solution*. John Wiley and Sons, 2010.
[3] J. H. Holland. *Adaptation in Natural and Artificial Systems: An Introductory Analysis with Applications to Biology, Control, and Artificial Intelligence*. MIT Press, 1992.
[4] E. Aarts and J. Korst. *Simulated Annealing and Boltzmann Machines: A Stochastic Approach to Combinatorial Optimization and Neural Computing*. John Wiley and Sons, Inc., 1989.
[5] M. Tsuzuki and T. DeCastro Martins. *Simulated Annealing: Strategies, Potential Uses and Advantages*. Nova Science Publishers, 2014.

10

Practicing Optimization—Larger Examples

10.1 Overview

This chapter covers the practice of optimization in diverse technical areas. What has been learned thus far is the practical knowledge that is needed to apply optimization to different types of designs or systems. We will explore optimizing systems/designs in the following disciplines: chemistry, mechanics, aerospace, automotive, mathematics (data fitting), nuclear, electrical, portfolio management, and business (Ref. [1]). Our ability to optimize these various designs provides us with the skills to apply optimization beyond the confines of this book or of the class you might be taking, and successfully venture into the real world!

10.2 Mechanical Engineering Example

10.2.1 Structural Example

Figure 10.1 represents a ten-bar truss. The members are connected to each other at six nodes, numbered 1 to 6. The truss is assumed to be planar, and every node has two degrees of freedom (*i.e.*, it is free to move both in the x and y directions). The left edge of the truss is fixed to the wall. The displacements u_1 to u_8 are specified at the specific non-fixed nodes. F_1 to F_8 represent loads applied to these nodes. The variables by x_1 to x_{10} denote the cross sectional areas of the truss members. The Young's modulus of the material is $E = 1 \times 10^6 \text{N/mm}^2$. The maximum allowable stress in each bar is $\sigma_{ult} = \pm100\text{N/mm}^2$ (*i.e.*, tension/compression), and the maximum allowable deflection is $\delta_{max} = \pm2$ mm. *We note that the unit of length in the figure and in this problem is mm, and that of force is N.* While the current example is commonly used in the literature, the reader is also encouraged to review the topology optimization survey by Rozvany in Ref. [2].

Imagine you are hired as the optimization expert for the ten-bar truss project. The structural engineer working on the project provides you with a program *blackbox_10bar.m*, which allows you to enter the loads at the nodes and the area of each member, and obtain the corresponding nodal displacements and stresses in the members. Figure 10.2 illustrates this task. The ten-bar truss is subjected to 10 different loading conditions given in Table 10.1.

Table 10.1. *Loading Conditions*

				Forces at Nodes$\times 10^5$N				
i	1	2	3	4	5	6	7	8
1	0.6154	0.0579	0.0153	0.8381	0.1934	0.4966	0.7271	0.7948
2	0.7919	0.3529	0.7468	0.0196	0.6822	0.8998	0.3093	0.9568
3	0.9218	0.8132	0.4451	0.6813	0.3028	0.8216	0.8385	0.5226
4	0.7382	0.0099	0.9318	0.3795	0.5417	0.6449	0.5681	0.8801
5	0.1763	0.1389	0.4660	0.8318	0.1509	0.8180	0.3704	0.1730
6	0.4057	0.2028	0.4186	0.5028	0.6979	0.6602	0.7027	0.9797
7	0.9355	0.1987	0.8462	0.7095	0.3784	0.3420	0.5466	0.2714
8	0.9169	0.6038	0.5252	0.4289	0.8600	0.2897	0.4449	0.2523
9	0.4103	0.2722	0.2026	0.3046	0.8537	0.3412	0.6946	0.8757
10	0.8936	0.1988	0.6721	0.1897	0.5936	0.5341	0.6213	0.7373

(Loading Conditions — row label for rows 1–10)

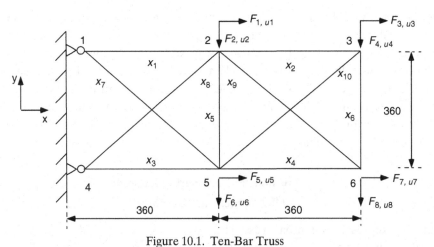

Figure 10.1. Ten-Bar Truss

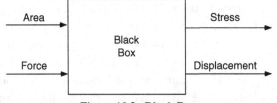

Figure 10.2. Black Box

Your task is to generate the most optimal design for the ten-bar truss by minimizing the mass of the configuration along with the maximum stress for a given set of conditions. The following steps will help you accomplish this optimization process.

1. Generate the row vector F that contains all the 8 forces:

$$F = [61{,}540\ 5{,}790\ 1{,}530\ 83{,}810\ 19{,}340\ 49{,}660\ 72{,}710\ 79{,}480]$$

2. Generate the row vector x that contains all the 10 areas:

$$x = [16.49 \; 4.67 \; 13.34 \; 0.40 \; 0.10 \; 0.10 \; 17.57 \; 11.36 \; 4.06 \; 16.19]$$

3. Use these values in the program *blackbox_10bar.m* to determine (1) stresses and (2) deflections.
4. Double the value of the forces and use these new values in the program *blackbox_10bar.m* to obtain the stresses and deflections.
5. Halve the areas and repeat the process to obtain the corresponding stresses and deflections.
6. Minimize the total material volume of the truss for the loading condition specified in Part 1 and the initial areas specified in Part 2. Note that to obtain the corresponding minimum mass, we would simply multiply by the density. (Hint: In forming the constraint function, use the black box program to determine your stresses and displacements, which should not exceed the maximum design conditions).
7. Minimize the total material volume of the truss for all ten loading conditions individually. (Hint: Change the set of loading conditions in each minimization subroutine. The different set of loading conditions are given in Table 10.1. In addition to the area vector provided in Part 2., randomly generate 9 row sets of area values to obtain the initial area matrix).
8. Minimize the maximum stress along with the total mass of the truss for the loading condition specified in Part 1 and the initial areas specified in Part 2. Plot the Pareto frontier for 11 solutions. (Hint: This is a bi-objective problem where the first objective is the combined mass of the truss, and the second objective is the maximum stress).
9. In this case, you are not so blessed as to have the Black Box simply provided to you. You started your job, and since you graduated in Mechanical Engineering (or Civil or Aerospace), your team simply expects that you will develop the Black Box yourself. Your task is to develop this Black Box. (i) Provide clear documentation and explain your assumptions in developing the analysis for this Black Box. Comment on how truss or frame structures assumptions apply here, and your pertinent decisions. (ii) Code your development in your own Black Box. (iii) Solve Item 8 above using your own Black Box; then using the one provided. Comment on the relative accuracies and computation times between the two Black Boxes, yours and the one provided. (*This part of the problem (9.) is geared towards graduate students in mechanical, aerospace, or civil engineering*).

10.2.2 Tolerance Allocation Problem

Consider the example of a two-part assembly shown in Fig. 10.3. The dimensions of the two parts, A and B, are given by x_1 through x_{12}. Parts A and B are mass produced by the same manufacturing company. To understand the concept of tolerance, assume that the design requirement for the dimension x_1 is 50 in. Consider a

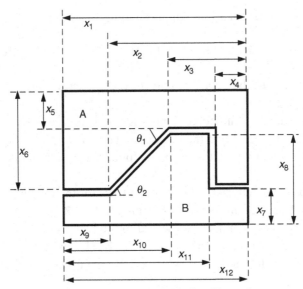

Figure 10.3. Tolerance Design Example

batch of 10,000 parts of A. If we measure the dimension x_1 of each part in the batch, it will not be exactly equal to 50 in. There is some variability in the measurement of x_1 in all the 10,000 parts. This variability can result from measurement errors or the manufacturing process itself. Each measurement of the dimension x_1 for the batch falls within a range $50 \pm tol$ inch, where *tol* is termed as the *tolerance* of the dimension.

It is desirable to have the tolerances be as close to zero as possible to ensure a high quality product. However, smaller tolerances require precise manufacturing processes, which, in turn, result in higher manufacturing costs. The range of $50 \pm tol$ inch mentioned above specifies the acceptable range of the dimension x_1. Parts that do not conform to the above specification are rejected, which are either scrapped or reworked, thus adding to the manufacturing costs. Setting a "good" tolerance value for a dimension involves understanding the tradeoff between cost and quality. Multiobjective optimization can be effectively applied in these problems.

To understand how to mathematically represent the variability in the part dimensions in the optimization formulation, the concept of a random variable and probability theory are used. A detailed description of probability theory can be found in probability textbooks (Ref. [3, 4]). In this problem, we present a brief summary of the basics of probability, normal distribution in particular.

Basic Concepts in Probability Theory

An event that can result in different outcomes, even though it is repeated in the same manner every time, is known as a random event. In the above example, each manufacturing event for the 10,000 parts in the batch results in a different length

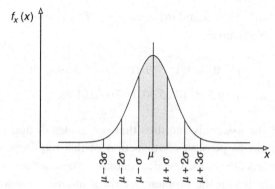

Figure 10.4. Normal Distribution

x_1. In probability theory, a variable referred to as a *random variable* is associated with the outcome of the random event. For the above example, the length of the dimension x_1 in each part of the batch is a random variable. Random variables are denoted by upper case letters. The dimensions of Parts A and B, x_1 through x_{12}, can be thought of as random variables, denoted X_1 through X_{12}.

A function that assigns probability values to a random variable is known as the probability density function (PDF). There are several random variables with standard PDF definitions that are typically used in engineering applications, such as uniform, Gaussian, and exponential distributions. In this example, we use Gaussian (or normal) distribution to represent the probabilities of the random variables X_1 through X_{12}. Figure 16.3 shows the PDF of a normal distribution.

The quantities μ and σ in Fig. 10.4 are referred to as the *mean* and the *standard deviation* of the normal distribution, respectively. The PDF curve provides information about the probabilities of random variables. Consider two points on the x-axis, such as $\mu + \sigma$ and $\mu - \sigma$ in Fig. 10.4. The probability that the random variable, X, lies between $\mu + \sigma$ and $\mu - \sigma$ is given by the area under the PDF curve, denoted by the shaded region in Fig. 10.4. The probabilities that X falls within the intervals $\mu \pm \sigma$, $\mu \pm 2\sigma$, and $\mu \pm 3\sigma$ (see Fig. 10.4) for a normal distribution are 0.6827, 0.9545, and 0.9973, respectively [4]. The area under the entire PDF curve is always equal to one.

The two quantities, mean and standard deviation, define the bell-shape of the normal distribution illustrated in Fig. 10.4. The normal distribution is symmetric about its mean. The smaller the value of σ, the narrower the bell-shaped PDF. In the manufacturing example, the closer the tolerance on the dimension X_1, the lower its variability. Thus, achieving closer tolerances requires reducing the standard deviation of the random variable X_1.

Finding Optimal Tolerances

To formulate the multiobjective problem that models the tradeoff between the tolerances and the manufacturing costs, the tolerances of the variables X_1 through X_{12} are assumed to be six times their standard deviations. The nominal or the mean values of X_1 through X_{12} are given as

$$X_n = [50.0\ 40.001,25\ 20.05\ 9.998,5\ 9.998,5\ 30.0,$$

$$10.0\ 30.0\ 10.05\ 30.0\ 40.0\ 50.0].$$

The quality of the assembly requires that the angles θ_1 and θ_2 in Fig. 10.3 be as close to each other as possible. Satisfaction of this requirement imposes tight tolerances on the parts dimensions, which increases manufacturing costs.

The design variables for the problem are the standard deviations of the dimensions. The first objective, J_1, is to minimize cost, which is given as a function of the standard deviations of the design variables as follows.

$$J_1 = 10 \sum_{i=1}^{12} \exp(-b_i * \sigma_{Xi}) \tag{10.1}$$

where $b = [50\ 50\ 50\ 50\ 50\ 50\ 50\ 50\ 50\ 50\ 50\ 50]$, and σ_{Xi} is the standard deviation of the i-th dimension. Note that the above cost function increases as the standard deviations decrease.

The second objective, J_2, ensures that the angles θ_1 and θ_2 are as close to each other as possible. Note that the tangent of the angles θ_1 and θ_2 can be expressed in terms of the dimensions X_6, X_5, X_2, X_3 and X_9, X_{10}, X_7, X_8, respectively. Since the above dimensions are random variables, the tangent of the angles θ_1 and θ_2 are also random. To ensure that θ_1 and θ_2 are as close to each other as possible, consider the random variable $\theta_1 - \theta_2$ and minimize its variation. One possible measure of the variation of $\theta_1 - \theta_2$ is its standard deviation, J_2, given as follows.

$$J_2 = \sqrt{V} \tag{10.2}$$

$$V = (X_n(2) - X_n(3))^2(\sigma_{X8}^2 + \sigma_{X7}^2)$$

$$+ (X_n(8) - X_n(7))^2(\sigma_{X2}^2 + \sigma_{X3}^2)$$

$$+ (X_n(6) - X_n(5))^2(\sigma_{X10}^2 + \sigma_{X9}^2)$$

$$+ (X_n(10) - X_n(9))^2(\sigma_{X6}^2 + \sigma_{X5}^2) \tag{10.3}$$

Where $X_n(i)$ is the i-th element of the mean vector of X (given as X_n); V is the variance of $\theta_1 - \theta_2$; and the square root of the variance yields the standard deviation.

The multiobjective optimization problem formed to obtain the optimal tolerances from a cost perspective is presented below.

$$\min_{\sigma_X}\{J_1, J_2\} \tag{10.4}$$

subject to

$$0.0001 \le \sigma_X \le 1 \tag{10.5}$$

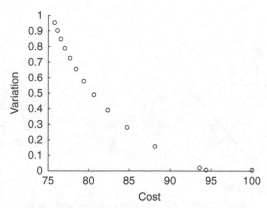

Figure 10.5. Pareto Frontier for the Tolerance Allocation Problem

$$J_1 \leq 100 \tag{10.6}$$

$$J_2 \leq 1 \tag{10.7}$$

$$-(X_n(6) - X_n(5)) + (X_n(8) - X_n(7))$$

$$+6\sqrt{\sigma_{X6}^2 + \sigma_{X5}^2 + \sigma_{X8}^2 + \sigma_{X7}^2} \leq 0 \tag{10.8}$$

$$-(X_n(3) - X_n(4)) + (X_n(11) - X_n(10))$$

$$+6\sqrt{\sigma_{X3}^2 + \sigma_{X4}^2 + \sigma_{X11}^2 + \sigma_{X10}^2} \leq 0 \tag{10.9}$$

Equation 10.5 represents the maximum and minimum allowable standard deviations for X_1 through X_{12}. Equations 10.8 and 10.9 ensure that the clearances between Parts A and B are positive, as shown in Fig. 10.3.

The Pareto frontier for the above bi-objective problem is illustrated in Fig. 10.5. The weighted sum method is used to obtain the Pareto frontier. Based on the relative preferences, any of the Pareto solutions may be chosen as the final design. From the design variable values at the chosen Pareto solution, the tolerances can be computed.

10.3 Aerospace Engineering Example

When designing an aircraft landing gear, the wheel track (the length between the left-most and the right-most wheels when looking at a front view) has several requirements that must be considered. These include (i) the ground lateral control, (ii) the ground lateral stability, and (iii) the structural integrity. The minimum allowable value for the wheel track must satisfy the ground lateral control and the ground lateral stability requirements. The maximum allowable value for the wheel track must satisfy the structural integrity requirements (Ref. [5]).

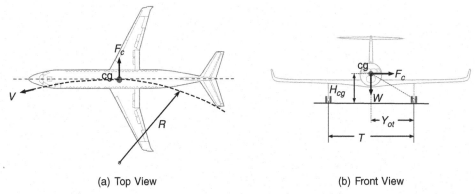

(a) Top View   (b) Front View

Figure 10.6. An Aircraft in a Ground Turn

10.3.1 Ground Controllability

The wheel track must be sufficiently wide so that the aircraft does not roll over while taxiing on the ground. The centrifugal force (F_c) during a turn, which may cause the aircraft to roll, is given by

$$F_c = m\frac{V^2}{R} \tag{10.10}$$

where m represents the aircraft mass, V denotes the aircraft ground speed, and R is the radius of the turn (see Fig. 10.6(b)). The sum of the two contributing moments (the moment of the centrifugal force (F_c) and the moment due to the aircraft weight, $W = mg$) about the gear is given by

$$\sum M_o = 0 \Rightarrow mg \cdot T/2 + F_c \cdot H_{cg} \tag{10.11}$$

where the parameter H_{cg} is the distance of the aircraft's center of gravity, cg, from the ground, and T is the length of the wheel track. The minimum constraint for the wheel track (T) is given by

$$T > 2\frac{F_c H_{cg}}{mg} \tag{10.12}$$

10.3.2 Ground Stability

While taxiing on the ground, wind affects the stability of the aircraft and, as such, must be considered during the design process. A cross wind (*i.e.*, perpendicular to the ground path or fuselage centerline of the aircraft) is the most important wind force when designing for ground stability.

The cross wind force (F_W) on an aircraft can be modeled as a drag force given by

$$F_w = \frac{1}{2}\rho V_W^2 A_S C_{D_s} \tag{10.13}$$

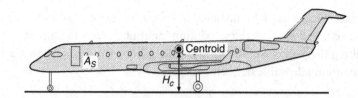

Figure 10.7. Aircraft Side Area and Its Centroid

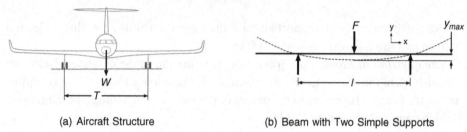

(a) Aircraft Structure   (b) Beam with Two Simple Supports

Figure 10.8. Front View of the Aircraft Structure Modeled as a Beam with Two Simple Supports

where ρ is the air density, V_W denotes the wind speed, and A_S is the aircraft side area (hatched area in Fig. 10.7). The parameter C_{D_s} is the side drag coefficient of the aircraft, and it varies from 0.3 to 0.8. To prevent an aircraft from overturning due to the cross wind, the wheel track (T) satisfies

$$T > 2\frac{F_W H_c}{mg} \qquad (10.14)$$

where H_c is the distance of the centroid from the ground.

10.3.3 Structural Integrity

When seen from the front, the aircraft structure can be modeled as a beam with a few simple supports (see Fig. 10.8). At the main gear station, the wing is the beam and the two main wheels are the simple supports. The wheel track is then the distance between these two supports.

The maximum deflection (y_{max}) in a beam (wing) is given by

$$y_{max} = -\frac{F_{m_{max}} T^3}{48EI} \qquad (10.15)$$

where $F_{m_{max}}$ is the maximum load on the main gear, E is the modulus of elasticity, and I is the beam area moment of inertia. The maximum static load, which is carried by the main gear, can also be estimated as

$$F_{m_{max}} = \frac{B_{n_{max}}}{B} mg \qquad (10.16)$$

where B is the wheel base (the distance between the nose gear and the main gear along the x-axis), and $B_{n_{max}}$ is the maximum length between the aircraft cg and the nose gear along the x-axis. Substituting Eq. 10.16 into Eq. 10.15, the wheel track in terms of maximum allowable deflection is given by

$$T = \left[\frac{48EIBy_{max}}{mgB_{n_{max}}} \right]^{\frac{1}{3}}$$

(10.17)

For a twin engine jet transport aircraft, the maximum allowable wing deflection is 0.03 m and the take-off mass is 50,000 kg.

The question at hand: Your job is to determine the range of wheel tracks that will satisfy all of the requirements discussed in this Aerospace Engineering Example. Explain the practical meaning of the nature of your answer, including an examination of which constraints are active of inactive.

Constants are given as

$$A_s = 150m^2$$

(10.18)

$$I = 0.003m^4$$

(10.19)

$$C_{D_s} = 0.8$$

(10.20)

$$E = 70Gpa$$

(10.21)

$$m = 50,000kg$$

(10.22)

$$y_{max} = 0.03m$$

(10.23)

$$H_{cg} = 3.5m$$

(10.24)

and constraints as

$$H_c = 3.6m$$

(10.25)

$$V = 20knot$$

(10.26)

$$40knot < V_w < 60knot \qquad (1knot = 0.5144m/s)$$

(10.27)

$$25m < R < 35m$$

(10.28)

$$12m < B_{n_{max}} < 14m$$

(10.29)

$$14m < B < 16m$$

(10.30)

$$1.1455kgm^{-3} < \rho < 1.422,4kgm^{-3}$$

(10.31)

10.4 Mathematical Example
10.4.1 Data Fitting

Data fitting is the process of constructing a mathematical function to fit a series of data points. Data points can be experiment results, simulation results, or observations of

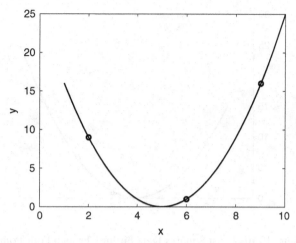

Figure 10.9. Data Fitting Through Three Points

n natural phenomena. There are many different methods for data fitting. This section will use the least squares approach, which requires the use of optimization.

Let us look at the three points marked as circles in Fig. 10.9. The coordinates of the three points are $(2, 9)$, $(6, 1)$, and $(9, 16)$.

We will use a quadratic function, expressed by $y = ax^2 + bx + c$, to fit the data. To fit a quadratic function that passes through all three points, the following system of equations must be solved. In the following system of equations, a, b, and c are unknowns and their values can be obtained exactly by solving the equations. The equations are satisfied by $a = 1$, $b = -10$, and $c = 25$.

$$2^2a + 2b + c = 9 \tag{10.32}$$

$$6^2a + 6b + c = 1 \tag{10.33}$$

$$9^2a + 9b + c = 16 \tag{10.34}$$

The fitted curve of the above example passes through all three points. If a quadratic function is required to fit more than three points, its curve can pass through all points only in special cases. The method of least squares is a popular and easy-to-implement approach to approximate the solution of overdetermined systems (sets of equations in which there are more equations than unknowns). The overall solution obtained by the least squares data fitting minimizes the sum of the squares of the errors when solving every single equation. It is used in Sec. 10.4.2 to fit a quadratic function to four points.

10.4.2 Least Squares Data Fitting

The least squares data fitting minimizes the sum of the squared residuals. The residuals are the differences between the recorded values and the fitted (or approximate) values provided by a model (Ref. [6]). In this example, there is one more point $(7.5, 6.5)$ in addition to the three points in Sec. 10.4.1. This new point is not on the

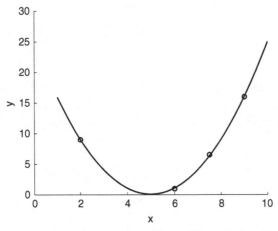

Figure 10.10. Least Squares Data Fitting Through Four Points

curve given by $y = x^2 - 10x + 25$. A fitted quadratic function cannot pass through all the four points. Instead, the least squares data fitting is used to fit a quadratic function that minimizes the sum of the errors between the fours points and the corresponding points on its curve. Let $\tilde{y}$ represent the fitted quadratic function. The sum of the errors is expressed as

$$\sum_{n=1}^{4} (\tilde{y} - y)^2 = (2^2 a + 2b + c - 9)^2 + (6^2 a + 6b + c - 1)^2$$

$$+ (9^2 a + 9b + c - 16)^2$$

$$+ (7.5^2 a + 7.5b + c - 6.5)^2 \tag{10.35}$$

Perform an unconstrained minimization using Eq. 10.35 as the objective function. After the minimization problem is solved using MATLAB, the corresponding values of the three parameters are $a = 0.9937$, $b = -9.9198$, and $c = 24.8524$. The sum of the squared residuals is 0.0383. The four points, as well as the curve given by the fitted quadratic function, are plotted in Fig. 10.10. Note that the points are very close to, but not exactly on, the fitted curve.

10.5 Civil Engineering Example

The company you work for is planning on building a new facility to account for recent growth. You have been chosen to design the exterior walls of the building. The indoor temperature is to remain at $T_i = 25°C$ for a comfortable indoor environment. During the winter months, the outdoor temperature is consistently at $T_o = 0°C$. Your objective is to minimize the heat loss during the winter months, while also being cost conscious.

In the building and construction industry, insulating materials are commonly described by their thermal resistance (R-value). The R-value being discussed is the

Table 10.2. *Material Properties*

Material	k (W/mK)	Cost ($/cm)
Concrete block	0.688	1
R-15 insulation board	0.04	15
Wood	0.212	3
Brick	1.24	0.50

unit thermal resistance. This is used for a unit value of any particular material. It is expressed as the thickness of the material divided by the thermal conductivity. For the thermal resistance of an entire section of material, the unit thermal resistance is divided by the thickness of the material. The larger the R-value, the better the building insulation's effectiveness (Ref. [7]).

Heat transfer through an insulating layer is analogous to electrical resistance. The heat transfer in a simple system can be solved by using electrical resistance in series with a fixed potential. The differences between electrical and thermal are: (i) the resistances of thermal systems are thermal resistances, and (ii) the potential is the difference in temperature from one side of the material to the other. The resistance of each material to heat transfer depends on the specific thermal resistance [R-value]/[unit thickness].

Assuming 1-D steady heat transfer, given by Eq. 10.36.

$$\frac{d^2 T}{dx^2} = 0 \tag{10.36}$$

The boundary conditions are given by

$$-k\frac{dT(0)}{dx} = h_0(T_0 - T(0)) \tag{10.37}$$

and

$$-k\frac{dT(L)}{dx} = h_i(T(L) - T_i) \tag{10.38}$$

where k is the thermal conduction coefficient, and h is the convection heat transfer coefficient. The convective heat transfer coefficients are $h_i = 2\ [\frac{W}{mK}]$ and $h_0 = 22\ [\frac{W}{mK}]$ for the inner and outer regions, respectively.

Your boss said that the wall must cost no more than $100 and be no thicker than $1m$. Instead of solving the differential equation by hand, you are provided with the file `steadyeqn.m` to calculate the inner and outer wall temperatures. Using the above equations along with the material properties in Table 10.2, minimize the heat transfer rate and minimize the cost per unit wall area.

1. Plot the Pareto frontier for each of the four materials on the same graph.
2. Plot the Pareto frontier for the entire range of materials and thicknesses.

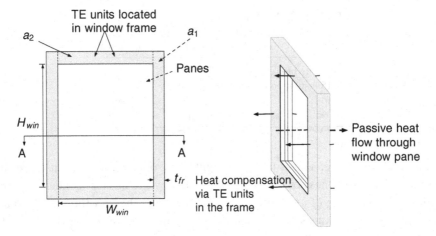

(a) Heat Transfer Process Taking Place through TE Window

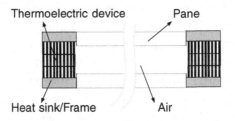

(b) Enlarged Broken View of Section A - A
Showing TE Integration in Window Frame.

Figure 10.11. Schematic of TE Window

10.6 Electrical Engineering Example

10.6.1 Introduction to Thermoelectric Window Design

The Thermoelectric (TE) window design presented in this example uses TE units to actively transfer heat to maintain the desired indoor temperatures. TE units are solid state devices that actively transfer heat in designated directions when supplied with electric power (see Refs. [8, 9, 10, 11]). They are sufficiently small to be integrated into a window. The simplified schematic of the TE window depicted in Fig. 10.11 indicates that the TE units are installed within the frame of the window, which, for practical purposes, also performs as a heat enhancer (heat sink). TE units in the window facilitate heat flow (solid lines) in the direction opposite that of the passive heat flow through the window panes (broken lines), as illustrated in Fig. 10.11(a). Though not shown in Fig. 10.11(a), the frame of the TE window will be equipped with fins to maximize heat transfer. With this design, it is possible to integrate TE units in any glazing system once the frame is modified to accommodate the TE units.

To achieve a high heat transfer rate through the TE units installed on the window frame, the connection of the TE units in their electric network should be optimized. Suppose the TE units are connected in N_p parallel circuits. In each of the parallel

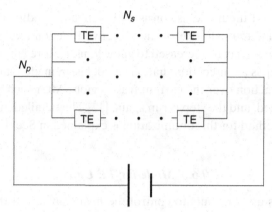

Figure 10.12. Electric Network

circuits, there are N_s TE units connected in series. The electric network is shown in Fig. 10.12.

The TE units shown in Fig. 10.12 are integrated into the window frames to transfer heat from inside to the outside in summer. Under this condition, the function of the TE units is to cool the air inside the room. The higher the rate of heat transfer through the TE units, the better the performance. The objective of this optimization is to maximize the total heat transfer rate through the TE units to the outside, expressed as $\dot{Q}_{cold}$. The numbers of TE units in series and in parallel, the total current, and the temperature difference across the TE units are the four parameters to be varied in the optimization. In this problem, the TE units are modeled based on their configuration and properties. The trust region method is used to generate an approximation for the optimization problem.

10.6.2 Brief Introduction to the Trust Region Method

The trust region method generates a series of intermediate steps with the help of a quadratic model of the objective function. This method defines a region around the current iterate within which it *trusts* the model to be an adequate representation of the objective function. It then defines the step leading to an approximate minimum of the model in this region. A detailed discussion of the theory and implementation of the method can be found in (Refs. [12, 13, 14, 15]).

The trust region method chooses the direction and length of the step simultaneously. If a step is not acceptable, it reduces the size of the region and finds a new minimum. In general, the direction of the step changes whenever the size of the trust region is altered. The size of the trust region is critical to the effectiveness of each step. If the region is too small, the algorithm misses an opportunity to take a substantial step that will move it much closer to the minimum of the objective function. If it is too large, the minimum of the model may be far from that of the objective function in the region, and the size of the region will need to be reduced. In practical algorithms, the size of the region is chosen according to the performance of the algorithm during previous iterations. If the model is consistently reliable, producing good steps and accurately predicting the behavior of the objective function along these steps, the size of the trust region may be increased to allow longer, more ambitious, steps to be taken. A failed step is an indication that our model is an inadequate representation of the objective function over the current trust region. After such a step, the size of the region is reduced, and the step is run again. [12] The detailed implementation of the trust region method for this optimization is explained in Sec. 10.6.4.

10.6.3 Modeling TE Units

The ATI windows use TE units to control the heat transferred through the inner panes (Ref. [9]). Each TE unit consists of thermocouples which, when supplied with electric current, induce heat flow in the direction of the current. This is known as the Peltier effect. Because of the thermocouples' electrical resistance, heat is generated. This is known as the Joules effect. As a result of the two conflicting effects, heat is absorbed on the cold side and released from the hot side. A temperature difference is created across the TE units. On the cold side of the TE units, the heat rate is predicted as

$$\dot{Q}_{cold} = 2N_{te}N\left[\alpha I_{te}T_c - \frac{I_{te}^2\rho}{2G} - \kappa\Delta T_{te}G\right] \tag{10.39}$$

where N_{te} is the number of TE units; N is the number of thermocouples in each TE unit; α is the Seebeck coefficient; I_{te} is the electric current; T_c is the cold side temperature; ρ is the resistivity; G is the geometry factor, which represents the area to thickness ratio of the thermocouple; κ is the thermal conductivity; and ΔT_{te} is the temperature difference across the thermocouple. For a given TE unit, α, ρ, and κ are temperature dependent properties, N and G are constants, and are all provided by the manufacturer. The remaining four variables, I_{te}, T_{te}, T_c, and N_{te}, are design variables for the problem presented in Sec. 10.6.4.

TE units are connected in an electrical network, and power is supplied to every TE unit. The TE units used for the TE window design are divided into several groups. The number of the groups is N_p. There is the same number of TE units in each group, which is N_s. Within each group, the TE units are connected in series. Then, all the groups are connected in parallel. The electric voltage supplied to each group is the same. The connection of the TE units is shown in Fig. 10.12. The total number of TE units is

$$N_{te} = N_sN_p \tag{10.40}$$

The voltage drop across the TE unit is given by

$$V_{te} = 2N\left[\frac{I_{te}\rho}{G} + \alpha\Delta T_{te}\right] \tag{10.41}$$

The heat released from the hot side of the TE unit is the combination of the heat absorbed by the TE and the heat generated by electric current, which is given by

$$\dot{Q}_{hot} = N_{te} I_{te} V_{te} + \dot{Q}_{cold} \tag{10.42}$$

The maximum allowable applied current, I_{max}, is given by

$$I_{max} = \frac{\kappa G}{\alpha} \left[\sqrt{(1 + 2ZT_h)} - 1 \right] \tag{10.43}$$

where Z is the figure-of-merit provided by the manufacturer, T_h is the hot side temperature, and the other variables are defined above.

The maximum allowable temperature difference, ΔT_{max}, is given by

$$\Delta T_{max} = T_h - \left[\frac{\sqrt{(1 + 2ZT_h)} - 1}{Z} \right] \tag{10.44}$$

10.6.4 Solving Optimization Problem

The optimization of the connection of the TE units can be solved as follows:

1. **Objective of Optimization**

   The TE units are integrated into the window frames to transfer the heat from the inside of a room to the outside in summer. In this condition, the function of the TE units is to cool the air inside the room. The objective of optimization is to maximize the overall heat transfer rate through the TE units to the outside, which is $\dot{Q}_{cold}$. The number of the TE units in series, N_s, the number of the TE units in parallel, N_p, the total electric current, I, and the temperature difference across the TE units, δT, are the four variables for optimization.

2. **Physical Constraints**

   The TE units are fixed on the frame of the window. The total area of all the TE units should not exceed the area of the frame, A_{frame}. The total area of the TE units is expressed as

   $$A_{all} = N_s N_p A_{te} \tag{10.45}$$

   The electric current in each set of TE units should not exceed the maximum allowable current.

   $$I_{te} \le I_{max} \tag{10.46}$$

   According to energy conservation, the outside temperature is less than the sum of the inside temperature and the temperature difference across the TE units.

   $$T_{out} \le T_{cold} + N_s \Delta T_{te} \tag{10.47}$$

Bounds on the total current are used to avoid excessively large power requirements or unreasonably small power consumption.

$$0.01 \leq N_p I_{te} \leq 100 \tag{10.48}$$

3. Modeling the Optimization Problem

Considering the objective of the optimization and the constraints, the optimization problem is as follows.

$$\min_{\{\Delta T, I, N_s, N_p\}} \dot{Q}_{cold}, \tag{10.49}$$

subject to

$$N_s N_p A_{te} \leq A_{frame}, \tag{10.50}$$

$$I = N_p I_{te}, \tag{10.51}$$

$$I_{te} \leq I_{max}, \tag{10.52}$$

$$T_{out} \leq T_{cold} + N_s \Delta T_{te}, \tag{10.53}$$

$$\Delta T_{te} \leq \Delta T_{max}, \tag{10.54}$$

$$0.01 \leq N_p I_{te} \leq 100. \tag{10.55}$$

4. Optimization Procedure

The trust region method is used for optimization. At the k iteration, the model of the TE units is approximated by a quadratic Taylor series expansion.

$$\dot{Q}_{cold_{approx}}(p) = \dot{Q}_{cold}(x_k) + G_k^T p + \frac{1}{2} p^T H_k p \tag{10.56}$$

Where $\dot{Q}_{cold_{approx}}(p)$ is the approximated objective function; p is the step length; $\dot{Q}_{cold}(x_k)$ is the optimal result from the last iteration and also the initial value of the objective function in the current iteration; G_k^T is the gradient at the starting point, and H_k is the Hessian at the starting point.

In each iteration, a ratio is defined to evaluate the agreement between the approximate objective function and the actual objective function, as given by

$$\rho_k = \frac{\dot{Q}_{cold}(x_k) - \dot{Q}_{cold}(x_k + p)}{\dot{Q}_{cold_{approx}}(0) - \dot{Q}_{cold_{approx}}(p)} \tag{10.57}$$

The approximate objective in Eq. 10.56 is optimized using MATLAB. In each iteration, the approximate model is optimized as follows.

Given a small number $\hat{\Delta} > 0$, $\gamma = 0.3$, $\Delta_0 = \gamma \hat{\Delta}$, $\alpha = 1$, and $\mu = \frac{1}{8}$;
for k=1, 2, ...
Solve optimization problem for the approximate model of Eq. 10.56 in MATLAB.

Check the improvement that $\Delta \dot{Q}_{cold}$ has on the value of the actual objective function.

if $\Delta \dot{Q}_{cold} \leq \alpha$
Stop optimization. The actual value of the objective
function $\dot{Q}_{cold}$ is the final optimal value.
else continue.
Evaluate ρ_k in Eq. 10.57
if $\rho_k \leq \frac{1}{4}$
$\quad \Delta_{k+1} = \mu \Delta_k$
else
$\quad$ if $\rho_k \geq \frac{3}{4}$ and $\|p_k\| = \Delta_k$
$\quad \quad \Delta_{k+1} = min(2\Delta_k, \hat{\Delta})$
$\quad$ else $\Delta_{k+1} = \Delta_k$
if $\rho_k \geq \mu$
$\quad x_{k+1} = x_k + p_k$
else
$\quad x_{k+1} = x_k$
end (for).

10.6.5 Results

The optimization of the problem in Sec. 10.6.4 is performed by MATLAB. The maximum value of the objective function is $Q_{cold} = 1924$. The optimal values of the design variables obtained from the optimization are $\Delta T = 6, I = 23.36, N_s = 16.8$, and $N_p = 6.6$. Since the numbers of the TE units in series and in parallel are integers, the values of N_s and N_p are rounded to the closest integers. The final optimal results are $\Delta T = 6, I = 23.36, N_s = 17$, and $N_p = 7$. The optimization stops only after four iterations to evaluate the approximate model.

10.7 Business Example

You are put in charge of rearranging the dining room at the restaurant you work at. The owner has told you that your salary will now be measured as a percentage of all food sold. Therefore, you have chosen to use optimization to maximize your paycheck.

The restaurant is a square with each side measuring 40 ft. You are free to select the size, position, and number of circular tables within the dining hall. In order for the kitchen staff to operate properly and the patrons to feel comfortable, the owner has decided that there must be a minimum distance of 3 ft between any two surfaces. This includes between tables, as well as away from the walls. In order to please the wait staff (whose salary is based on tips), you have chosen a minimum of two seats per table. The minimum radius of these two person tables is 1 ft.

For every 2.5 ft. of additional circumference, you may add a place setting. If you assume that your salary is directly proportional to the number of seats you can fit (neglecting atmosphere, quality of food, sanitary requirements, fire regulations, etc.), formulate and solve a problem to maximize your paycheck.

10.8 Summary

In order to fully appreciate the imperative role of optimization in designing systems - from simple everyday products and complex engineering systems, to non-engineering systems - it is important to begin the application of optimization to practical problems. With this perspective in mind, this chapter provided the opportunity to apply optimization to diverse practical problems. These included (i) designing a truss structure, (ii) designing an aircraft landing gear, (iii) data fitting, (iv) designing a thermoelectric window, and (v) arranging the seating pattern in a restaurant. Pertinent mathematical models were also provided for each problem, such that additional references were not required to formulate and solve each optimization problem.

10.9 Problems

Warm-up Problems

10.1   Prove the results of Sec. 10.4.1. Briefly discuss the issues as you see them.

Intermediate Problems

10.2   A structural engineering example is presented in Sec. 10.2.1, solve Items 1. through 5. Explain your work throughout.

10.3   A structural engineering example is presented in Sec. 10.2.1, solve Items 1. through 8. Explain your work throughout.

10.4   Generate the Pareto frontier for the tolerance allocation problem shown in Fig. 10.5 with 20 points. Prepare a five-page PowerPoint presentation for a customer who wants to understand the tradeoffs involved in the problem. Also submit a concise report that explains how you developed your presentation results. The customer will give your report to her technical group leader who will pass judgement on the credibility of your findings.

10.5   In Section 10.3, an aerospace engineering example is provided. You are to determine the range of wheel track values that will satisfy all of the requirements discussed in this Aerospace Engineering Example. Explain the practical meaning of the nature of your answer, including an examination of which constraints are active or inactive, and any pertinent implications. Discover any tradeoffs that result from the range provided in your answer. Prepare a five-page PowerPoint presentation of your findings, and a concise report that supports your results.

10.6   Solve the problem in Section 10.4.2, using computational optimization.

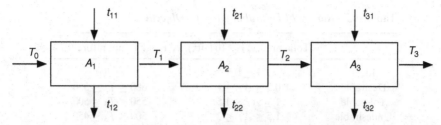

Figure 10.13. Heat Exchanger Network Design Problem

10.7 Solve the problem in Section 10.4.2, using analytical optimization. That is, find the necessary conditions, and solve them using linear algebra.

10.8 (i) Solve the problem in Section 10.5. (ii) Prepare a five-page PowerPoint presentation of your findings. (iii) Submit a brief report that supports your analytical developments.

10.9 HEAT EXCHANGER: In this problem, we are interested in designing a three-stage heat exchanger network (Refs. [16, 17]). A schematic of the heat exchanger network is shown in Fig. 10.13. A cold stream of fluid of a given flow rate, W, and specific heat, C_p, is heated from temperature T_0 °F to T_3 °F using three heat exchangers arranged in three stages. At each stage, the cold fluid is heated by a hot fluid of the same flow rate, W, and specific heat, C_p, as for the cold fluid. The temperatures of the hot fluid entering Stages 1, 2, and 3 are t_{11}, t_{12}, and t_{13}, respectively. The overall heat transfer coefficients U_1, U_2, and U_3 of the heat exchangers are known.

Our focus is to minimize the sum of the areas of the individual heat exchangers, while maximizing the final temperature of the cold fluid, T_3. We assume that the initial temperature of the cold fluid, T_0, is given.

A total of six inequality constraints are imposed in this design. The first three arise from the fact that the rate of heat transferred to the cold fluid is less than or equal to the rate of heat lost by the hot fluid. For the i-th heat exchanger, this constraint can be expressed as

$$WC_p(T_i - T_{i-1}) \le WC_p(t_{i1} - t_{i2}) \qquad i = \{1,2,3\} \qquad (10.58)$$

The remaining inequality constraints reflect that the heat gained by the cold fluid is less than or equal to the heat lost by the heat exchanger. For the i-th heat exchanger, this constraint can be expressed in a simplified form as

$$WC_p(T_i - T_{i-1}) \le U_i A_i(t_{i2} - T_{i-1}) \qquad i = \{1,2,3\} \qquad (10.59)$$

The areas and temperatures are given by the following upper and lower bounds.

$$100 \text{ ft}^2 \le A_i \le 10,000 \text{ ft}^2 \qquad i = \{1,2,3\} \qquad (10.60)$$

Table 10.3. *Ranges of Desirability for the Objectives*

	Total area, $A_T(\times 104 \ \text{ft}^2)$	Final Temperature (°F)
Ideal	$A_T \leq 0.5$	≥ 600
Desirable	$0.5 \leq A_T \leq 1$	$600 \leq T_3 \leq 550$
Tolerable	$1 \leq A_T \leq 1.5$	$550 \leq T_3 \leq 500$
Undesirable	$1.5 \leq A_T \leq 2$	$500 \leq T_3 \leq 450$
Unacceptable	$A_T \geq 2$	$T_3 \leq 450$

$$10\,°F \leq T_i \leq 1,000\,°F \qquad\qquad\qquad i = \{1,2\} \qquad (10.61)$$

$$10\,°F \leq t_{i2} \leq 1,000\,°F \qquad\qquad\qquad i = \{1,2,3\} \qquad (10.62)$$

The given design parameters are $T_0 = 100$ °F, $WC_p = 105$ BTU/Hr-°F, $t_{i1} = \{300, 400, 600\}$ °F, and $U_i = \{120, 80, 40\}$ BTU/hr-sq.ft-°F, where $i = \{1, 2, 3\}$. Given the above information, solve the following problems.

(a) Assume that the initial temperature of the cold fluid is 100 °F, and the desired final temperature is 500 °F. Minimize the total area of the heat exchanger network that accomplishes the desired heating of the cold fluid, subject to the constraints discussed above.

(b) Assume that the initial temperature of the cold fluid is still 100 °F. Find a design that maximizes the final temperature of the cold fluid, subject to the given constraints.

(c) We now wish to simultaneously minimize the total area, and maximize the final temperature of the cold fluid, subject to the above constraints. Formulate the multiobjective problem, and solve it using the weighted sum method. With supporting information, comment on how easy/difficult it was to obtain the Pareto frontier using the weighted sum method.

(d) Now that we have obtained the Pareto frontier, use it as a design tool to choose the "optimum" design for the heat exchanger. You are given the ranges of desirability for each of the objectives (see Table 10.3). On the Pareto frontier plot, mark each of these ranges for the two objectives. A grid of different possibilities in the design space (in terms of ranges of differing desirabilities) is obtained. As an example, a region in the design space could have total area as "unacceptable" and the final temperature as "desirable."

Is it possible for both of the objectives to be in the "ideal" or "desirable" range of desirability? Explain.

(e) Our task now is to choose a final design that achieves the "best of both worlds" from the above generated grid of possibilites. Make sure that your final design does not fall into "undesirable" or "unacceptable" ranges for either objectives. Discuss your thought process behind your choice of final design (note that there is generally no single *correct* answer for such problems).

Advanced Problems

10.10 UNIVERSAL MOTOR: Refer to the universal motor problem given in Sec. 7.7. Use the information in that example to complete the following tasks.

1. Use the following values for the design variables, $N_a = 1236.8, N_f = 53.23, A_{wf} = 0.2596 \times 10^{-6}, A_{wa} = 0.2601 \times 10^{-6}, I = 6, r_o = 0.025448, t = 0.007184$, and $L = 0.024894$. Input the information as a row vector and use the function umotor.m to determine the mass, power, efficiency, torque, and magnetizing intensity of the motor from the design variables.

2. Minimize the mass of the motor using the values in Part 1 as initial guesses.

3. Minimize the mass and maximize the efficiency of the motor, for both objective functions bearing the same weight (*i.e.*, $w = 0.5$ in $f = w(\mu_1) + (1 - w)(\mu_2)$). This is a bi-objective problem where μ_1 is the mass and μ_2 is efficiency. What is the result? Discuss possible reasons for your results.

4. Implement the scaling of the design variables discussed in the chapter. Minimize the mass and maximize the efficiency of the motor for both objective functions bearing the same weight. (Remember to specify appropriate upper and lower bounds.)

5. Plot the Pareto frontier for the bi-objective problem. Maximize efficiency and minimize mass. Obtain 100 Pareto points.

6. Design the motor for a mass target of 0.5kg and an efficiency target of 0.7.

10.11 WIND Energy: You have installed a small wind turbine at Syracuse University. In the next 36 hours, it can be operated for only 4 hours 30 minutes (4.5 hours). The turbine can be switched on and off twice over the concerned 36 hours (*i.e.*, it can be operated only twice). It cannot be operated when *the chance of precipitation (C)* is 50% or more. The power generated (*P*) by the turbine in watts depends on the wind speed (*U*), as given by

$$P = 3.0 \times U^3 \tag{10.63}$$

The energy generated (*E*) over a period of *T* minutes can be estimated as:

$$E = \sum_{i=1}^{T} P_i \times 60 \tag{10.64}$$

where P_i is the power generated by the turbine in the ith minute.

1. Model the variations of wind speed (*U*) and chance of precipitation (*C*) with time (*T*) over the concerned 36 hours. You may use quadratic or cubic polynomial functions for this purpose.

2. Formulate, model, and solve an optimization problem to maximize the total energy produced (E_T) by the turbine over the next 36 hours, starting from a point of data recording. The design variables are the two starting times, T_{T1} and T_{T2}, and the two stopping times, T_{P1} and T_{P2}. All four design variables are to be expressed in minutes.

The hourly (and 15-min interval) weather data (*i.e.*, wind speed (U) and chance of precipitation (C)) can be obtained from the website: www.weather.com.

10.12 Solve the business problem presented in Sec. 10.7. Provide brief and helpful pertinent discussions.

Graduate Level Problems

10.13 A structural engineering example is presented in Sec. 10.2.1. Solve Items 1. through 9. (*This problem is geared towards graduate students in mechanical, civil, or aerospace engineering*). This problem may require some independent research for some students.

10.14 We are given the means, μ_X, and standard deviations, σ_X, of a set of independent random variables, $X = \{X_1, X_2, ... X_{n_x}\}$. The mean and standard deviation of a function of X, say $g(X)$, can be found using a Taylor series approximation, given as follows [18].

$$\mu_g = g(\mu_X) \tag{10.65}$$

$$V_g = \sum_{i=1}^{n_x} \left[\left. \frac{\partial g}{\partial X_i} \right|_{X=\mu_X} \sigma_{X_i} \right]^2 \tag{10.66}$$

$$\sigma_g = \sqrt{V} \tag{10.67}$$

where μ_g, V_g, and σ_g are the mean, variance, and standard deviation of g, respectively. Using the above equation, derive the expression for the variance and standard deviation of $\theta_1 - \theta_2$ (Eqs. 10.2 and 10.3) for the tolerance allocation problem.

BIBLIOGRAPHY OF CHAPTER 10

[1] K. Deb. *Optimization for Engineering Design: Algorithms and Examples*. PHI Learning, 2012.

[2] G. I. Rozvany. A critical review of established methods of structural topology optimization. *Structural and Multidisciplinary Optimization*, 37(3):217–237, 2009.

[3] S. M. Ross. *A First Course in Probability*. Pearson, 9th edition, 2014.

[4] D. C. Montgomery and G. C. Runger. *Applied Statistics and Probability for Engineers*. John Wiley and Sons, 6th edition, 2014.

[5] N. S. Currey. *Aircraft Landing Gear Design: Principles and Practices*. AIAA, 1988.

[6] P. C. Hansen, V. Pereyra, and G. Scherer. *Least Squares Data Fitting with Applications.* Johns Hopkins University Press, 2012.

[7] Y. A. Cengel and A. J. Ghajar. *Heat and Mass Transfer: Fundamentals and Applications.* McGraw-Hill, 2011.

[8] K. Miettinen. *Nonlinear Multiobjective Optimization.* Springer, 1998.

[9] R. Khire. *Selection-Integrated Optimization (SIO) Methodology for Adaptive Systems and Product Family Optimization.* PhD thesis, Rensselaer Polytechnic Institute, Troy, New York, September 2006.

[10] A. Messac, R. S. Birthright, T. Harren-Lewis, and S. Rangavajhala. Optimizing thermoelectric cascades to increase the efficiency of thermoelectric windows. In *4th AIAA Multidisciplinary design optimization specialist conference. Schaumburg, IL,* 2008.

[11] T. Harren-Lewis, S. Rangavajhala, A. Messac, and Junqiang Zhang. Optimization-based feasibility study of an active thermal insulator. *Building and Environment,* 53:7–15, 2012.

[12] J. Nocedal and S. J. Wright. *Numerical Optimization.* Springer, 2nd edition, 2006.

[13] W. Sun and Y. Yuan. *Optimization Theory and Methods: Nonlinear Programming.* Springer, 1st edition, 2006.

[14] N. M. Alexandrov, Jr. J. E. Dennis, R. M. Lewis, and V. Torczon. A trust-region framework for managing the use of approximation models in optimization. *Structural Optimization,* 15(1):16–23, 1998.

[15] J. Sobieszczanski-Sobieski, C. L. Bloebaum, and P. Hajela. Sensitivity of control-augmented structure obtained by a system decomposition method. *AIAA Journal,* 29(2):264–270, 1991.

[16] M. Avriel and A. C. Williams. An extension of geometric programing with applications in engineering optimization. *Journal of Engineering Mathematics,* 5(2):187–194, 1971.

[17] C. A. Floudas, P. M. Pardalos, C. S. Adjiman, W. R. Esposito, Z. H. Gumus, S. T. Harding, J. L. Klepeis, C. A. Meyer, and C. A. Schweiger. *Handbook of Test Problems in Local and Global Optimization.* Kluwer Academic Publishers, The Netherlands, 1999.

[18] A. Haldar and S. Mahadevan. *Probability, Reliability, and Statistical Methods in Engineering Design.* John Wiley and Sons, 1999.

PART 4

GOING DEEPER: INSIDE THE CODES AND THEORETICAL ASPECTS

Part IV of the book explains what is inside the code and how it works. This knowledge will make it possible to use the optimization code with more confidence, and more reliably. It will also help you know what to do when things do not work. The material presented will also be great preparation for further studies in optimization. With this more advanced knowledge, it will be possible to understand such things as the code error messages. The material examined in this part of the book will reinforce and advance the knowledge of optimization that was previously learned.

Metaphorically, the material up to this point taught us how to *drive the design* from a bad state to an optimal state. We did so without understanding how the *car we are driving* works, how to fix it, design it, or actually build it. However, we have sufficient practical knowledge to *drive* from a bad to a good design. The following material will teach us how the car works, how to fix it, how to make it perform very well, and even prepare us venture into the advanced world of designing cars (if that is our interest). That is, building optimization codes and algorithms. Interestingly, most of us want to use the power of optimization, without having to become expert in the intricate details of the inner-workings of the optimization code. This book provides the breadth of presentation to suit these diverse objectives. More advanced topics will be presented beyond Part IV.

Specifically, Part IV presents three elementary topics:

11. Linear Programming

12. Nonlinear Programming with No Constraints

13. Nonlinear Programming with Constraints

11

Linear Programming

11.1 Overview

Linear programming (LP) is a technique for optimizing a linear objective function, subject to linear equality and linear inequality constraints. Linear programming is an important field of optimization for several reasons (Refs. [1, 2]). Many practical problems in engineering and operations research can be expressed as linear programming problems. Linear programming problems can be solved in an easy, fast, and reliable manner.

Linear programming was first used in the field of economics, and still remains popular in the areas of management, economics, finance, and engineering. During World War II, George Dantzig of the U.S. Air Force used LP techniques for planning problems. He invented the Simplex method, which is one of the most popular methods for solving LP problems.

Linear programming is a well developed subject in operations research, and several references that focus on linear programming alone are available [3, 4, 5, 6]. The reader is encouraged to consult these references for a more detailed discussion of linear programming.

In this chapter, the basics of linear programming problems will be studied. First, an introduction of the basic terminology and important concepts in LP problems will be presented in Sec. 11.2. In Sec. 11.3, the graphical approach for solving LP problems will be discussed, and four types of possible solutions for an LP problem will be introduced. Details on the MATLAB linear programming solver are presented in Sec. 11.4. Sections 11.5 and 11.6 discuss the various concepts involved in the Simplex method. In Sec. 11.7, the notion of duality and interior point methods are briefly discussed. Section 11.8 concludes the chapter with a summary.

11.2 Basics of Linear Programming

In this section, some basic terminology and definitions in linear programming will be discussed.

A generic linear programming problem consisting of linear equality and linear inequality constraints is given as

$$\min_{x} \quad z = c_1 x_1 + c_2 x_2 + \ldots + c_n x_n \tag{11.1}$$

subject to

$$a_{11} x_1 + a_{12} x_2 + \ldots + a_{1n} x_n \leq b_1 \tag{11.2}$$

$$\vdots \qquad\qquad \vdots$$

$$a_{m1} x_1 + a_{m2} x_2 + \ldots + a_{mn} x_n \leq b_m \tag{11.3}$$

$$a_{eq_{11}} x_1 + a_{eq_{12}} x_2 + \ldots + a_{eq_{1n}} x_n = b_{eq_1} \tag{11.4}$$

$$\vdots \qquad\qquad \vdots$$

$$a_{eq_{p1}} x_1 + a_{eq_{p2}} x_2 + \ldots + a_{eq_{pn}} x_n = b_{eq_p} \tag{11.5}$$

$$x_{1-lb} \leq x_1 \leq x_{1-ub} \tag{11.6}$$

$$\vdots \qquad\qquad \vdots$$

$$x_{n-lb} \leq x_n \leq x_{n-ub} \tag{11.7}$$

where z is the linear objective function to be minimized; c represents the vector of coefficients for each of the n design variables; A and b represent the matrix and vector of coefficients for m inequality constraints, respectively; A_{eq} and b_{eq} represent the matrix and vector of coefficients for p equality constraints, respectively; and x_{i-lb} and x_{i-ub} are the lower and upper bounds on the i-th design variable, respectively. Note that the above problem is *not* defined in the so-called standard form, which is commonly used in some LP solvers. The standard form will be discussed later.

In a matrix notation, the above formulation can be written as

$$\min_{x} \quad z = c^T x \tag{11.8}$$

subject to

$$Ax \leq b \tag{11.9}$$

$$A_{eq} x = b_{eq} \tag{11.10}$$

$$x_{lb} \leq x \leq x_{ub} \tag{11.11}$$

where x_{lb} and x_{ub} are vectors of the lower and upper bounds on the design variables, respectively. Note that in some references, the objective function, z, is called the cost function. The coefficients $c_1, \ldots, c_n$ are also known as *cost coefficients*. Some references pose the LP problem as a maximization problem, which is different from the convention in the present chapter, a minimization problem.

11.3 Graphical Solution Approach: Types of LP Solutions

The graphical approach is a simple and easy technique to solve LP problems. The procedure involves plotting the contours of the objective function and the constraints. The feasible region of the constraints and the optimal solution is then identified graphically. Note that the graphical approach to solving LP problems may not be possible as the number of variables increases.

There are four possible types of solutions in a generic LP problem: (1) unique solution, (2) segment solution, (3) no solution, and (4) solution at infinity. These four types of solutions will be studied using examples.

11.3.1 The Unique Solution

Consider the following LP problem.

$$\min_x \ x_1 - 2x_2 \tag{11.12}$$

subject to

$$-4x_1 + 6x_2 \leq 9 \tag{11.13}$$

$$x_1 + x_2 \leq 4 \tag{11.14}$$

$$x_1, x_2 \geq 0 \tag{11.15}$$

Plot the constraints and the objective function contours, as shown in Fig. 11.1(a). There is a unique solution for this problem, where the objective function contour has the least value while remaining in the feasible region.

Here is an important point to keep in mind. If a constant value is added to the objective function in the above problem, for example, $x_1 - 2x_2 + 5$, and the constraints remain unchanged, how would it affect the optimal solution? The optimal values of the design variables will not change because of the added constant. This is because the additional constant in the objective function does not change the slope of the function contours. However, the optimal value of the objective function will change when compared to the original objective function.

11.3.2 The Segment Solution

Consider the LP problem with infinitely many solutions in the form of a segment solution. Examine the following example:

$$\min_x \ x_1 - 2x_2 \tag{11.16}$$

subject to

$$-0.5x_1 + x_2 \leq 10 \tag{11.17}$$

$$x_1 + x_2 \leq 4 \tag{11.18}$$

$$x_2 \geq 0 \tag{11.19}$$

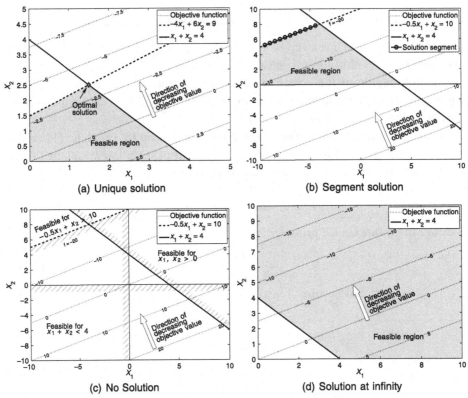

Figure 11.1. Types of LP Solutions

From Figure 11.1(b), the slope of the objective function and the slope of the constraint function, $-0.5x_1 + x_2 \leq 10$, are the same. Upon optimization, the objective function coincides with the constraint function, resulting in infinitely many solutions along the segment shown in Fig. 11.1(b).

A discussion of the third type of LP solutions is presented next.

11.3.3 No Solution

In this case, we will study LP problems that do not have a feasible solution. Consider the following example.

$$\min_x \quad x_1 - 2x_2 \tag{11.20}$$

subject to

$$-0.5x_1 + x_2 \geq 10 \tag{11.21}$$

$$x_1 + x_2 \leq 4 \tag{11.22}$$

$$x_1, x_2 \geq 0 \tag{11.23}$$

Observe the feasible region of this problem from Fig. 11.1(c). The respective feasible regions of the inequality constraints do not intersect. There is no solution that satisfies all the constraints. Thus, there is no solution to this problem.

11.3.4 The Solution at Infinity

Consider the following problem.

$$\min_{x} \; x_1 - 2x_2 \tag{11.24}$$

subject to

$$x_1 + x_2 \geq 4 \tag{11.25}$$

$$x_1, x_2 \geq 0 \tag{11.26}$$

The above problem leads to a solution at infinity. The feasible region, as illustrated in Fig. 11.1(d), is not bounded to yield a finite optimum value. The solution for this problem lies at infinity.

An unconstrained linear programming problem has a solution at infinity. These problems are rarely encountered in practice. On the contrary, unconstrained nonlinear programming problems are fairly common in practice. Nonlinear optimization problems do not require the presence of a constraint in order to yield a finite optimum.

Thus far, four possible types of LP solutions have been described. The graphical approach to solving LP problems is feasible only for small scale problems. For large scale problems, software implementations of LP solution algorithms are used. However, note that graphical visualization means can be used to examine the solution obtained for a large scale LP problem.

11.4 Solving LP Problems Using MATLAB

In this section, the command linprog in MATLAB is revisited. The MATLAB LP solvers use variations of the numerical methods that will be learned later in this chapter. The command linprog employs different solution strategies, such as the Simplex method and the interior point methods based on the size of the problem. The command allows for linear equality and linear inequality constraints and bounds on the design variables.

Different software codes follow their own respective formulations. Before using the solver, the problem is transformed into the formulation specified by the solver. The default problem formulation for linprog is given as

$$\min_{x} \; f^T x \tag{11.27}$$

subject to

$$Ax \leq b \tag{11.28}$$

$$A_{eq}x = b_{eq} \tag{11.29}$$

$$x_{lb} \leq x \leq x_{ub} \tag{11.30}$$

In the above formulation, f is the vector of coefficients of the objective function; A and A_{eq} represent the matrices of the left-hand-side coefficients of the linear inequality and linear equality constraints, respectively; b and b_{eq} represent the vectors of the right-hand-side values of the linear inequality and equality constraints, respectively; and x_{lb} and x_{ub} are the lower and upper bounds on the design variables, respectively.

Example: Consider the following example to demonstrate the use of the `linprog` command.

$$\min_{x} \; x_1 - 2x_2 \tag{11.31}$$

subject to

$$-9 - 4x_1 + 6x_2 \leq 0 \tag{11.32}$$

$$x_1 + x_2 - 4 \leq 0 \tag{11.33}$$

$$x_1, x_2 \geq 0 \tag{11.34}$$

The constraints can be rewritten as $-4x_1 + 6x_2 \leq 9$ and $x_1 + x_2 \leq 4$, according to the MATLAB standard formulation. The MATLAB code that solves the above problem is summarized below.

```
f = [1;-2]  % Defining Objective
A= [-4 6; 1  1];   % LHS inequalities
b = [9; 4]   % RHS inequalities
Aeq = [];    % No equalities
beq = [];    % No equalities
lb = [0;0]   % Lower bounds
ub = []      % No upper bounds
x0 = [1;1]   % Initial guess


x = linprog(f,A,B,Aeq,beq,lb,ub,x0)
```

The output generated by MATLAB is given below.

```
Warning: Large scale (interior point) method uses a
built-in starting point;
ignoring user-supplied X0.
> In linprog at 235
Optimization terminated.

x =

    1.5000
    2.5000
```

The warning displayed above informs the reader that the interior point algorithm used by the MATLAB solver does not need a starting point. The starting point provided has been ignored by the solver. These messages are warnings that often help the user understand the solver algorithm and should not be mistaken for error messages.

Thus far, we have studied the graphical approach to solving LP problems, followed by the use of the MATLAB software options. Next, we present a popular numerical technique for solving LP problems that forms the basis of many commercial solvers.

11.5 Simplex Method Basics

This section introduces the Simplex method. Before discussing the algorithm of the Simplex method, some terminology and basic definitions associated with the Simplex method are presented.

11.5.1 The Standard Form

In order to apply the Simplex method, the problem must be posed in standard form. The standard form of an LP problem for the Simplex method is given below.

$$\min_{x} \quad z = c_1 x_1 + c_2 x_2 + \dots + c_n x_n \tag{11.35}$$

subject to

$$a_{11} x_1 + a_{12} x_2 + \dots + a_{1n} x_n = b_1 \tag{11.36}$$

$$a_{21} x_1 + a_{22} x_2 + \dots + a_{2n} x_n = b_2 \tag{11.37}$$

$$\vdots \qquad\qquad \vdots \tag{11.38}$$

$$a_{m1} x_1 + a_{m2} x_2 + \dots + a_{mn} x_n = b_m \tag{11.39}$$

$$x_1, x_2, \dots, x_n \geq 0 \tag{11.40}$$

The standard formulation does not contain inequality constraints. In a matrix notation, the standard formulation can be written as follows.

$$\min_{x} \quad z = c^T x \tag{11.41}$$

subject to

$$Ax = b \tag{11.42}$$

$$x \geq 0 \tag{11.43}$$

where $c = [c_1, \dots, c_n]$ is the vector of the cost coefficients for the objective function; A is an $m \times n$ matrix of the coefficients for the linear equality constraints; and b is the $m \times 1$ vector of the right-hand-side values.

The feasible region of the standard LP problem is a convex polygon. The optimal solution of the LP problem lies at one of the vertices of the polygon. In the Simplex method, the solution process moves from one vertex of the polygon to the next along the boundary of the feasible region.

11.5.2 Transforming into Standard Form

In the standard definition of LP problems, there are no inequality constraints. All design variables have non-negativity constraints. If a given problem is not in this standard form, certain operations are performed to transform the problem into the Standard Form. How the operations are performed for inequality constraints and for unbounded variables is discussed below.

Inequality Constraints

If the given problem formulation contains inequality constraints of the form $g(x) \leq 0$, they are transformed into equality constraints by using *slack* variables. For constraints of the form $g(x) \leq 0$, *add* a non-negative slack variable, s_1, to the left-hand-side of the constraint, yielding $g(x) + s_1 = 0$. The variable s_1 is called a slack variable because it represents the slack between the left-hand-side and the right-hand-side of the inequality.

For inequalities of the form $g(x) \geq 0$, they are transformed into equality constraints by using *surplus* variables. *Subtract* a non-negative surplus variable, s_2, from the left-hand-side of the constraint, yielding $g(x) - s_2 = 0$. The variable s_2 represents the surplus between the left-hand-side and the right-hand-side of the inequality.

The slack/surplus variables are unknowns, and will be determined as part of the LP solution process.

Example: Consider the following linear programming problem.

$$\min_x \; x_1 - 2x_2 \tag{11.44}$$

subject to

$$-4x_1 + 6x_2 \leq 9 \tag{11.45}$$

$$x_1 + x_2 \leq 4 \tag{11.46}$$

$$x_1, x_2 \geq 0 \tag{11.47}$$

The standard form for the above formulation can be given as

$$\min_x \; x_1 - 2x_2 \tag{11.48}$$

subject to

$$-4x_1 + 6x_2 + s_1 = 9 \tag{11.49}$$

$$x_1 + x_2 + s_2 = 4 \tag{11.50}$$

$$x_1, x_2, s_1, s_2 \geq 0 \tag{11.51}$$

The standard form for the LP formulation requires the design variables to be non-negative.

Unbounded Design Variables

In the standard form, the design variables should be non-negative, $x \geq 0$. However, in practice, the bounds on the design variable could also be of the form $x \leq 0$. In some problems, the design variables may be indefinite (*i.e.*, no bounds may be specified). In these cases, how are the design variables put into a standard form?

If a design variable, x_i, does not have bounds imposed in the problem, the following technique is used. Let $x_i = s_1 - s_2$, where $s_1, s_2 \geq 0$. In the standard form, the variable x_i is then replaced by $s_1 - s_2$, and the additional constraints $s_1, s_2 \geq 0$ are added to the problem. In other words, the unbounded design variable is rewritten as the difference between two non-negative additional variables. The additional variables will be determined as part of the LP solution process.

Example: Consider the following formulation.

$$\min_{x} \ x_1 - 2x_2 + 3x_3 \tag{11.52}$$

subject to

$$-4x_1 + 6x_2 + x_3 \leq 9 \tag{11.53}$$

$$x_1 + x_2 - 2x_3 \leq 4 \tag{11.54}$$

$$x_1, x_2 \geq 0 \tag{11.55}$$

Note that the variable x_3 is unbounded in the above formulation. Assume that $x_3 = s_1 - s_2$ and $s_1, s_2 \geq 0$. The standard formulation can be written as follows.

$$\min_{x} \ x_1 - 2x_2 + 3(s_1 - s_2) \tag{11.56}$$

subject to

$$-4x_1 + 6x_2 + (s_1 - s_2) + s_3 = 9 \tag{11.57}$$

$$x_1 + x_2 - 2(s_1 - s_2) + s_4 = 4 \tag{11.58}$$

$$x_1, x_2, s_1, s_2, s_3, s_4 \geq 0 \tag{11.59}$$

Next, we introduce the Gauss Jordan method, which forms an important component of the Simplex method.

11.5.3 Gauss Jordan Elimination

Note that the number of variables (including the n design variables and the p slack variables) is not necessarily equal to the number of equations, m. If the number of variables is equal to the number of equality constraints, then the solution is uniquely defined. In most LP problems, there exists more variables than equations. This results in an under-determined system of equations, resulting in infinitely many feasible solutions for the equality constraint set. The optimization problem then lies in determining which feasible solution(s) results in the minimization of the objective function, while satisfying the non-negativity constraints for the design variables.

Example: Consider the following under-determined system of equations:

$$x_1 + x_2 - x_3 + 3x_4 = 2 \tag{11.60}$$

$$-x_1 + 3x_2 - 5x_3 - 2x_4 = 5 \tag{11.61}$$

$$x_1 + 2x_2 - x_4 = 6 \tag{11.62}$$

The above set of equations have more variables than equations. Therefore, there are infinitely many solutions for this case.

To efficiently work with the constraint set of the LP problem, reduce the constraint set into a special form. The set of equations in the special form is said to be in a *canonical form*. The original constraint set and the canonical form are *equivalent* (*i.e.*, they have the same set of solutions). By transforming the LP constraint set into a canonical form (which is easier to solve), the solutions can be found more efficiently. To solve LP problems, use a canonical form known as the *reduced row echelon form*. This approach of using a reduced row echelon form to solve a set of linear equations is known as the *Gauss Jordan elimination*.

How can a given constraint be reduced into a row echelon form? A canonical form is usually defined with respect to a set of dependent or *basic variables*, which are defined in terms of a set of independent or *non-basic variables*. The choice of basic variables is arbitrary. A system of m equations and n variables is said to be in a reduced row echelon form with respect to a set of *basic variables*, $x_1, ..., x_m$, if all the basic variables have a coefficient of one in only one equation, and have a zero coefficient in all the other equations. A generic matrix based representation of a set of equations in a reduced row echelon form with m basic variables and p non-basic variables is given as

$$
\begin{bmatrix}
1 & 0 & \cdots & 0 & \vdots & d_{11} & \cdots & d_{1p} \\
0 & 1 & \cdots & 0 & \vdots & d_{21} & \cdots & d_{2p} \\
\vdots & \vdots & \ddots & \vdots & \vdots & \vdots & \ddots & \vdots \\
0 & 0 & \cdots & 1 & \vdots & d_{m1} & \cdots & d_{mp}
\end{bmatrix}
\begin{bmatrix}
x_{b1} \\ x_{b2} \\ \vdots \\ x_{bm} \\ \cdots \\ x_{nb1} \\ \vdots \\ x_{nbp}
\end{bmatrix}
=
\begin{bmatrix}
b_1 \\ b_2 \\ \vdots \\ b_m
\end{bmatrix}
\tag{11.63}
$$

where d is the matrix of coefficients for the non-basic variables; x_b is the set of basic variables; and x_{nb} is the set of non-basic variables. The number of basic variables is equal to the number of equations.

The reduced row echelon form is illustrated with the help of the following example.

Example: The following equations are in a reduced row echelon form with respect to the variables x_1, x_2, x_3, and x_4.

$$x_1 + x_6 = 5 \tag{11.64}$$

$$x_2 - 3x_5 + 4x_6 = 10 \tag{11.65}$$

$$x_3 + 3x_5 = 2 \tag{11.66}$$

$$x_4 + 2x_5 - 5x_6 = 7 \tag{11.67}$$

Writing the above equation set in a matrix form allows us to obtain the following.

$$\begin{bmatrix} 1 & 0 & 0 & 0 & 0 & 1 \\ 0 & 1 & 0 & 0 & -1 & 4 \\ 0 & 0 & 1 & 0 & 3 & 0 \\ 0 & 0 & 0 & 1 & 2 & -5 \end{bmatrix} \begin{bmatrix} x_1 \\ x_2 \\ x_3 \\ x_4 \\ x_5 \\ x_6 \end{bmatrix} = \begin{bmatrix} 5 \\ 10 \\ 2 \\ 7 \end{bmatrix} \tag{11.68}$$

Notice the matrix of coefficients on the left-hand-side. Each row of the matrix represents one constraint. We note that the coefficients corresponding to the basic variables x_1, x_2, x_3, and x_4 are equal to one in only one equation, and zero elsewhere.

The given set of constraints of an LP problem is not usually given in the row echelon form. A series of row operations must be performed to transform it into the standard form. The details are discussed next.

11.5.4 Reducing to a Row Echelon Form

A *pivot operation* consists of a series of elementary row operations to make a particular variable a basic variable (*i.e.*, reduce the coefficient of the variable to unity in only one equation, and to zeroes in the other equations). A particular variable to be made basic, x_{bi}, could exist in some or in all of the m equations. We need to decide which coefficient of the basic variable should be made *one* (*i.e.*, in which equation). By definition, each equation is to have only one basic variable with unit coefficient. The choice as to which equation corresponds to which basic variable is arbitrary, or is based on algebraic convenience for the reduced row echelon form. There will be further restrictions on this issue when the Simplex method is discussed.

There are two types of elementary row operations that can be performed to reduce a set of equations into a reduced row echelon form: (1) Multiply both sides of an equation with the same non-zero number. (2) Replace one equation by a linear combination of other equations.

Example: Reduce the following set of equations into a row echelon form.

$$R_1 \equiv x_1 + x_2 - x_3 + 3x_4 = 2 \tag{11.69}$$

$$R_2 \equiv -x_1 + 3x_2 - 7x_3 + x_4 = 6 \tag{11.70}$$

$$R_3 \equiv x_1 + 2x_2 \qquad - 2x_4 = 7 \tag{11.71}$$

Choose x_1, x_2, and x_3 as basic variables. In order to obtain a row echelon form, perform operations such that the variables x_1, x_2, and x_3 appear in only one equation with a unit coefficient, and do not appear in the other equations. Use the following representation, where the first four columns represent the coefficients of each variable, and the last column represents the right-hand-side of the equation.

$$\begin{bmatrix} & x_1 & x_2 & x_3 & x_4 & b \\ R_1 & 1 & 1 & -1 & 3 & 2 \\ R_2 & -1 & 3 & -7 & 1 & 6 \\ R_3 & 1 & 2 & 0 & -2 & 7 \end{bmatrix} \tag{11.72}$$

First, make x_1 a basic variable. Choose x_1 to have a unit coefficient in R_1, and zero coefficients in R_2 and R_3. Replace R_2 by $R_2 + R_1$ and R_3 by $R_3 - R_1$ to obtain

$$\begin{bmatrix} & x_1 & x_2 & x_3 & x_4 & b \\ R_1 & 1 & 1 & -1 & 3 & 2 \\ R_2 & 0 & 4 & -8 & 4 & 8 \\ R_3 & 0 & 1 & 1 & -5 & 5 \end{bmatrix} \tag{11.73}$$

With the above transformations, x_1 is made a basic variable (appears only in R_1). Now make x_2 a basic variable. That is, make its coefficient one in R_2 and zeroes in the other rows. First, divide R_2 by 4 to obtain a unit coefficient in R_2.

$$\begin{bmatrix} & x_1 & x_2 & x_3 & x_4 & b \\ R_1 & 1 & 1 & -1 & 3 & 2 \\ R_2 & 0 & 1 & -2 & 1 & 2 \\ R_3 & 0 & 1 & 1 & -5 & 5 \end{bmatrix} \tag{11.74}$$

Now replace R_1 by $R_1 - R_2$ and R_3 by $R_3 - R_2$ to make the coefficients of x_2 zeroes in the other equations to obtain

$$\begin{bmatrix} & x_1 & x_2 & x_3 & x_4 & b \\ R_1 & 1 & 0 & 1 & 2 & 0 \\ R_2 & 0 & 1 & -2 & 1 & 2 \\ R_3 & 0 & 0 & 3 & -6 & 3 \end{bmatrix} \tag{11.75}$$

Next, make x_3 a basic variable by making its coefficient unity in R_3 and zeroes in the other equations. Divide R_3 by 3 to obtain the following.

$$\begin{bmatrix} & x_1 & x_2 & x_3 & x_4 & b \\ R_1 & 1 & 0 & 1 & 2 & 0 \\ R_2 & 0 & 1 & -2 & 1 & 2 \\ R_3 & 0 & 0 & 1 & -2 & 1 \end{bmatrix} \tag{11.76}$$

Replace R_2 by $R_2 + 2R_3$ and R_1 by $R_1 - R_3$ to make the coefficients of x_3 zeroes in the other equations.

$$\begin{bmatrix} & x_1 & x_2 & x_3 & x_4 & b \\ R_1 & 1 & 0 & 0 & 4 & -1 \\ R_2 & 0 & 1 & 0 & -3 & 4 \\ R_3 & 0 & 0 & 1 & -2 & 1 \end{bmatrix} \tag{11.77}$$

The above set of equations are in the reduced row echelon form with respect to the basic variables x_1, x_2, and x_3. Note that the particular choice of making x_1, x_2, and x_3 the basic variables is arbitrary.

11.5.5 The Basic Solution

Thus far, our discussion has dealt with how to reduce a given set of equations into a canonical form. The next step is to find the solution for the set of equations in the canonical form. A *basic solution* is obtained from the canonical form by setting the non-basic (or independent) variables to zero. A *basic feasible solution* is a basic solution in which the values of the basic variables are non-negative. In other words, the basic feasible solution is feasible for the standard LP formulation, whereas a basic solution is not always feasible.

Example: Consider the following canonical form derived previously.

$$\begin{bmatrix} & x_1 & x_2 & x_3 & x_4 & b \\ R_1 & 1 & 0 & 0 & 4 & -1 \\ R_2 & 0 & 1 & 0 & -3 & 4 \\ R_3 & 0 & 0 & 1 & -2 & 1 \end{bmatrix} \tag{11.78}$$

Set the non-basic variable, x_4, to zero. Then, obtain the basic solution by solving for x_1, x_2, and x_3. The basic solution can then be written as $x_1 = -1, x_2 = 4, x_3 = 1$, and $x_4 = 0$. Note that this basic solution is not a basic feasible solution for the standard LP formulation since $x_1 < 0$.

The choice of the set of basic variables is not unique, and can be decided according to computational convenience. In the example considered above, x_2, x_3, and x_4 could have been chosen as basic variables instead of x_1, x_2, and x_3. For a generic problem, any set of m variables (recall that there are m equations) from the possible n variables can be chosen as basic variables. This implies that the number of basic solutions for a generic standard LP problem with m constraints and n variables is given as

$$C_m^n = \frac{n!}{m!(n-m)!} \tag{11.79}$$

Example: Consider the following set of equations from the previous example.

$$R_1 \equiv x_1 + x_2 - x_3 + 3x_4 = 2 \tag{11.80}$$

$$R_2 \equiv -x_1 + 3x_2 - 7x_3 + x_4 = 6 \tag{11.81}$$

$$R_3 \equiv x_1 + 2x_2 \qquad - 2x_4 = 7 \tag{11.82}$$

Here, $m = 3$ and $n = 4$. Therefore, $C_3^4 = \frac{4!}{3!(4-3)!} = 4$ basic solutions.

As mentioned earlier, the feasible region of the standard LP problem is a convex polygon. Each vertex of this convex polygon corresponds to a basic feasible solution of the constraint set. The optimal solution of the LP problem, which is the basic feasible solution with the minimum objective function value, lies at one of the vertices of the polygon. In the Simplex method, the solution process moves from one vertex of the polygon to the next along the boundary of the feasible region. Now that the supporting mathematics of the Simplex method has been presented, we may proceed to discuss the Simplex algorithm.

11.6 Simplex Algorithm
11.6.1 Basic Algorithm

The optimal solution of the LP problem lies at one of the vertices of the feasible convex polygon. A cumbersome approach to solve an LP problem would be to list all the basic feasible solutions from which the optimal solution could be chosen. In the Simplex method, the solution process *efficiently* moves from one basic feasible solution to the next, while ensuring objective function reduction at each iteration (Ref. [7]).

Before discussing the details of the Simplex algorithm, please note the following. The standard LP problem was posed as a minimization problem. The rules of the following algorithm apply to minimization problems only. Some other textbooks may treat the standard LP problem as a maximization problem, and the rules of the algorithm would differ accordingly.

The Simplex method algorithm is presented in 6 generic steps:

1. Transformation into the Standard LP Problem
2. Formation of the Simplex Tableau
3. Choice of the Variable that Enters the Basis – Identify the Pivotal Column
4. Application of the Minimum Ratio Rule – Identify the Pivotal Row
5. Reduction to Canonical Form
6. Checking for Optimality

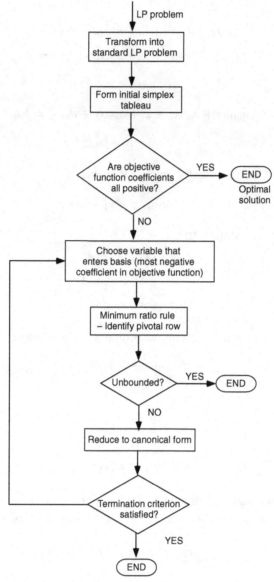

Figure 11.2. Simplex Algorithm

The presentation of these steps follows. The algorithm of the Simplex method is summarized in Fig. 11.2.

1. **Transform into Standard LP Problem:** Transform the given problem into the standard LP formulation by adding slack/surplus variables. Consider a generic formulation with n design variables, m inequality constraints, and m slack variables. The standard LP formulation is given as

$$\min_x \quad z = c_1 x_1 + c_2 x_2 + \dots + c_n x_n \tag{11.83}$$

subject to

$$a_{11}x_1 + a_{12}x_2 + \ldots + a_{1n}x_n + d_1s_1 = b_1 \tag{11.84}$$

$$a_{21}x_1 + a_{22}x_2 + \ldots + a_{2n}x_n + d_2s_2 = b_2 \tag{11.85}$$

$$\vdots \qquad\qquad \vdots$$

$$a_{m1}x_1 + a_{m2}x_2 + \ldots + a_{mn}x_n + d_ms_m = b_m \tag{11.86}$$

$$x_1, \ldots, x_n, s_1, \ldots, s_m \geq 0 \tag{11.87}$$

Example: Consider the following LP problem.

$$\min_x \ z = x_1 - 2x_2 \tag{11.88}$$

subject to

$$x_1 + x_2 \leq 4 \tag{11.89}$$

$$-x_1 + x_2 \leq 3 \tag{11.90}$$

$$x_1, x_2 \geq 0 \tag{11.91}$$

The standard form for the above problem is given below.

$$\min_x \ z = x_1 - 2x_2 \tag{11.92}$$

subject to

$$x_1 + x_2 + s_1 = 4 \tag{11.93}$$

$$-x_1 + x_2 + s_2 = 3 \tag{11.94}$$

$$x_1, x_2, s_1, s_2 \geq 0 \tag{11.95}$$

2. **Form the Initial Simplex Tableau:** List the constraints and the objective function coefficients in the form of a table, known as the Simplex tableau, shown below.

$$
\begin{bmatrix}
 & x_1 & \cdots & x_n & s_1 & \cdots & s_p & b \\
\hline
\text{Constraint 1} & a_{11} & \cdots & a_{1n} & 1 & \cdots & 0 & b_1 \\
\vdots & \vdots & \vdots & \vdots & \vdots & \vdots & \vdots & \\
\hline
\text{Constraint } m & a_{m1} & \cdots & a_{mn} & 0 & \cdots & 1 & b_m \\
\hline
\text{Objective} & c_1 & \cdots & c_n & 0 & 0 & 0 & f
\end{bmatrix}
\tag{11.96}
$$

Example: The Simplex tableau for the example is shown in Table 11.1.
The basic feasible solution for the initial Simplex tableau with s_1 and s_2 as basic variables is $x_1 = 0, x_2 = 0, s_1 = 4$, and $s_2 = 3$, and the function value is $f = 0$ (see Fig. 11.3).

Table 11.1. *Initial Simplex Tableau*

	x_1	x_2	s_1	s_2	b
R_1	1	1	1	0	4
R_2	−1	1	0	1	3
R_3	1	−2	0	0	f

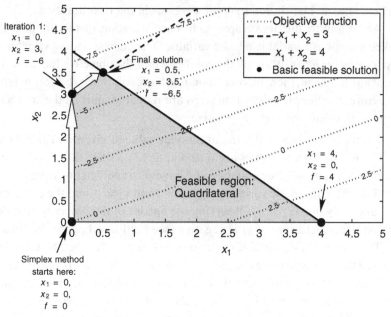

Figure 11.3. Simplex Method Example

3. **Choose the Variable that Enters the Basis – Identify the Pivotal Column:** The above set of equations is in the reduced row echelon form with respect to the slack variables, $s_1, ... s_p$. The Simplex algorithm begins with this initial basic feasible solution. By observing the coefficients of the objective function row in the initial Simplex tableau, the algorithm moves to the *adjacent basic feasible solution that reduces the objective function*. The other adjacent basic solution(s) with objective function value(s) higher than the current solution are ignored.

   In order to move to the next basic feasible solution, the algorithm proceeds by making an existing basic variable a non-basic variable. In addition, an existing non-basic variable is made into a basic variable. How do we choose which non-basic variable becomes a basic variable? Similarly, how do we choose which will be the next basic variable?

   The non-basic variable with the highest negative coefficient in the objective function is selected to become the basic variable in the next iteration. This choice is driven by our interest in minimizing the objective function value. The variable with the highest negative coefficient has the potential to reduce the objective function value to the maximum extent, when compared to the other variables. In other words, this choice can incur the maximum improvement to the objective value per unit of increase of the variable. In contrast, the coefficient of a non-basic variable would not have the opportunity to play an active role in the minimization process, since the value of all non-basic variables are set to zero – in determining the basic feasible solutions. As the non-basic variable enters the basis (in the next iteration), it will directly impact the function minimization process. Similarly, in a function maximization process, the non-basic variable with the highest positive coefficient would be chosen to enter the basis in the next iteration.

According to the above approach, each iteration of the Simplex method makes a non-basic variable a basic variable. The corresponding variable is then said to *enter the basis*. When all the coefficients of the objective function are positive at any iteration, then the corresponding basic solution is the optimal solution. Therefore, the Simplex algorithm steps are repeated until all the coefficients in the objective function row are positive.

> **Example:**   Consider the initial Simplex tableau given in Table 11.1. Currently, the basic variables are $s_1 = 4$ and $s_2 = 3$ since they appear with unit coefficient in only one equation and have zero coefficients in the other equations. The non-basic variables are $x_1 = 0$ and $x_2 = 0$, and the corresponding function value is $f = 0$. Observing the entries of the objective function row (R_3), the coefficient of x_2 is negative (see bold-faced entries in Table 11.2). The current basic variables, s_1 and s_2, are not part of the objective function (see the corresponding coefficients in R_3 in Table 11.2). If x_2 is made a basic variable, the objective function value can be reduced from its current value. Therefore, the variable x_2 *enters the basis*, as per Table 11.2. (Note that enters the basis means *will enter the basis in the next iteration.*

4. **Minimum Ratio Rule – Identify the Pivotal Row:**   In the previous step, the variable entering the basis was identified. However, the variable may appear in all the equations in the constraint set. In other words, the previous step identified the column of interest, or the *pivotal column*. For a variable to be basic, recall that it must have unit coefficient in one equation only, and zero coefficients in all other equations. Which equation will have the unit coefficient? In other words, how is the row of interest determined, or the *pivotal row*?

The number of basic variables in a set of equations cannot be greater than the number of equations. Since one variable was added to the basic variable set in the previous step, an existing basic variable must be made non-basic. In order to determine which variable is eliminated from the basic variable set, use the *minimum ratio rule*. The variable that is selected using this rule is said to be *leaving the basis*. To select the variable that leaves the basis, compute the following ratio for the selected column in the previous step corresponding to the variable that enters the basis, say x_j.

$$\min_{j,\text{ for all } a_{ij}>0} \frac{b_i}{a_{ij}} \tag{11.97}$$

Table 11.2. *Simplex Method Example: Identifying the Pivotal Column*

	x_1	x_2	s_1	s_2	b
R_1	1	**1**	1	0	4
R_2	−1	**1**	0	1	3
R_3	1	**−2**	0	0	f

The row that satisfies the above minimum ratio rule is then selected as the pivotal row.

In the previous step, the entering basic variable was chosen based on its potential to reduce the objective function value. However, the allowable reduction of the objective function depends on the condition that the constraints are not violated. The minimum ratio rule above provides the largest increase in the objective function value possible while the constraints are not violated.

Special Cases:
(1) If all the coefficients of the column corresponding to the chosen basic variable, a_{ij}, are negative, the minimum ratio cannot be computed using Eq. 11.97. In these cases, the LP problem has an *unbounded solution*.
(2) If two or more rows have the same minimum ratio as computed from Eq. 11.97, any of these can be chosen as the pivotal row.
(3) When one or more of the basic variables have zero values, the solution is said to be *degenerate*. This can happen when the right-hand-side value b_i is zero and, consequently, the minimum ratio in Eq. 11.97 is zero. This usually implies that adding a new variable to the basic variable set may not reduce the objective function value. This may result in *cycling*, where the Simplex algorithm may enter an infinite loop. Fortunately, in most practical problems, this situation usually does not arise.

> **Example:** After applying the minimum ratio rule, the pivotal row has been identified as shown in Table 11.3.

5. **Reduce to Canonical Form:** Once the pivotal row and the pivotal column have been chosen based on the above rules, the pivotal element can be identified. The constraint set is then transformed into a reduced row echelon form with respect to the newly identified incoming basic variable.

> **Example:** By performing elementary row operations as discussed in Sec. 11.5.3, the initial Simplex tableau is reduced into the canonical form with respect to the newly added basic variable, x_2. The coefficient of x_2 is reduced to one in R_2, and is reduced to zero in R_1 and R_3. Table 11.4 provides the updated Simplex tableau in the canonical form with respect to x_2.

Table 11.3. *Simplex Method Example:*
Identifying the Pivotal Row

	x_1	x_2	s_1	s_2	b	b_i/a_{ij}
R_1	1	1	1	0	4	$4/1 = 4$
R_2	−1	1	0	1	3	$3/1 = 3$
R_3	1	−2	0	0	f	

The basic solution for the above Simplex tableau is $x_1 = 0, x_2 = 3, s_1 = 1$, and $s_2 = 0$, while the function value is $f = -6$ (see Fig. 11.3). The variable x_2 has *entered the basis* and s_1 has *left the basis*. The value of x_2 has increased from zero in the initial Simplex tableau to three in the current iteration, and the function value has been *reduced* from $f = 0$ to $f = -6$.

As shown in Fig. 11.3, the Simplex method moves from one basic feasible solution to an adjacent one with a reduced objective function value. The other adjacent basic feasible solution at $x_1 = 4, x_2 = 0$ is ignored since the objective function value is higher ($f = 4$) than the initial basic feasible solution ($f = 0$).

6. **Check for Optimality:** If the coefficients of the objective function are all positive in the current Simplex tableau, the optimum has been reached and the algorithm can be terminated. If not, repeat the Simplex algorithm until the above termination criterion is met.

   Example: In Table 11.4, the coefficient of x_1 in the objective function row is negative. Therefore, choose x_1 as the variable entering the basis. The minimum ratio rule in the last column indicates that R_1 is the pivotal row (see Table 11.5).

   By reducing Table 11.5 to a canonical form with respect to x_1, the Simplex tableau provided in Table 11.6 is obtained. The basic feasible solution for Table 11.6 is $x_1 = \frac{1}{2}, x_2 = \frac{7}{2}, s_1 = 0$, and $s_2 = 0$, and the function value is $f = -\frac{13}{2}$. The coefficients in the objective function in Table 11.6 are all positive. Therefore, the current basic feasible solution is the optimal solution.

   Figure 11.3 illustrates the progression of the iterations of the Simplex method. The solid black circles represent the four basic feasible solutions. The feasible region of the standard LP problem is a polygon which, in this case, is a quadrilateral. The hollow block arrows illustrate the progression of the Simplex method along the boundary of the feasible quadrilateral. The initial Simplex tableau has a basic feasible solution at $x_1 = 0, x_2 = 0$. Using the rules of the Simplex algorithm, the method proceeds to the adjacent vertex of the quadrilateral *that reduces* the objective function value.

The above example concludes our discussion of the Simplex algorithm. Next, we briefly examine how to deal with cases where the Simplex method cannot be applied.

Table 11.4. *Simplex Method Example:*
Tableau in Canonical Form

	x_1	x_2	s_1	s_2	b
R_1	2	**0**	1	−1	1
R_2	−1	**1**	0	1	3
R_3	−1	**0**	0	2	$f+6$

Table 11.5. *Simplex Method Example:*
The Second Iteration

	x_1	x_2	s_1	s_2	b	b_i/a_{ij}
R_1	**2**	**0**	**1**	**−1**	**1**	**1/2**
R_2	**−1**	1	0	1	3	3
R_3	**−1**	0	0	2	$f+6$	

Table 11.6. *Simplex Method*
Example: Final Simplex Tableau

	x_1	x_2	s_1	s_2	b
R_1	1	0	1/2	−1/2	1/2
R_2	0	1	1/2	1/2	7/2
R_3	0	0	1/2	3/2	$f+13/2$

11.6.2 Special Cases

The discussion of the Simplex method in the previous subsection assumes that the LP problem can be reduced into the standard formulation. In some cases, it will not be directly possible to do so. One requirement of the standard LP problem is that the variables must be non-negative, and the right-hand-side constants for the constraints must be non-negative. In some problems, obtaining a standard LP problem is not straightforward. In these cases, *artificial variables* are used. Variations of the Simplex method known as the *two-phase Simplex method* and the *dual Simplex method* are used. A detailed discussion of the theory and implementation of these methods can be found in [3, 8]. Next, we explore some advanced concepts in the area of linear programming.

11.7 Advanced Concepts

Two important topics will be studied briefly in this section: duality and interior point methods. While a detailed discussion of these topics is outside the scope of this chapter, the following discussion provides the reader with a basic understanding of the theory underlying these advanced concepts.

11.7.1 Duality

Duality is an important concept in linear programming problems. Each LP problem, known as the *primal*, has another corresponding LP problem, known as the *dual*, associated with it. The dual and the primal problems share some common features, but are arranged differently. How is this concept useful? As will be discussed later, the solutions of the primal and dual problems have interesting relationships. In some cases, solving a dual problem might be computationally more convenient than solving the primal problem. The concept of duality is of great importance in various numerical optimization algorithms and LP solvers. This section introduces the basic concept of duality in LP problems.

Consider the following matrix representation of the LP problem.

$$\min_{x} \; z = c^T x \tag{11.98}$$

subject to

$$Ax \le b \tag{11.99}$$

$$x \ge 0 \tag{11.100}$$

For the above primal problem, the corresponding dual problem can be defined as follows.

$$\max_{y} \; z_d = b^T y \tag{11.101}$$

subject to

$$A^T y \ge c \tag{11.102}$$

$$y \ge 0 \tag{11.103}$$

If the primal is a minimization problem, the dual will be a maximization problem, and vice versa.

Example: Consider the following LP problem.

$$\min_{x} \; z = x_1 - x_2 - 2x_3 + 4x_4 \tag{11.104}$$

subject to

$$x_1 + 5x_2 - 2x_3 + 3x_4 \le 1 \tag{11.105}$$

$$5x_1 + x_2 + 3x_3 + 8x_4 \le 5 \tag{11.106}$$

$$x_1 + 2x_2 + 3x_3 + 5x_4 \le 3 \tag{11.107}$$

$$x_1, x_2, x_3, x_4 \ge 0 \tag{11.108}$$

In the above problem, c is a 4×1 column vector given as $c = [1, -1, -2, 4]^T$; the left-hand-side coefficients form a 3×4 matrix

$$A = \begin{bmatrix} 1 & 5 & -2 & 3 \\ 5 & 1 & 3 & 8 \\ 1 & 2 & 3 & 5 \end{bmatrix} \qquad (11.109)$$

and the right-hand-side values of the inequality constraints form a 3×1 column vector given as $b = [1, 5, 3]^T$.

The dual of the above problem is given as

$$\max_{y} \; z_d = y_1 + 5y_2 + 3y_3 \qquad (11.110)$$

subject to

$$y_1 + 5y_2 + y_3 \geq 1 \qquad (11.111)$$

$$5y_1 + y_2 + 2y_3 \geq -1 \qquad (11.112)$$

$$-2y_1 + 3y_2 + 3y_3 \geq -2 \qquad (11.113)$$

$$3y_1 + 8y_2 + 5y_3 \geq 4 \qquad (11.114)$$

$$y_1, y_2, y_3 \geq 0 \qquad (11.115)$$

Note the following relationships between the primal and the dual problem:

1. The primal problem had three inequality constraints, while the dual problem has three design variables.
2. The primal problem had four design variables, while the dual problem has four inequality constraints.
3. The primal is a minimization problem, while the dual is a maximization problem.
4. The primal inequalities are of the form ≤ 0, whereas the dual constraints are of the form ≥ 0.

Next, we examine the relationships between the primal and the dual problems.

11.7.2 Primal-Dual Relationships

The feasible sets and the solutions of the primal and the dual problems are related. *Duality theory* presents the relationship between the two problems. The following are the important relationships between the primal and the dual problems.

1. The objective function value of the dual problem evaluated at any feasible solution provides a lower bound on the objective function value of the primal problem evaluated at any feasible solution. This is known as the *weak duality theorem*.
2. Every feasible solution for the dual problem provides a lower bound on every feasible solution of the primal.
3. The dual of a dual problem is the primal problem.

274

4. If the primal solution has a feasible solution and the dual problem has a feasible solution, then there exists optimal feasible solutions such that the objective function values of the primal and the dual are the same. This is known as the *strong duality theorem*.

5. If the primal problem is unbounded, the dual problem is infeasible.

The second topic in this section, interior point methods, is introduced next.

11.7.3 Interior Point Methods

The Simplex algorithm moves along the boundary of the feasible region to find the optimal solution (see Fig. 11.3). For large scale problems, the Simplex method can become cumbersome and computationally expensive. There exists another class of solution approaches for LP problems known as the interior point methods. These algorithms are known to converge faster than the Simplex method. Unlike the Simplex method, which shifts from one vertex of the feasible polygon to the other, interior point methods move *across* the feasible region based on pre-defined criteria, such as the feasibility of the constraints or the objective function value.

Some variations of the interior point method employ the concept of duality. An important component of the interior point method, similar to most numerical optimization algorithms, is determining the search direction and the step size. The search direction idea here is similar to those discussed in Chapter 12, such as the steepest descent direction and the Newton's search direction. Comprehensive presentations of the interior point method can be found in Refs. [5] and [9].

11.7.4 Solution Sensitivity

In linear programming (LP), it is crucial to understand the impact of the coefficients in the objective and the constraint functions on the optimum solution. In other words, for LP problems, the optimum solution is highly sensitive to the slope (or gradient) of the objective function and the constraint functions. Therefore, a careful observation of the slope of these functions provides important insight into the behavior of the LP problem even prior to optimization. Two important characteristics of the solution sensitivity are pointed out in this section. In order to discuss these characteristics, we revisit the LP problem defined by Eqs. (11.12 - 11.15) and illustrated in Fig. 11.4.

The more readily evident characteristics are the direct proportionality of the function to the coefficients of the linear terms. For example, in Eq. 11.12, the objective function value increases at the same rate at which the variable x_1 increases, and decreases at twice the rate at which variable x_2 increases. The impact of x_2 on the objective function is double that of x_1. However, when we consider the relative slopes of the objective function and of the constraint functions, potentially more significant impacts become evident. A minor change in the slope (or in the linear term coefficients) can switch the optimum from one vertex of the feasible region (or feasible polytope) to another adjacent one. To explain this scenario, we will call the

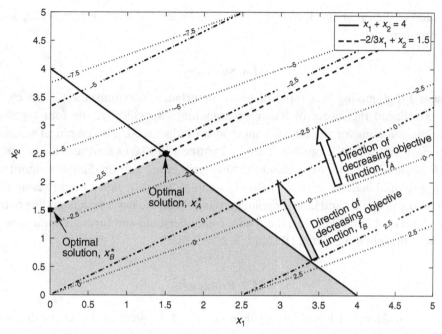

Figure 11.4. Sensitivity of Optimum to Slope of Linear Objective Function

objective function given by Eq. 11.12 as f_A, and consider an alternative objective function, f_B. These two objective functions are described below:

$$f_A = x_1 - 2x_2 \qquad (11.116)$$

$$f_A = x_1 - 1.4x_2 \qquad (11.117)$$

The direction in which objective functions f_A and f_B decrease are provided in Fig. 11.4, represented by the dash-dot line and the dashed line, respectively. Note that, with the same feasible region, the optimum for f_A lies at the vertex $x_A^* = (1.5, 2.5)$, while the optimum for f_b lies at the vertex $x_B^* = (0, 1.5)$. The constraint given by Eq. 11.13 is the active constraint in both cases. Now, if we carefully compare the slopes of the objective functions (f_A and f_B) with that of the active constraint given by Eq. 11.13, we find that the slope of f_A is slightly greater than that of the constraint, while the slope of f_B is slightly smaller. Thus, as a result of a minor change in the slope of the objective function (with respect to the active constraint), the location of the optimum can change drastically.

In the context of practical application of LP, the following comment can be made. Although mathematically, the sensitivity of the objective function to the design variables remain the same across the design space for an LP problem, practically, the location of the optimum becomes more critical (or sensitive) to the slope of the objective function when this slope is close to that of an active constraint. Under these scenarios, when formulating a practical optimization problem, if the objective function coefficients are changed slightly, it may lead to a substantial change in the location of the optimum. Extreme care should be exercised when formulating such LP problems.

11.8 Summary

Linear Programming is a topic of great importance in optimization. This chapter introduced the basics of linear programming. Specifically, the four types of possible solutions for the Simplex method were discussed. The graphical solution approach and software options, such as `linprog` in MATLAB, were also discussed with examples. The theory behind the Simplex method and the Simplex algorithm were explored in detail with the help of examples. We provided a brief overview of the theory of duality and interior point methods. With the basic understanding of the topic provided in this chapter, the reader can understand and further explore linear programming in more detail (Refs. [1, 3, 2, 4, 5, 6].

11.9 Problems

11.1   Consider the LP problem given in Sec. 11.3.1. Reproduce the results shown in Fig. 11.1(a) using MATLAB. How does the optimal solution change if a constant is added to the objective function? Run your MATLAB code with five different starting points. Do you observe a change in the optimal results?

11.2   Consider the LP problem given in Sec. 11.3.2. Reproduce the results shown in Fig. 11.1(b) using MATLAB. Run your MATLAB code with five different starting points. Do you observe a change in the optimal results?

11.3   You are given the following optimization problem. Solve the following problem graphically.

$$\min_{x} \quad 8x_1 + 10x_2 + 4 \tag{11.118}$$

subject to

$$x_1 - x_2 \geq -4 \tag{11.119}$$

$$x_1 + x_2 \leq 6 \tag{11.120}$$

$$x_1, x_2 \geq 0 \tag{11.121}$$

Plot the objective function and the constraint equations. Identify the feasible design space and the optimal solution.

11.4   Transform the following problem into the standard LP formulation.

$$\min_{x} \quad z = x_1 + 5x_2 + 3x_3 \tag{11.122}$$

subject to

$$x_1 - 5x_2 + x_3 \geq 1 \tag{11.123}$$

$$5x_1 + x_2 + 2x_3 \leq 5 \tag{11.124}$$

$$-2x_1 + 3x_2 + 3x_3 \leq 4 \tag{11.125}$$

$$3x_1 + 8x_2 + 5x_3 \leq 3 \tag{11.126}$$

$$x_1, x_2 \geq 0, x_3 \text{ unbounded} \tag{11.127}$$

11.5 Consider the following set of equations.
(1) How many basic solutions are possible for this set of constraints?
(2) Transform them into the reduced row echelon form with respect to the basic variables x_1, x_2, and x_3.

$$x_1 + 2x_2 + 2x_3 = 4 \tag{11.128}$$

$$x_1 + x_2 + x_3 - x_4 = 1 \tag{11.129}$$

$$3x_2 + x_3 + 2x_4 = 6 \tag{11.130}$$

11.6 Solve the LP formulation given in Problem 11.4 using linprog.

11.7 Consider the following LP problem from Sec. 11.3.1. Solve it using the Simplex method.

$$\min_{x} \quad x_1 - 2x_2 \tag{11.131}$$

subject to

$$-4x_1 + 6x_2 \leq 9 \tag{11.132}$$

$$x_1 + x_2 \leq 4 \tag{11.133}$$

$$x_1, x_2 \geq 0 \tag{11.134}$$

11.8 Solve the following problem using the Simplex method. Verify the correctness of your solution using linprog.

$$\min_{x} \quad x_1 + 2x_2 - 7x_3 \tag{11.135}$$

subject to

$$2x_1 + x_2 + x_3 \leq 15 \tag{11.136}$$

$$-x_1 + 2x_2 - x_3 \leq 7 \tag{11.137}$$

$$x_1 + 5x_2 + 5x_3 \leq 25 \tag{11.138}$$

$$x_1, x_2, x_3 \geq 0 \tag{11.139}$$

BIBLIOGRAPHY OF CHAPTER 11

[1] D. G. Luenberger and Y. Ye. *Linear and Nonlinear Programming.* Springer, 2008.
[2] S. I. Gass. *Linear Programming: Methods and Applications.* Dover, 5th edition, 2010.
[3] G. B. Dantzig and M. N. Thapa. *Linear Programming 1: Introduction.* Springer, 1997.
[4] V. Chvatal. *Linear Programming.* W. H. Freeman and Company, New York, 1983.

[5] S. J. Wright. *Primal-Dual Interior-Point Methods*. Society of Industrial and Applied Mathematics, Philadelphia, 1997.

[6] R. J. Vanderbei. *Linear Programing: Foundations and Extensions*. Springer, 2014.

[7] G. Hurlber. *Linear Optimization: The Simplex Workbook*. Undergraduate Texts in Mathematics. Springer New York, 2010.

[8] A. Ravindran, G. V. Reklaitis, and K. M. Ragsdell. *Engineering Optimization: Methods and Applications*. John Wiley and Sons, 2006.

[9] C. Roos, T. Terlaky, and J.-Ph. Vial. *Interior Point Methods for Linear Optimization*. Springer, 2nd edition, 2010.

12

Nonlinear Programming with No Constraints

12.1 Overview

Nonlinear programming is a technique used to solve optimization problems that involves nonlinear mathematical functions (*e.g.*, objective functions or constraints, or both). The methods presented in this chapter are capable of solving unconstrained optimization problems with unimodal objectives. The procedures used for these methods are illustrated via minimization problems. Maximization problems can be converted to minimization problems by multiplying the objective functions by -1.

Unconstrained nonlinear programming has many practical applications. Many real life design problems have nonlinear objectives due to the nonlinearities in nature. Some design problems do not have constraints, or the constraints are negligible. Constrained optimization problems can be reformulated as unconstrained problems. The methods used to solve unconstrained nonlinear problems are powerful tools for solving optimization problems.

12.2 Necessary and Sufficient Conditions

Chapter 2 provides mathematical knowledge concerning the first and second derivatives of functions. Local optima can be determined by their derivatives. This chapter provides the conditions for local minima. These conditions involve the gradients (first derivative vectors) and Hessian matrices (second derivative matrices) for the objective functions. These conditions are the foundation for several of the algorithms described in this chapter (see Refs. [1, 2]).

Assume a function, $f(x)$, is continuous and has continuous first and second derivatives. The function can be expressed as a Taylor expansion.

$$f(x^* + \Delta x) = f(x^*) + \nabla f(x^*)^T \Delta x + \frac{1}{2} \Delta x^T \nabla^2 f(x^*) \Delta x + O_3(\Delta x) \qquad (12.1)$$

Neglecting the higher order terms, $O_3(\Delta x)$, the objective function change can be expressed as

$$\Delta f(x) = f(x^* + \Delta x) - f(x^*) = \nabla f(x^*)^T \Delta x + \frac{1}{2} \Delta x^T \nabla^2 f(x^*) \Delta x \qquad (12.2)$$

If x^* is a local minimum, any other points in its neighborhood should produce a greater objective value, which can be expressed as

$$\Delta f(x) = f(x^* + \Delta x) - f(x^*) \geq 0 \tag{12.3}$$

From Eq. 12.2, in order for the sign of $\Delta f(x)$ to be known (*e.g.*, positive for a minimum at x^*) for arbitrary values of Δx, the first derivative of $f(x)$ should be zero. Otherwise, $\Delta f(x)$ can be forced to be positive or negative by changing the sign of Δx, and Eq. 12.3 cannot be satisfied. Therefore, a local minimum, x^*, should satisfy the following first-order necessary conditions.

Theorem (First-Order Necessary Conditions)
If x^* is a local minimum of $f(x)$ and $f(x)$ is continuously differentiable in an open neighborhood of x^*, then the gradient $\nabla f(x^*) = 0$.
  Since $\nabla f(x^*) = 0$ at the local minimum, Eq. 12.2 becomes

$$\Delta f(x) = f(x^* + \Delta x) - f(x^*) = \frac{1}{2}\Delta x^T \nabla^2 f(x^*)\Delta x \tag{12.4}$$

It should satisfy $\Delta f(x) \geq 0$ since x^* is a local minimum. Then, $\nabla^2 f(x^*)$ should be positive semidefinite, which is stated as the second-order necessary conditions as follows.

Theorem (Second-Order Necessary Conditions)
If x^* is a local minimum of $f(x)$ and $\nabla^2 f(x)$ exists and is continuous in an open neighborhood of x^*, then $\nabla f(x^*) = 0$ and $\nabla^2 f(x^*)$ is positive semidefinite.
  The second-order necessary conditions do not guarantee a local minimum. At $x = 0$, the function, $f(x) = x^3$, satisfies the first-order and second-order necessary conditions. However, $x = 0$ is not a local minimum of $f(x) = x^3$.
  If the Hessian of $f(x)$ is positive definite at the point where $\nabla f(x) = 0$, then

$$\Delta f(x) = f(x^* + \Delta x) - f(x^*) = \frac{1}{2}\Delta x^T \nabla^2 f(x^*)\Delta x > 0 \tag{12.5}$$

Equation 12.5 guarantees the point, x^*, is a strict local minimum. It is stated as the second-order sufficient condition. The second-order sufficient conditions are stronger than the second-order necessary conditions.

Theorem (Second-Order Sufficient Conditions)
Given $f(x)$, if $\nabla^2 f(x)$ is continuous in an open neighborhood of x^* and $\nabla f(x^*) = 0$; if $\nabla^2 f(x^*)$ is positive definite, then x^* is a strict local minimum of $f(x)$.

12.3 Single Variable Optimization

Some unconstrained nonlinear optimization problems may only have a single variable. In this section, the methods for solving single variable nonlinear optimization problems are presented. We parenthetically note that the Bisection and the Golden search methods require bounds which, technically, are constraints. However, these bounds need not have physical meanings, and can be arbitrarily chosen for the sole purpose of implementing these algorithms.

12.3.1 Interval Reduction Methods

Bisection

The interval reduction methods are applicable to single variable nonlinear optimization problems whose variables have lower bounds and upper bounds.

The *Bisection method* (also known as the *Interval Halving method*) finds an extreme point in a bounded region for a unimodal single variable function. The function and its first derivative are assumed continuous. The Bisection method successively halves the interval, and decides in which one of the two half intervals the extreme point exists. The Bisection method uses the first derivative of the function to determine in which half the extreme point lies. Figure 12.1 illustrates how to update the interval according to the sign of the first derivative. Suppose the variable x of a unimodal convex function $f(x)$ is inside the interval $[a, b]$. The function $f(x)$ and its derivative $f'(x)$ are continuous over this interval. The first derivatives at the two ends satisfy the condition $f'(a)f'(b) < 0$. This condition implies that there is a minimum in the interval $[a, b]$ (with 0 derivative). The Bisection method procedure for finding the minimum is outlined as follows.

1. Specify the convergence tolerance for an interval of length $\epsilon > 0$. Specify the convergence tolerance of the gradient $\gamma > 0$. Set the iteration number, k, to 0. The bounds are $l_0 = a$ and $r_0 = b$.
2. If $r_k - l_k < \epsilon$, stop. The midpoint $x_k = \frac{l_k + r_k}{2}$ is taken as the minimum, x^*, and the corresponding function value $f(x^*)$ is the optimal solution.
3. Evaluate the gradient of the function at the midpoint $f'(x_k) = f'(\frac{l_k + r_k}{2})$.
4. If $\|f'(x_k)\| < \gamma$, stop. The corresponsing x_k is considered the minimum, x^*, and the corresponding function value, $f(x^*)$, is the optimal solution.
5. Evaluate the derivative $f'(x_k)$. If it is positive, let $l_{k+1} = l_k$ and $r_{k+1} = x_k$. If it is negative, let $l_{k+1} = x_k$ and $r_{k+1} = r_k$.
6. $k = k + 1$. Go to Step 2.

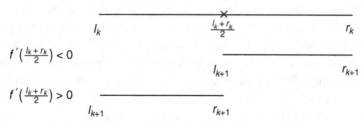

Figure 12.1. Interval Updates of the Bisection Method

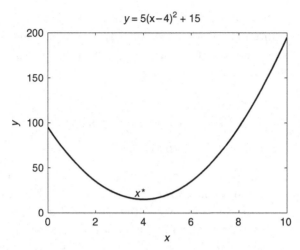

Figure 12.2. Plot of the Objective Function

Table 12.1. *The End Points and*
the Derivatives at Each Midpoint

k	a_k	b_k	Derivative
1	0.0000	5.0000	-15.0000
2	2.5000	5.0000	-2.5000
3	3.7500	5.0000	3.7500
4	3.7500	4.3750	0.6250
5	3.7500	4.0625	-0.9375
6	3.9063	4.0625	-0.1563
7	3.9844	4.0625	0.2344
8	3.9844	4.0234	0.0391
9	3.9844	4.0039	-0.0586
10	3.9941	4.0039	-0.0098

Example: The following optimization problem is used to illustrate the Bisection method.

$$\min_x f(x) = 5(x - 4)^2 + 15 \qquad 0 \le x \le 10 \qquad (12.6)$$

Figure 12.2 presents the function curve in the defined region.
The convergence tolerance of the interval length, ϵ, is set as 0.001. The convergence tolerance of the gradient at the midpoints, γ, is set as 0.01. Table 12.1 provides the end points, a_k and b_k, and the derivatives at each midpoint. The Bisection method stops after 10 iterations. At the 10th iteration, the distance between the two points, 0.0098, is larger than the convergence tolerance of the interval length, ϵ. The absolute value of the derivative at the midpoint is less than the convergence tolerance of the gradient at the midpoint, γ. It is determined that the optimal value is at the midpoint 3.9990. The corresponding function value is $f(x^*) = 15.0000$.

Golden Section Search

The Golden Section Search is a method to minimize or maximize a unimodal function of one variable. The algorithm maintains the function values for triples of points whose distances form a golden ratio. This method does not require information about the derivatives. If the minimum is known to exist inside a region of the variable, this method successively narrows the range of function values inside which the minimum exists.

In mathematics, two quantities are in the golden ratio if the ratio of the sum of the quantities to the larger quantity is equal to the ratio of the larger quantity to the smaller quantity. Figure 12.3 illustrates the geometric relationship that defines the golden ratio. The total length is l_{ac}, the larger segment is l_{ab}, and the smaller segment is l_{bc}. From Fig. 12.3, the golden ratio can be expressed as

$$\varphi \equiv \frac{l_{ab} + l_{bc}}{l_{ab}} = \frac{l_{ab}}{l_{bc}} \tag{12.7}$$

Solving Eq. 12.7, the value of φ is

$$\varphi = \frac{1 + \sqrt{5}}{2} \approx 1.618 \tag{12.8}$$

The value of its conjugate is

$$\tau \equiv \frac{1}{\varphi} = \frac{-1 + \sqrt{5}}{2} \approx 0.618 \tag{12.9}$$

Suppose the variable of a unimodal function $f(x)$ is inside the interval $[a, b]$, and $f(x)$ is continuous over this interval. The minimum exists inside this interval. The steps of the Golden Section Search are illustrated as follows.

1. Select $\epsilon > 0$ as the convergence tolerance of the interval. Set the iteration number, k, to 0. The bounds are $a_0 = a$ and $b_0 = b$. Set $l_0 = b_0 - \tau(b_0 - a_0)$ and $r_0 = a_0 + \tau(b_0 - a_0)$.
2. If $b_k - a_k < \epsilon$, stop. The midpoint, $x_k = \frac{a_k + b_k}{2}$, is considered the optimal value, x^*; and the corresponding function value, $f(x^*)$, is the optimal solution.

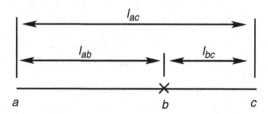

Figure 12.3. Line Segments Divided According to the Golden Ratio

3. If $f(l_k) > f(r_k)$, the parameters are updated as shown in the $f(l_k) > f(r_k)$ case in Fig. 12.4.

$$a_{k+1} = l_k$$
$$b_{k+1} = b_k$$
$$l_{k+1} = r_k$$
$$r_{k+1} = a_{k+1} + \tau(b_{k+1} - a_{k+1})$$

4. If $f(l_k) < f(r_k)$, the parameters are updated as the $f(l_k) < f(r_k)$ case in Fig. 12.4.

$$a_{k+1} = a_k$$
$$b_{k+1} = r_k$$
$$r_{k+1} = l_k$$
$$l_{k+1} = b_{k+1} - \tau(b_{k+1} - a_{k+1})$$

5. $k = k + 1$, go to Step 2.

Inside each iteration, depending on $f(l_k) > f(r_k)$ or $f(l_k) < f(r_k)$, the objective function is evaluated at different new points. The two bounds of the interval are updated and the interval becomes shorter. At Step 2 of each iteration, the bounds are $[a_k, b_k]$. If $f(l_k) < f(r_k)$, the right bound, b_k, is updated in Step 4. The length of the new interval is

$$b_{k+1} - a_{k+1} = r_k - a_k = [a_k + \tau(b_k - a_k)] - a_k = \tau(b_k - a_k) \qquad (12.10)$$

The length of the interval is reduced by a factor of τ at each iteration. The length reduction is similar to when $f(l_k) > f(r_k)$.

The following observations are noteworthy. The Golden Section Search method requires one evaluation of the objective function at each iteration to determine the interval where the minimum lies; while the Bisection method requires one evaluation of the first derivative at each iteration.

Example: The same optimization example used for the Bisection method is again used to illustrate the Golden Section Search. The convergence tolerance rate, ϵ, is set as 0.1 for this example, which is different from the previous example

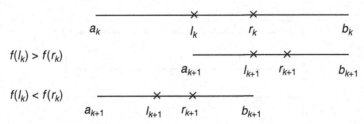

Figure 12.4. Interval Updates of the Golden Section Search

Table 12.2. *Endpoints, Golden Section Points and Function Values at Each Iteration*

k	a_k	b_k	$f(a_k)$	$f(b_k)$	l_k	r_k	$f(l_k)$	$f(r_k)$
1	0.00	10.00	95.00	195.00	3.82	6.18	15.16	38.77
2	0.00	6.18	95.00	38.77	2.36	3.82	28.44	15.16
3	2.36	6.18	28.44	38.77	3.82	4.72	15.16	17.60
4	2.36	4.72	28.44	17.60	3.26	3.82	17.72	15.16
5	3.26	4.72	17.72	17.60	3.82	4.16	15.16	15.13
6	3.82	4.72	15.16	17.60	4.16	4.38	15.13	15.71
7	3.82	4.38	15.16	15.71	4.03	4.16	15.01	15.13
8	3.82	4.16	15.16	15.13	3.95	4.03	15.01	15.01
9	3.95	4.16	15.01	15.13	4.03	4.08	15.01	15.03
10	3.95	4.08	15.01	15.03	4.00	4.03	15.00	15.01

for the Bisection method.

$$\min_x f(x) = 5(x - 4)^2 + 15, \qquad 0 \le x \le 10 \qquad (12.11)$$

Table 12.2 provides the four points, a_k, b_k, l_k, and r_k, and their function values. The Golden Section Search stops after 10 iterations. At the 10th iteration, the values of the objective function are compared at two points, $l_{10} = 4.00$ and $r_{10} = 4.03$. It is determined that the optimal value is between the points, 3.95 and 4.03. The distance between the two points is less than the convergence tolerance, ϵ. Finally, the point $x^* = (3.95 + 4.03)/2 = 3.99$ is considered the optimal value, and the corresponding function value is $f(x^*) = 15.00$.

12.3.2 Polynomial Approximations: Quadratic Approximation

If we assume an objective function is unimodal and continuous inside an interval, then it can be approximated by a polynomial. A is the simplest interpolation of the objective function (see Ref. [3]). The quadratic approximation method consists of a sequence of reducing intervals, and iterative approximations in the reduced intervals. As the approximation approaches the actual minimum, its accuracy increases. When the error between the approximation and the actual function is less than a predefined tolerance, terminate the iteration.

A quadratic approximating function can be constructed given three consecutive points, x_1, x_2, and x_3, and their corresponding function values, $f(x_1), f(x_2)$, and $f(x_3)$. The quadratic function is expressed as follows.

$$\tilde{f}(x) = c_0 + c_1(x - x_1) + c_2(x - x_1)(x - x_2) \qquad (12.12)$$

At the three points, x_1, x_2, and $x_3, \tilde{f}(x) = f(x)$. The three constants, c_0, c_1, and c_2, can be determined as follows.

At the point x_1, since $f(x_1) = \tilde{f}(x_1) = c_0$, the first constant is $c_0 = f(x_1)$. At the point $x_2, f(x_2) = \tilde{f}(x_2) = c_0 + c_1(x_2 - x_1)$. The second constant, c_1, can be evaluated as follows.

$$c_1 = \frac{f(x_2) - c_0}{x_2 - x_1} = \frac{f(x_2) - f(x_1)}{x_2 - x_1} \tag{12.13}$$

At the point x_3, $f(x_3) = \tilde{f}(x_3) = c_0 + c_1(x_3 - x_1) + c_2(x_3 - x_1)(x_3 - x_2)$. The third constant, c_2, is obtained from the following equation.

$$
\begin{aligned}
c_2 &= \frac{f(x_3) - c_0 - c_1(x_3 - x_1)}{(x_3 - x_1)(x_3 - x_2)} \\
&= \frac{1}{x_3 - x_2} \left(\frac{f(x_3) - f(x_1)}{x_3 - x_1} - \frac{f(x_2) - f(x_1)}{x_2 - x_1} \right)
\end{aligned} \tag{12.14}
$$

Using the above quadratic approximation, the minimum point, x^*, should satisfy the first-order necessary conditions.

$$\frac{d\tilde{f}(x)}{dx} = c_1 + c_2(x - x_1) + c_2(x - x_2) = 0 \tag{12.15}$$

Then, the minimum point, x^*, can be expressed as

$$x^* = \frac{x_1 + x_2}{2} - \frac{c_1}{2c_2} \tag{12.16}$$

The quadratic approximation can be implemented in successively reduced intervals. A successive quadratic approximation method is outlined as follows.

1. Define the tolerance of the function successive-values difference, $\gamma > 0$, and the tolerance of the variable, $\epsilon > 0$. Set Δx. Set the iteration number, k, to 1.

2. Set the initial point, x_1^1. Compute $x_2^1 = x_1^1 + \Delta x$. Evaluate $f(x_1^1)$ and $f(x_2^1)$. If $f(x_1^1) > f(x_2^1)$, then $x_3^1 = x_2^1 + \Delta x$. If $f(x_1^1) < f(x_2^1)$, then $x_3^1 = x_1^1 - \Delta x$. Evaluate $f(x_3^1)$.

3. Compare the function values at the three points, x_1^k, x_2^k, and x_3^k. Find the minimum, $f_{min}^k = min\{f(x_1^k), f(x_2^k), f(x_3^k)\}$, and the corresponding point, x_{min}^k.

4. Using the three points, x_1^k, x_2^k, and x_3^k, construct a quadratic approximation. Compute the minimum point, x_k^*, using Eq. 12.16. Evaluate $f(x_k^*)$.

5. If $\|f(x_k^*) - f_{min}^k\| < \gamma$ and $\|x_k^* - x_{min}^k\| < \epsilon$, take x_k^* as the minimum. Terminate the iteration.

6. Take the current best point x_k^* and the two points bracketing it as the three points for the next quadratic approximation. $k = k + 1$. Go to Step 3.

Example: Use the quadratic approximation method to solve the following optimization problem. The variable, x, is inside $(1.001, 10)$.

$$\text{Minimize } f(x) = x^2 - 2x + \frac{8}{x - 1} + 6 \tag{12.17}$$

Table 12.3. *The Iterations for the Quadratic*
Approximation Method

	x_1^k	$f(x_1^k)$	x_2^k	$f(x_2^k)$	x_3^k	$f(x_3^k)$
k=1	2	14	3	13	4	16.67
	x_{min}^k	f_{min}^k	x_k^*	$f(x_k^*)$	δf^k	δx^k
	3	13	2.71	12.61	0.03	0.11
	x_1^k	$f(x_1^k)$	x_2^k	$f(x_2^k)$	x_3^k	$f(x_3^k)$
k=2	2	14	2.71	12.61	3	13
	x_{min}^k	f_{min}^k	x_k^*	$f(x_k^*)$	δf^k	δx^k
	2.71	12.61	2.65	12.57	0.002	0.02

Define the tolerance of the function successive-values difference as $0.01\|f(x_k^*)\|$ and the tolerance of the variable as $0.05\|x_k^*\|$. Set $\Delta x = 1$. Set the initial point, $x_1^1 = 2$.
Define δf^k and δx^k as

$$\delta f^k = \frac{\|f(x_k^*) - f_{min}^k\|}{\|f(x_k^*)\|} \tag{12.18}$$

$$\delta x^k = \frac{\|x_k^* - x_{min}^k\|}{\|x_k^*\|} \tag{12.19}$$

The results for all the iterations are listed in Table 12.3.

At the end of the second iteration, the errors satisfy the tolerances. The iterations are terminated. The optimal point is 2.65 and the optimal function value is 12.57

12.4 Multivariable Optimization

Most practical unconstrained nonlinear optimization problems have multiple design variables. In this section, the methods for solving multivariable nonlinear optimization problems are illustrated.

12.4.1 Zeroth-Order Methods

Simplex Search Method

The Simplex Search method is a direct-search method that does not require function derivatives (see Ref. [4]). Although the name of this method includes the word simplex, it has no relationship to the Simplex method of linear programming.

In geometry, a Simplex is a triangle in two-dimensional space and a tetrahedron in three-dimensional space. In n-dimensional space, it is an n-dimensional polytope. The main idea of the Simplex method is that, at each iteration, based on the comparison of function values at all the vertices, one vertex is projected through the centroid of the other vertices at a suitable distance. At the beginning, the Simplex Search method sets up a regular Simplex in the space of the variables. The objective function is evaluated at each vertex. For minimization problems, the vertex with the highest function value is the one that is reflected through the centroid of the other vertices to generate a new point. The point and the remaining vertices construct the next Simplex.

At the beginning of the algorithm, it requires the first Simplex to be generated given a base point, x^0, and an appropriate scale factor, α. Assuming the variable vector has N dimensions, two increments, δ_1 and δ_2, are defined by the following two expressions.

$$\delta_1 = \left[\frac{(N+1)^{1/2} - 1}{N\sqrt{2}} \right] \alpha \tag{12.20}$$

$$\delta_2 = \left[\frac{(N+1)^{1/2} + N - 1}{N\sqrt{2}} \right] \alpha \tag{12.21}$$

Using the two increments, the jth dimension for the ith vertex can be calculated using the following expression.

$$x_j^{(i)} = \begin{cases} x_j^{(0)} + \delta_1 & \text{if } j \neq i \\ x_j^{(0)} + \delta_2 & \text{if } j = i \end{cases} \tag{12.22}$$

At each iteration, the vertex with the highest function value is selected as x_{high}. It is reflected through the centroid of the remaining points. The centroid is evaluated by the following expression.

$$x_{centroid} = \frac{1}{N} \sum_{\substack{i=0 \\ i \neq j}}^{N} x^{(i)} \tag{12.23}$$

The line passing through x_{high} and $x_{centroid}$ is given by the following expression.

$$x = x_{high} + \lambda(x_{centroid} - x_{high}) \tag{12.24}$$

The selection of λ can yield the desired point on the line. If $\lambda = 0$, it yields the point, x_{high}. If $\lambda = 1$, the result is the point, $x_{centroid}$. In order to generate a symmetric reflection, λ is set as 2. The reflected point, $x_{reflected}$, is evaluated as

$$x_{reflected} = 2x_{centroid} - x_{high} \tag{12.25}$$

The reflection process is illustrated in Fig. 12.5. The point, x^2, has the highest function value, and it is reflected through the centroid of x^1 and x^3.

During the optimization, the following two situations may occur.

1. At the current iteration, the vertex with the highest function value is the reflected point generated in the last iteration. In this situation, choose instead the vertex with the next highest function value and generate a reflected point.

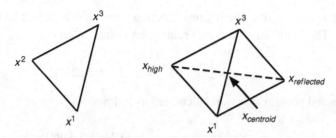

Figure 12.5. Reflection of the Vertex with the Highest Function Value

2. The iterations can cycle between two or more Simplexes. If a vertex remains in the Simplex for more than M iterations, set up a new Simplex with the lowest point as the base point and reduce the size of the Simplex. The number of the dimension for the variable is N. The value of M can be estimated by $M = 1.65N + 0.05N^2$ and M is rounded to the nearest integer.

The Simplex Search method is terminated when the size of the Simplex is sufficiently small or other termination criteria are met. The vertex with the lowest function value is taken as the minimum.

The Simplex construction and point reflection are illustrated in the following example.

Example: Use the Simplex Search method to minimize the following function.

$$\text{Minimize } f(x) = (x_1 - 2)^2 + (x_2 - 3)^2 \tag{12.26}$$

The number of the dimensions is 2. Each Simplex has 3 vertices. Take $x^0 = [1, 1]^T$ as the starting point, and set $\alpha = 2$. The increments, δ_1 and δ_2, can be evaluated as follows.

$$\delta_1 = \left[\frac{(2+1)^{1/2} - 1}{2\sqrt{2}} \right] \times 2 = 1.9318 \tag{12.27}$$

$$\delta_2 = \left[\frac{(2+1)^{1/2} + 2 - 1}{2\sqrt{2}} \right] \times 2 = 0.5176 \tag{12.28}$$

The coordinates of the other two vertices are calculated as follows.

$$x^1 = [1 + 1.9318, 1 + 0.5176]^T = [2.9318, 1.5176]^T \tag{12.29}$$

$$x^2 = [1 + 0.5176, 1 + 1.9318]^T = [1.5176, 2.9318]^T \tag{12.30}$$

The function values at the three points, x^0, x^1, and x^2, are as follows.

$$f(x^0) = 5 \tag{12.31}$$

$$f(x^1) = 3.0658 \tag{12.32}$$

$$f(x^2) = 0.2374 \tag{12.33}$$

Since the vertex, x^0, has the highest function value, it is reflected to form a new Simplex. The centroid of the two remaining points, x^1 and x^2, is calculated as follows.

$$x_{centroid} = \frac{1}{2}(x^1 + x^2) = [2.2247, 2.2247]^T \tag{12.34}$$

The reflected point, $x_{reflected}$, is calculated as follows.

$$x_{reflected} = 2x_{centroid} - x^0 = [3.4494, 3.4494]^T \tag{12.35}$$

At the point, $x_{reflected}$, $f(x_{reflected}) = 2.3027$. The vertex with the highest function value in this new Simplex is x^1. The algorithm will continue by reflecting the x^1, and the iterations proceed following the reflection rule until the optimal point, $[2, 3]^T$, is found.

Pattern Search Methods

The pattern search methods are a family of derivative-free numerical optimization methods. They can be applied to not only the optimization of continuous differentiable functions, but also to the optimization of discrete or nondifferentiable functions. The pattern search methods look along certain specified directions, and evaluate the objective function at a given step length along each of these directions. These points form a frame around the current iteration. Depending on whether any point within the pattern has a lower objective function value than the current point, the frame shrinks or expands in the next iteration. The search stops after a minimum pattern size is reached.

An important decision in the pattern search methods is to choose the search direction set, D_k, at each iteration k. A key condition is that at least one direction in this set should give a descent direction for the objective function whenever the current point is not a stationary point. If a search direction, p, satisfies the following inequality, it can be a direction of descent for the objective function.

$$\cos \theta = \frac{-\nabla f_k^T p}{\|\nabla f_k\| \|p\|} \geq \delta \tag{12.36}$$

where δ is a positive constant.

If the search direction, p, satisfies Eq. 12.36 at each iteration, it is a descent direction. Choose the search direction set, D_k, so that at least one direction, $p \in D_k$, will satisfy $\cos \theta > \delta$, regardless of the value of ∇f_k.

A second condition on D_k is that the lengths of the vectors in this set are all similar. The diameter of the frame formed by this set is captured adequately by the step length. This condition can be expressed as

$$\beta_{min} \leq \|p\| \leq \beta_{max}, \text{ for all } p \in D_k \tag{12.37}$$

where β_{min} and β_{max} are positive constants.

For the pattern search methods in n-dimensional space, examples of sets D_k that satisfy the above two conditions in Eqs. 12.36 and 12.37 include the coordinate direction set

$$e_1, e_2, ..., e_n, -e_1, -e_2, ..., -e_n \qquad (12.38)$$

and the set of $n + 1$ vectors defined by

$$p_i = \frac{1}{2n}e - e_i, i = 1, 2, ..., n; p_{n+1} = \frac{1}{2n}e \qquad (12.39)$$

where $e = (1, 1, ..., 1)^T$.

Another important decision in the pattern search method is how to choose a sufficient decrease function $\rho(l)$. The sufficient decrease function is used to determine the convergence of the results. It is a function of the step length and its domain is $[0, \infty)$. It is an increasing function of l and the function values are positive. As l approaches 0, the limit of $\rho(t)/t$ is 0. An appropriate choice of the sufficient decrease function is $Mt^{3/2}$, where M is some positive constant. If the decrease of the objective function value is less than the value of the sufficient decrease function, the pattern search is converged to the minimum.

At each iteration, the search direction set is D_k and the step length is l_k. The frame that consists of the points at the next iteration is $x_k + l_k p_k$ for all $p_k \in D_k$. When one of the points in the frame yields a significant decrease of the objective function, it becomes the next searching point, x_{k+1}, and the next step length, l_{k+1}, is increased. If none of the points in the frame has a significantly smaller function value than f_k, the next step length, l_{k+1}, will be reduced. The procedure of the pattern search algorithm is illustrated as follows.

1. Define the sufficient decrease function $\rho(l)$. Choose convergence tolerance l_{tol}, contraction parameter θ_{max}, and search direction set D_k. Choose initial point x_0, initial step length l_0, and initial direction set D_0. Set the iteration number, k, to 1.
2. Evaluate the objective function value $f(x_k)$.
3. If the step length is $l_k \leq l_{tol}$, take x_k as the optimal point, x^*. Stop.
4. If $f(x_k + l_k p_k) < f(x_k) - \rho(l_k)$ for some $p_k \in D_k$, set x_{k+1} as $x_k + l_k p_k$ for some $p_k \in D_k$, and increase the step length for the next iteration.
5. If $f(x_k + l_k p_k) \geq f(x_k) - \rho(l_k)$ for all $p_k \in D_k$, use the same point x_k as the point x_{k+1} for the next iteration. Reduce the step length for the next iteration, l_{k+1}. Set l_{k+1} as $\theta_k l_k$, where $0 < \theta_k \leq \theta_{max} < 1$.
6. Set $k = k + 1$. Go to Step 2.

MATLAB has the function, `patternsearch`, that finds the minimum of a function using the pattern search method. It can handle optimization problems with nonlinear, linear, and bound constraints, and does not require functions to be differentiable or continuous. An example is provided below to illustrate how to solve a nonlinear optimization problem without constraints using `patternsearch`.

Example: Use the pattern search method to solve the following three-dimensional quadratic problem.

$$f(x) = \frac{1}{2}x^T Q x - c^T x \qquad (12.40)$$

$$Q = \begin{bmatrix} 2 & 0 & 0 \\ 0 & 3 & 0 \\ 0 & 0 & 5 \end{bmatrix} \qquad (12.41)$$

$$c = \begin{bmatrix} -8 \\ -9 \\ -8 \end{bmatrix} \qquad (12.42)$$

The sufficient decrease function is set as $\rho(l) = 10l^{3/2}$. The contraction factor is 0.5. The convergence tolerance is set as 0.001. The contraction factor, θ_{max}, is set as 0.5. The initial step length, l_0, is set as 0.5. The starting point, x_0, is $[0, 0, 0]^T$. The search directions set, D_k, is the coordinate direction set, which is as follows.

$$\left\{ \begin{bmatrix} 1 \\ 0 \\ 0 \end{bmatrix}, \begin{bmatrix} 0 \\ 1 \\ 0 \end{bmatrix}, \begin{bmatrix} 0 \\ 0 \\ 1 \end{bmatrix}, \begin{bmatrix} -1 \\ 0 \\ 0 \end{bmatrix}, \begin{bmatrix} 0 \\ -1 \\ 0 \end{bmatrix}, \begin{bmatrix} 0 \\ 0 \\ -1 \end{bmatrix} \right\} \qquad (12.43)$$

At the initial point, x_0, the function value, $f(x_0)$, is 0. At the first iteration, the function values at $x_0 + l_0 p_k, (k = 1, ..., 6)$ are 9, 10.5, 10.5, -7, -7.5, and -5.5. The value of sufficient decrease function is as follows.

$$\rho(l_0) = \rho(0.5) = 10 \times 0.5^{3/2} = 3.53 \qquad (12.44)$$

For the six points $x_0 + l_0 p_k$, three of them ($k = 4, 5, 6$) satisfy the equation, $f(x_k + l_k p_k) < f(x_k) - \tilde{N}(l_k)$. Set the point, $x_0 + l_0 p_5$, as x_1, and increase the step length. The iterations continue. The minimum point is obtained as follows.

$$x^* = \begin{bmatrix} -4 \\ -3 \\ -1.6000 \end{bmatrix} \qquad (12.45)$$

The optimal function value $f(x^*) = -35.9$.

This problem can also be solved by MATLAB function `patternsearch`. The result is the same. The main file is as follows.

```
clc
clear
X0 = [0; 0; 0];
[x,fopt]=patternsearch(@obj_fun,X0)
```

The file for the objective function is as follows.

```
function f = obj_fun(x)
```

```
Q = [2 0 0; 0 3 0; 0 0 5];
c = [-8; -9; -8];
f = 0.5*x'*Q*x-c'*x;
```

12.4.2 First-Order Methods

Steepest Descent

The steepest descent method is a first-order optimization algorithm. In each iteration, it takes the negative direction of the gradient as the descent direction of the objective function. This method takes steps along the descent direction to find a local minimum. The procedure is as follows.

1. Choose a starting point x_0. Set $k = 0$.
2. Check the conditions of $f(x_k)$. If x_k is the minimum, stop.
3. Calculate the descent direction, p_k, as follows.

$$p_k = -\nabla f(x_k) \tag{12.46}$$

4. Evaluate the step length, α_k. The step length, α_k, should be an acceptable solution to the following minimization problem.

$$\min_{\alpha_k > 0} f(x_k + \alpha_k p_k) \tag{12.47}$$

5. Update the value of the variable, x_{k+1}, using the descent direction and the step length.

$$x_{k+1} = x_k + \alpha_k p_k \tag{12.48}$$

6. Set $k = k + 1$ and go to Step 2.

The steepest descent method is the simplest Newton-type method for nonlinear optimization. The search direction is the opposite of the gradient. It does not require the computation of second derivatives, and it does not require matrix storage. The computational cost per iteration is lower compared to other Newton-type methods. However, this method is very inefficient at solving most problems. It converges only at a linear rate. This means it requires more iterations. As a result, although the computational cost per iteration is low, the overall costs are high. However, the steepest descent method is theoretically useful in proving the convergence of other methods; and it provides a lower bound on the performance of other algorithms.

The procedure to solve an unconstrained nonlinear problem using the steepest descent method is illustrated with the following example. Only the first two iterations are illustrated in this example.

Example: Use the steepest descent method to solve the following three-dimensional quadratic problem.

$$f(x) = \frac{1}{2}x^T Q x - c^T x \tag{12.49}$$

$$Q = \begin{bmatrix} 1 & 0 & 0 \\ 0 & 5 & 0 \\ 0 & 0 & 10 \end{bmatrix} \tag{12.50}$$

$$c = \begin{bmatrix} -1 \\ -1 \\ -1 \end{bmatrix} \tag{12.51}$$

In this problem, the steepest descent direction is

$$p_k = -\nabla f(x_k) = -(Qx_k - c) \tag{12.52}$$

The step length used for exact line search $x_{k+1} = x_k + \alpha_k p_k$ is

$$\alpha_k = -\frac{\nabla f(x_k)^T p_k}{p_k^T Q p_k} \tag{12.53}$$

Choose an initial point $x_0 = (0,0,0)^T$. At x_0,

$$f(x_0) = 0 \tag{12.54}$$

$$\nabla f(x_0) = \begin{bmatrix} 1 \\ 1 \\ 1 \end{bmatrix} \tag{12.55}$$

and

$$\|\nabla f(x_0)\| = 1.7321. \tag{12.56}$$

Since $\|\nabla f(x_0)\| > 0$, go to the next iteration. The step length is $\alpha_0 = 0.1875$. The next point is

$$x_1 = \begin{bmatrix} -0.1875 \\ -0.1875 \\ -0.1875 \end{bmatrix} \tag{12.57}$$

At point x_1, the function value, the gradient, and the norm of the gradient are as follows.

$$f(x_1) = -0.2813 \tag{12.58}$$

$$\nabla f(x_1) = \begin{bmatrix} 0.8125 \\ 0.0625 \\ -0.8750 \end{bmatrix} \tag{12.59}$$

and

$$\|\nabla f(x_1)\| = 1.1957 \tag{12.60}$$

The next step length is $\alpha_1 = 0.1715$ and the new point is

$$x_2 = \begin{bmatrix} -0.3269 \\ -0.1982 \\ -0.0374 \end{bmatrix} \tag{12.61}$$

At point x_2, the function value, the gradient, and the norm of the gradient are as follows.

$$f(x_2) = -0.4039 \tag{12.62}$$

$$\nabla f(x_2) = \begin{bmatrix} 0.6731 \\ 0.0089 \\ 0.6257 \end{bmatrix} \tag{12.63}$$

and

$$\|\nabla f(x_2)\| = 0.9191 \tag{12.64}$$

The above are the first two iterations to solve this problem. It takes 36 iterations before the norm of the gradient is less than 0.001. The optimal solution is

$$x^* = \begin{bmatrix} -0.9993 \\ -0.2000 \\ -0.0999 \end{bmatrix} \tag{12.65}$$

The corresponding optimal function value is $f(x^*) = -0.65$.

Conjugate Gradient Methods

In addition to the Gaussian elimination method, iterative methods are also a valuable tool for solving linear equations in the form of Eq. 12.66. One of the many iterative methods is the conjugate gradient method, which is designed to solve Eq. 12.66 when the matrix A is symmetric and positive definite.

$$Ax = b \tag{12.66}$$

Equation 12.67 is an optimization problem that has a quadratic objective function.

$$\min_x f(x) = \frac{1}{2} x^T A x - b^T x \tag{12.67}$$

The matrix A is symmetric and positive definite. The first-order necessary conditions require the gradient of $f(x)$ to be equal to zero. The gradient can be expressed as Eq. 12.68.

$$\nabla f(x) = Ax - b \tag{12.68}$$

Since the Hessian matrix, $\nabla^2 f(x) = A$, is positive definite, the sufficient conditions for a local minimum are satisfied. This optimization problem is equivalent to the problem of linear equations as shown by Eq. 12.66.

If a set of vectors, p_i, satisfies Eq. 12.69, the vectors are said to be conjugate with respect to the matrix A. Any set of conjugate vectors is also linearly independent.

$$p_i^T A p_j = 0, \text{ if } i \neq j \tag{12.69}$$

The residual of Eq. 12.68 is

$$r = Ax - b \tag{12.70}$$

In Eq. 12.67, if the matrix A is diagonal, the contours of the function $f(x)$ are ellipses. The axes of the ellipses are aligned with coordinate directions. The conjugate directions are along the coordinate directions. Figure 12.6 illustrates the conjugate directions in a two dimensional space. The minimum of the function can be found by performing successive one-dimensional minimizations along the conjugate directions.

In Eq. 12.67, if Matrix A is non-diagonal, the contours of the function $f(x)$ are also ellipses. However, the axes of the ellipses are not aligned with the coordinate directions. The conjugate directions are not along the coordinate directions. Figure 12.7 shows the conjugate directions for a non-diagonal matrix in a two dimensional space.

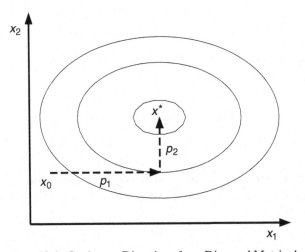

Figure 12.6. Conjugate Directions for a Diagonal Matrix A

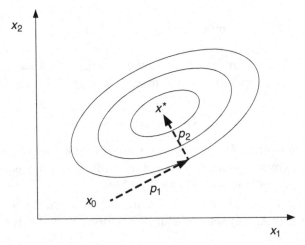

Figure 12.7. Conjugate Directions for a Non-diagonal Matrix A

The conjugate gradient methods used to solve the quadratic optimization problem (Eq. 12.67) exist in different versions of the algorithm. The steps for one of these versions are illustrated as follows.

1. Set an initial guess x_0. The residual is $r_0 = Ax_0 - b$. Set a vector with the same dimension of the conjugate vector, $p_0 = -r_0$. Specify the convergence tolerance, $\epsilon > 0$. Initialize the iteration counter, $i = 0$.
2. Check the value of the residual $\|r_i\|$. If $\|r_i\| < \epsilon$, stop.
3. Set $\alpha_i = r_i^T r_i / p_i^T A p_i$.
4. Evaluate the next point, $x_{i+1} = x_i + \alpha_i p_i$.
5. Evaluate the residual, $r_{i+1} = r_i + \alpha_i A p_i$.
6. Set $\beta_{i+1} = r_{i+1}^T r_{i+1} / r_i^T r_i$.
7. Set the conjugate vector, $p_{i+1} = -r_{i+1} + \beta_{i+1} p_i$.
8. Set $i = i + 1$. Go to Step 2.

Example: Use the above algorithm for conjugate gradient methods to solve the following three-dimensional quadratic problem.

$$f(x) = \frac{1}{2} x^T Q x - c^T x \tag{12.71}$$

$$Q = \begin{bmatrix} 2 & 0 & 0 \\ 0 & 3 & 0 \\ 0 & 0 & 5 \end{bmatrix} \tag{12.72}$$

$$c = \begin{bmatrix} -8 \\ -9 \\ -8 \end{bmatrix} \tag{12.73}$$

The convergence tolerance, ϵ, is set as 0.01. Choose an initial point $x_0 = (0, 0, 0)^T$. At this point, the residual is $r_0 = (8, 9, 8)^T$. This point is not optimal. The algorithm performs several iterations in search of the optimal solution. Table 12.4 provides the results for all the iterations.

At the initial point x_0, $\|r_0\| = 14.46$. In the first iteration, $\|r_1\| = 5.24$. In the second iteration, $\|r_2\| = 1.22$. In the third iteration, $\|r_3\| = 0$. Since $\|r_3\| \leq \epsilon$, the point $x_3 = (-4.00, -3.00, -1.60)^T$ is the optimal solution to this example, and the optimal function value is $f(x_3) = -35.9$.

Different versions of the conjugate gradient method can be applied to quadratic optimization problems and general unconstrained nonlinear optimization problems. The Fletcher-Reeves method extends the conjugate gradient method to nonlinear optimization problems by making two changes. First, the one-dimensional minimum along each conjugate vector is identified by a line search. Second, the residual is replaced by the gradient of the nonlinear function. The steps of the Fletcher-Reeves method are as follows.

Table 12.4. *The Iterations for the Conjugate Gradient Method*

k	α_k	x_k^T	r_k^T	β_k	p_k^T
0		0, 0, 0	8, 9, 8		
1	0.30	−2.42, −2.72, −2.42	3.16, 0.83, −4.10	0.13	−4.21, −2.02, 3.05
2	0.29	−3.65, −3.31, −1.53	0.70, −0.93, 0.35	0.05	−0.93, 0.82, −0.19
3	0.38	−4.00, −3.00, −1.60	0, 0, 0	0	0, 0, 0

1. Set an initial guess x_0. At x_0, evaluate the function, $f(x_0)$, and its gradient, $\nabla f(x_0)$. Set $p_0 = -\nabla f(x_0)$. Specify the convergence tolerance of the residual, $\epsilon > 0$. Set the iteration number, i, to 0.
2. Check the norm of the gradient $\|\nabla f(x_i)\|$. If $\|\nabla f(x_i)\| < \epsilon$, stop.
3. Use a line search along the direction of the conjugate vector to determine the minimum $x_{i+1} = x_i + \alpha_i p_i$.
4. Evaluate the gradient $\nabla f(x_{i+1})$.
5. Set $\beta_{i+1} = \nabla f(x_{i+1})^T \nabla f(x_{i+1}) / \nabla f(x_i)^T \nabla f(x_i)$.
6. Set the conjugate vector $p_{i+1} = -\nabla f(x_{i+1}) + \beta_{i+1} p_i$.
7. Set $i = i + 1$. Go to Step 2.

12.4.3 Second-Order Methods

Newton Method

Newton's method approximates the function $f(x)$ with a quadratic function at each iteration. The approximated quadratic function is minimized exactly, and a descent direction is evaluated.

In each iteration, the function is approximated as follows.

$$\tilde{f}(x) \approx f(x_k) + \nabla f(x_k)^T (x - x_k)$$
$$+ \frac{1}{2}(x - x_k)^T \nabla^2 f(x_k)(x - x_k) \tag{12.74}$$

According to the first-order necessary condition, the first derivative of $\tilde{f}(x)$ should satisfy $\nabla \tilde{f}(x) = 0$. This condition can be expressed as

$$\nabla \tilde{f}(x) = \nabla f(x_k) + \nabla^2 f(x_k)(x - x_k) = 0 \tag{12.75}$$

Solving the above equation,

$$\nabla^2 f(x_k)(x - x_k) = -\nabla f(x_k) \tag{12.76}$$

Assuming the inverse of the Hessian exists, then

$$x = x_k - [\nabla^2 f(x_k)]^{-1} \nabla f(x_k) \tag{12.77}$$

The second term in Eq. 12.77 is used as the reasonable direction to estimate x_{k+1} in the next iteration.

$$x_{k+1} = x_k - [\nabla^2 f(x_k)]^{-1} \nabla f(x_k) \tag{12.78}$$

The descent direction, p_k, can be expressed as

$$p_k = [\nabla^2 f(x_k)]^{-1} \nabla f(x_k). \tag{12.79}$$

If $[\nabla^2 f(x_k)]^{-1}$ exists, p_k can be evaluated using Eq. 12.79. The other way to evaluate the descent direction, p_k, is by solving the following equation.

$$\nabla^2 f(x_k) p_k = -\nabla f(x_k) \tag{12.80}$$

Equation 12.80 can be solved by Gaussian Elimination, Cholesky Factorization, or other suitable methods.

The expression of the descent direction is simple in Newton's method. However, Newton's method has certain shortcomings. (1) Newton's method does not always find a minimum, and might only find a stationary (not minimum) point. (2) It converges only if one starts sufficiently near an optimal solution. (3) The computational cost to evaluate the inverse of the Hessian is high, especially for high-dimensional optimization problems.

The steps to solve an unconstrained nonlinear problem are illustrated with the following example.

Example: Use Newton's method to minimize the following function.

$$f(x) = 4x_1^2 + 2x_1 x_2 + 2.5x_2^2 \tag{12.81}$$

The gradient of f(x) is as follows.

$$\nabla f(x) = \begin{bmatrix} 8x_1 + 2x_2 \\ 2x_1 + 5x_2 \end{bmatrix} \tag{12.82}$$

The Hessian of f(x) is as follows.

$$\nabla^2 f(x) = \begin{bmatrix} 8 & 2 \\ 2 & 5 \end{bmatrix} \tag{12.83}$$

The inverse of $\nabla^2 f(x)$ is as follows

$$[\nabla^2 f(x)]^{-1} = \begin{bmatrix} \frac{5}{36} & -\frac{1}{18} \\ -\frac{1}{18} & \frac{2}{9} \end{bmatrix} \tag{12.84}$$

Choose an initial point, $x_0 = (10, 10)^T$. The next point, x_1, is evaluated as follows.

$$x_1 = x_0 - [\nabla^2 f(x_0)]^{-1} \nabla f(x_0) = (0, 0)^T \tag{12.85}$$

The point $x_1 = (0,0)^T$ is the optimal solution for this example, and the optimal function value is $f(x_1) = 0$.

Quasi-Newton Methods

Quasi-Newton methods are among the most widely used methods for nonlinear optimization. When the Hessian is difficult to compute, quasi-Newton methods are effective for solving these kinds of problems. There are different quasi-Newton methods, but they are all based on approximations of the Hessian by another matrix to reduce the computational cost. In Sec. 12.4.3, Newton's method estimates the search direction, p_k, that satisfies the following equation.

$$\nabla^2 f(x_k) p_k = -\nabla f(x_k) \tag{12.86}$$

Quasi-Newton methods use an approximation of the Hessian, B_k, to obtain the search direction, p_k. The search direction is obtained by solving

$$B_k p_k = -\nabla f(x_k) \tag{12.87}$$

For one-dimensional nonlinear optimization problems, quasi-Newton methods are generalizations of the secant method. The secant method approximates the Hessian using

$$f''(x_k) \approx \frac{f'(x_k) - f'(x_{k-1})}{x_k - x_{k-1}} \tag{12.88}$$

where $f'(x_k)$ can be approximated as

$$f'(x_k) \approx \frac{f(x_k) - f(x_{k-1})}{x_k - x_{k-1}} \tag{12.89}$$

For multidimensional nonlinear optimization problems, the above condition is rewritten as

$$\nabla^2 f(x_k)(x_k - x_{k-1}) \approx \nabla f(x_k) - \nabla f(x_{k-1}) \tag{12.90}$$

From the above Eq. 12.90, the condition for the approximation matrix B_k is derived, which can be expressed as

$$B_k(x_k - x_{k-1}) = \nabla f(x_k) - \nabla f(x_{k-1}) \tag{12.91}$$

The matrix B_k has n^2 entries and the secant condition only has a set of n equations. This condition is insufficient to define B_k uniquely. The matrix B_k does not always have a unique solution for the secant condition.

There are several ways to approximate and update the matrix B_k. Define two vectors, s_k and y_k, as follows.

$$s_k = x_{k+1} - x_k \tag{12.92}$$

$$y_k = \nabla f(x_{k+1}) - \nabla f(x_k) \tag{12.93}$$

As one of the approximation formulae, the symmetric rank-one update formula can be expressed as

$$B_{k+1} = B_k + \frac{(y_k - B_k s_k)(y_k - B_k s_k)^T}{(y_k - B_k s_k)^T s_k} \tag{12.94}$$

The B_k's updated by Eq. 12.94 are symmetric. However, they are not necessarily positive definite.

As one of the most widely used formulae, the BFGS update formula can be expressed as

$$B_{k+1} = B_k - \frac{(B_k s_k)(B_k s_k)^T}{s_k^T B_k s_k} + \frac{y_k y_k^T}{y_k^T s_k} \tag{12.95}$$

Upon selecting the appropriate update formula of B_k, the quasi-Newton algorithm is illustrated as follows.

1. Choose a starting point x_0. Choose an initial Hessian approximation B_0, which can be the diagonal matrix, I. Set $k = 0$.
2. If x_k is optimal, stop.
3. Determine the search direction, p_k, by solving $B_k p_k = -\nabla f(x_k)$.
4. Find the step length, α_k, by a line search to minimize $f(x_k + \alpha_k p_k)$.
5. Set $x_{k+1} = x_k + \alpha_k p_k$.
6. Update B_{k+1} using the selected update formula.
7. Set $k = k + 1$ and go to Step 2.

The steps to solve an unconstrained nonlinear problem are illustrated with the following example.

Example: Use a quasi-Newton method with the symmetric rank-one B_k update formula to solve the following three-dimensional quadratic problem.

$$f(x) = \frac{1}{2}x^T Q x - c^T x \tag{12.96}$$

$$Q = \begin{bmatrix} 2 & 0 & 0 \\ 0 & 3 & 0 \\ 0 & 0 & 5 \end{bmatrix} \tag{12.97}$$

$$c = \begin{bmatrix} -8 \\ -9 \\ -8 \end{bmatrix} \tag{12.98}$$

Choose an initial point $x_0 = (0, 0, 0)^T$. The initial guess is $B_0 = I$. At this point, x_0, $\|\nabla f(x_0)\| = \| - c\| = 14.4568$, so this point is not optimal. From $B_0 p = -\nabla f(x_0)$, the first search direction is derived as

$$p_0 = \begin{bmatrix} -8 \\ -9 \\ -8 \end{bmatrix} \tag{12.99}$$

The line search formula gives $\alpha_0 = 0.3025$. The new estimated point is

$$x_1 = \begin{bmatrix} -2.4197 \\ -2.7221 \\ -2.4197 \end{bmatrix} \tag{12.100}$$

The gradient at the new point x_1 is

$$\nabla f(x_1) = \begin{bmatrix} 3.1606 \\ 0.8336 \\ -4.0984 \end{bmatrix} \tag{12.101}$$

To update B_1, first evaluate s_0 and y_0.

$$s_0 = x_1 - x_0 = \begin{bmatrix} -2.4197 \\ -2.7221 \\ -2.4197 \end{bmatrix} \tag{12.102}$$

$$y_0 = \nabla f(x_1) - \nabla f(x_0) = \begin{bmatrix} -4.8394 \\ -8.1664 \\ -12.0984 \end{bmatrix} \tag{12.103}$$

$$\begin{aligned} B_1 &= I + \frac{(y_0 - Is_0)(y_0 - Is_0)^T}{(y_0 - Is_0)^T s_0} \\ &= \begin{bmatrix} 1.1328 & 0.2988 & 0.5311 \\ 0.2988 & 1.6722 & 1.1950 \\ 0.5311 & 1.1950 & 3.1245 \end{bmatrix} \end{aligned} \tag{12.104}$$

At this new point x_1, $\|\nabla f(x_1)\| = 5.2423$. It is not optimal. The search direction is

$$p_1 = \begin{bmatrix} -3.5444 \\ -1.6971 \\ 2.5633 \end{bmatrix} \tag{12.105}$$

The line search step length is $\alpha_1 = 0.3471$. At the new point, x_2, the new estimates of the solution, the gradient, and the Hessian are

$$x_2 = \begin{bmatrix} -3.6499 \\ -3.3112 \\ -1.5300 \end{bmatrix} \tag{12.106}$$

$$\nabla f(x_2) = \begin{bmatrix} 0.7002 \\ -0.9335 \\ 0.3501 \end{bmatrix} \tag{12.107}$$

$$B_2 = \begin{bmatrix} 1.4875 & 0.6833 & -0.2562 \\ 0.6833 & 2.0890 & 0.3416 \\ -0.2562 & 0.3416 & 4.8719 \end{bmatrix} \tag{12.108}$$

At the point x_2, $\|\nabla f(x_2)\| = 0.8397$. It is not optimal. The new search direction is

$$p_2 = \begin{bmatrix} -0.4851 \\ 0.5749 \\ -0.2426 \end{bmatrix} \tag{12.109}$$

and the step length $\alpha_2 = 0.4145$. The line search yields the point

$$x_3 = \begin{bmatrix} -4 \\ -3 \\ -1.6000 \end{bmatrix} \tag{12.110}$$

At point x_3,

$$\nabla f(x_3) = \begin{bmatrix} 0 \\ 0 \\ 0 \end{bmatrix} \tag{12.111}$$

The point x_3 is the optimal solution to this example, and the optimal function value is $f(x_3) = -35.9$.

12.5 Comparison of Computational Issues in the Algorithms

12.5.1 Rate of Convergence

The optimization algorithms discussed in this chapter are iterative methods, and are used to evaluate a sequence of approximate solutions. The rate of convergence is a measure of efficiency, which describes how quickly the estimates of the solution approach the exact solution. Efficient algorithms reduce the computational cost. Assume a sequence of points x_k converges to a solution x^*. The error at the kth iteration can be defined as

$$e_k = x_k - x^* \tag{12.112}$$

As the sequence approaches the solution x^*, the limit of e_k is 0.

If Eq. 12.113 holds, the sequence x_k is said to converge to x^* with rate r and rate constant C. The constant $C < \infty$.

$$\lim_{k \to \infty} \frac{\|e_{k+1}\|}{\|e_k\|^r} = C \tag{12.113}$$

For a sequence of errors, when $r = 1$, the convergence is said to be *linear*, as Eq. 12.114 indicates. Note that if $0 < C < 1$, then the norm of the error is reduced at every iteration. If $C > 1$, then the sequence diverges.

$$\|e_{k+1}\| = C\|e_k\| \tag{12.114}$$

If $r = 1$ and $C = 0$, the rate of the convergence is said to be *superlinear*. For any $r > 1$, if

$$\lim_{k \to \infty} \frac{\|e_{k+1}\|}{\|e_k\|^r} = C < \infty \qquad (12.115)$$

then Eq. 12.116 holds. Any convergence with $r > 1$ has a *superlinear* rate of convergence.

$$\lim_{k \to \infty} \frac{\|e_{k+1}\|}{\|e_k\|} = \lim_{k \to \infty} \frac{\|e_{k+1}\|}{\|e_k\|^r} \|e_k\|^{r-1} = C \times \lim_{k \to \infty} \|e_{k+1}\|^{r-1} = 0 \qquad (12.116)$$

When $r > 2$, the convergence is called *quadratic*. The rates of convergence for some optimization algorithms in this chapter are listed in Table 12.5.

12.5.2 Line Search Methods

In the above sections, at each iteration, the steepest descent method, the conjugate gradient method, the Newton method and the quasi-Newton methods first evaluate a search direction p_k, and then do a line search along the direction to determine an appropriate step length, α_k. The iteration is given by

$$x_{k+1} = x_k + \alpha_k p_k \qquad (12.117)$$

Most line search algorithms require the search direction, p_k, to be a descent direction, and it should satisfy

$$p_k^T \nabla_k f(x_k) < 0 \qquad (12.118)$$

The step length, α_k, should generate a substantive reduction of $f(x)$ along the descent direction. The minimum of Eq. 12.119 is chosen as a function of α_k.

$$\phi(\alpha_k) = f(x_k + \alpha_k p_k), \alpha_k > 0 \qquad (12.119)$$

In general, it is not computationally practical to determine that exact minimum. Some practical strategies perform an inexact line search to identify a step length with adequate reduction in $f(x)$. These strategies reduce computational cost.

A popular inexact line search condition requires that α_k give a sufficient decrease in the objective function, $f(x)$, as measured by the following inequality:

$$f(x_k + \alpha_k p_k) \leq f(x_k) + c_1 \alpha_k \nabla f_k^T p_k \qquad (12.120)$$

where the constant $c_1 \in (0, 1)$. This inequality is called the Armijo condition.

The sufficient-decrease condition is not, by itself, adequate to ensure that the algorithm makes reasonable progress. To rule out unacceptable short steps, the curvature condition is introduced as the following inequality.

$$\nabla f(x_k + \alpha_k p_k)^T p_k \geq c_2 \nabla f_k^T p_k \qquad (12.121)$$

Table 12.5. *Comparison of Optimization Methods for Unconstrained Nonlinear Problems*

Method	Converge Proof	Convergence Rate	Computation Cost	Problem Scale
Bisection	yes	linear	very high	1-dimensional
Golden Section	yes	linear	high	1-dimensional
Quadratic Appro	yes	linear	very high	1-dimensional
Pattern Search	yes		high	small
Steepest Descent	yes	linear	high	medium
Conjugate Grad	yes	linear-quadratic	medium high	large
Newton	no	quadratic	medium - high	small-medium
Quasi-Newton	yes	superlinear	low - medium	small - mid

where the constant $c_2 \in (c_1, 1)$. The sufficient-decrease condition and the curvature condition are collectively known as the Wolfe conditions.

12.5.3 Comparison of Different Methods

The optimization methods presented in this chapter include the Bisection method, the Golden Section Search method, the quadratic approximation method, the pattern search methods, the steepest descent method, the conjugate gradient methods, the Newton method, and the quasi-Newton methods.

All of these methods have applications in different fields. Unconstrained nonlinear problems can be classified as small, mid, or large scale problems. Computers used to run the codes of the algorithms can have different computational capacity. The appropriate method can be selected, depending on the application, the computation capability, and the storage requirements. Note that computation capacity, in this regard, and storage requirements have become less critical issues in recent years.

Table 12.5 provides a comparison of the different methods. The comparison includes the following properties: convergence guarantee, convergence rate, computation cost, and problem scale. It is important to keep in mind that the overall computational cost is a combination of (i) the number of iterations, and (ii) the cost of each iteration. While a method may converge in fewer iterations, it might require more costly computation at each iteration. For example, the Newton method may often converge with fewer iterations than the Conjugate Gradient method. However, at each iteration, Conjugate Gradient only requires gradient computation, while Newton requires a very costly Hessian computation.

The Conjugate Gradient method may exhibit linear to quadratic convergence, depending on the particular implementation. As standalone algorithms, Bisection, Golden Section, and Quadratic Approximation methods are capable of performing only univariate optimization or line search. However, they can be applied as techniques to perform line search along a direction of improvement (direction vector) within the context of multivariate optimization. We note that some of the above comments are somewhat subjective, and that much depends on the particular problem and the computing resources at hand.

This chapter provides the reader with sufficient information to understand the basics of unconstrained nonlinear programming. This background makes it possible to explore more theoretical aspects of the topic [5].

12.6 Summary

Since most engineering design problems tend to be nonlinear in nature, the topic of nonlinear programming (NLP) is of paramount importance in learning optimization. This chapter introduced the basics of unconstrained NLP. The chapter began with a description of the necessary and sufficient conditions for optimality in unconstrained NLP problems. This description was followed by the introduction of the major techniques/algorithms used for solving uni-vartiate and multivariate NLP problems. Specifically, two major classes of uni-variate methods were presented: (i) interval reduction methods and (ii) polynomial approximation methods. This was followed by a description of the three major classes of multivariate unconstrained NLP methods. These are: the Zeroth order methods (e.g., line search and pattern search), the first order methods (e.g., steepest descent and conjugate gradient), and the second order methods (e.g., Newton's and Quasi Newton methods). The chapter concluded with a comparative discussion of the rate of convergence and the computational cost of these different classes of methods for solving unconstrained NLP problems.

12.7 Problems

Warm-up Problems

12.1   Consider the function:

$$f(x) = 8x_1^2 + 4x_1x_2 + 5x_2^2$$

1. Calculate the Hessian matrix of the function.
2. Calculate the Hessian at (i)$[1,1]$, (ii)$[2,2]$, and (iii)$[0,0]$.
3. Use optimality conditions to classify the points in Question (2) as minimum, maximum, or saddle points.

Intermediate Problems

12.2   Consider the following two dimensional function:

$$f(x) = 8x_1^2 + 4x_1x_2 + 5x_2^2$$

1. Calculate the gradient vector of the above function by hand.
2. Calculate the gradient at the following points by hand: (i)$[1,1]$, (ii)$[2,2]$, and (iii)$[0,0]$.

3. Write a small program in MATLAB that evaluates the gradient at each point in a two-dimensional grid in the space $-5 \le x_1 \le 5$. Choose an appropriate grid spacing (at least 100 points).
4. Find a way in MATLAB to plot each gradient vector of Question (3) as a small arrow (pointing in the correct direction) at its x_1, x_2 location; that is, if you evaluated 1,000 gradient vectors in Question 3, then your plot should contain 1,000 arrows. In addition, plot the contours of f on top of the gradient plot.
5. Interpret your plot and comment on the likely location of the minimum point by observing the gradient vectors and the contours of f.

12.3 Consider the function:

$$f(x) = (x_1 - 4)^2 + (x_2 - 4)^2 + 10(5 - x_1 - x_2)^2$$

1. Write a program to estimate the minimum of the function using the steepest descent method. Use the starting point $[10, 1]$.
2. Compare your optimum results with the results of `fmincon`. How many function evaluations does your code need? How many function evaluations does `fmincon` need?
3. On the contour plot of the function, plot the sequence of points obtained during the optimization iterations using your steepest descent code.

Advanced Problems

12.4 Consider the following function:

$$f(x) = -3x^3 + \frac{12}{x^2} + 2e^{x^2}, \quad \frac{1}{2} \le x \le 2$$

We wish to find the minimum of the above function in the given range of x.
1. Using the following listed methods, estimate the minimum of the above function using up to three iterations. Solve Parts (a) through (c) by hand, and please show your complete work.
   (a) Interval Halving Method (*i.e.*, Bisection method)
   (b) Newton-Raphson Method using $x_0 = 1$ as the starting point
   (c) Golden Section Method
   (d) Plot the function $f(x)$ using MATLAB, and identify the optimum solution in the range of x. For Parts (a) through (c), calculate the percentage error between the optimum x value and the value of x after three iterations.
   Present your results in a tabular form, as shown below, for methods in Parts (a), (b), and (c).

2. Write a MATLAB code for the Newton-Raphson method that evaluates the minimum of the function in Problem 1. Use an appropriate stopping criterion.

Table 12.6. *Results Summary for Given Search Methods*

Iteration	(a)		(b)		(c)	
	x	$f(x)$	x	$f(x)$	x	$f(x)$
1						
2						
3						

Table 12.7. *Results Summary for Given Search Methods*

Iteration	(a)		(b)		(c)	
	x	$f(x)$	x	$f(x)$	x	$f(x)$
1						
2						
3						

3. Verify your results using `fmincon`. Make sure your code gives you the same solution as that given by `fmincon`. You might have to change your stopping criterion.
4. How many function evaluations does your code need? How many function evaluations does `fmincon` need? Use the same starting point for your code and `fmincon`.

12.5 Consider the following function:

$$f(x) = -3x^3 + \frac{12}{x^2} + 2e^{x^2}, \quad \frac{1}{2} \le x \le 2$$

We want to find the minimum of the above function in the given range of x.
1. Using the following methods, estimate the minimum of the above function by implementing up to three iterations. Solve Parts (a) through (c) by hand, and please show your complete work.
   (a) Bisection Method
   (b) Secant Method
   (c) Successive quadratic estimation
       Present your results in a tabular form, as shown below, for methods in Parts (a), (b), and (c).
2. Write a MATLAB code for the above three methods for the given function. Use an appropriate stopping criterion.

BIBLIOGRAPHY OF CHAPTER 12

[1] M. S. Bazaraa, H. D. Sherali, and C. M. Shetty. *Nonlinear Programming: Theory and Algorithms*. John Wiley and Sons, 3rd edition, 2013.

[2] A. Ruszczynski. *Nonlinear Optimization*. Princeton University Press, 2011.

[3] B. Zwicknagl and R. Schaback. Interpolation and approximation in taylor spaces. *Journal of Approximation Theory*, 171:65–83, 2013.

[4] A. R. Conn, K. Scheinberg, and L. N. Vicente. *Introduction to Derivative-free Optimization*. SIAM, 2009.

[5] R. J. Vanderbei. *Linear Programing: Foundations and Extensions*. Springer, 2014.

13

Nonlinear Programming with Constraints

13.1 Overview

Practical engineering optimization problems often involve constraints. Since nonlinearities are pervasive in practical constraints and objective functions, it is necessary to explicitly study the methods used to solve nonlinear optimization problems with constraints [1, 2, 3].

In the previous chapter (Chapter 12), the methods used to solve nonlinear programming problems without constraints were discussed. The present chapter presents a representative set of methods used to solve nonlinear programming problems with constraints. These include: the elimination method, the penalty method, the Karush-Kuhn Tuker (KKT) condition that defines optimality, sequential linear programming, and sequential quadratic programming.

13.2 Structure of Constrained Optimization

The general formulation of optimization problems has been defined as

$$\min_{x} \quad f(x) \tag{13.1}$$

subject to

$$g(x) \leq 0 \tag{13.2}$$

$$h(x) = 0 \tag{13.3}$$

$$x_l \leq x \leq x_u \tag{13.4}$$

The function $f(x)$ represents the objective function. The constraints include the inequality constraint function $g(x)$, the equality constraint function $h(x)$, and the side constraints. An optimization problem is classified as a constrained nonlinear programming problem when (i) it involves at least one constraint, and (ii) when at least one of the functions (among its objective function or its constraint functions) is nonlinear. An example of constrained nonlinear optimization problems is as follows.

310

Example:

$$\min_{x} \quad f(x) = x_1 + x_2 \tag{13.5}$$

subject to

$$g(x) = x_1^2 + x_2^2 - 1 \le 0 \tag{13.6}$$

$$h(x) = x_1 + \frac{1}{2} = 0 \tag{13.7}$$

This example involves one nonlinear function: the inequality constraint $g(x)$. The feasible region is depicted in Fig. 13.1. The inequality constraint, $g(x)$, constrains the feasible region inside the circle. The equality constraint function, $h(x)$, constrains the feasible region to the single straight line segment inside the circle. Therefore, the feasible region is the line segment between the two points, a and b, as given by the dashed line in the figure.

This optimization problem can be solved using graphical approaches. The location of the minimum is point b. (In this chapter, more advanced numerical methods are introduced to solve constrained nonlinear optimization problems.)

Note that it is acceptable to include the side constraints (Eq. 13.4) into the set of behavioral inequality constraints (Eq. 13.2), so that no side constraints explicitly appear in the formulation. This inclusion of side constraints into the behavioral constraints is done in Sec. 13.5 in the development of the Karush-Kuhn-Tucker Conditions. Note that, in software application, this inclusion can have negative computational consequences. This is because optimization software treat these constraints differently for computational efficiency purposes. We combine them here only for theoretical convenience. In this case, the problem formulation takes the form

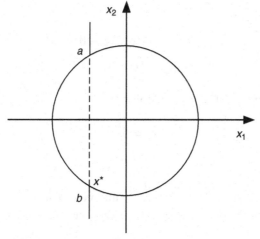

Figure 13.1. An Example of a Constrained Nonlinear Optimization Problem

$$\min_{x} \quad f(x) \tag{13.8}$$

subject to

$$g_i(x) \le 0, i = 1, ..., m \tag{13.9}$$

$$h_j(x) = 0, j = 1, ..., n \tag{13.10}$$

13.3 Elimination Method

If an optimization problem only has equality constraints, it can be solved as an unconstrained problem by eliminating variables. This class of problems is defined as

$$\min \quad f(x_1, x_2, ..., x_N) \tag{13.11}$$

subject to

$$h_j(x_1, x_2, ..., x_N) = 0, j = 1, ..., n \tag{13.12}$$

In the above problem definition, N represents the number of design variables, and n represents the number of equality constraints. Suppose $N > n$ and the n equality constraints are mutually independent. In this case, the equality constraints can be solved, and the expressions of n variables can be substituted into the objective function. This process reduces the number of variables from N to $N - n$. In addition, the constrained optimization problem becomes unconstrained. Methods used to solve unconstrained optimization problems can then be used to search for the optimum solution. The following example is used to illustrate this elimination method for solving constrained nonlinear programming problems involving only equality constraints.

Example: Use the elimination method to solve the following optimization problem with equality constraints.

$$\min f(x) = x_1 x_2 - x_3 - 3 \tag{13.13}$$

subject to

$$x_2 - 2x_1 - x_3 - 2 = 0 \tag{13.14}$$

$$x_3 - 4x_1 = 0 \tag{13.15}$$

The equality constraints can be simplified to

$$x_2 = 6x_1 + 2 \tag{13.16}$$

$$x_3 = 4x_1 \tag{13.17}$$

Substituting the above expressions for x_2 and x_3 into the objective function of the constrained optimization problem yields the following unconstrained problem:

$$\min f(x) = 6x_1^2 - 2x_1 - 3 \tag{13.18}$$

Solving the above unconstrained optimization problem, the minimum point is found to be $x^* = 0.1667$, and the minimum value of the objective function is -3.1667.

13.4 Penalty Methods

The Penalty method is used to replace the original optimization problem with a sequence of subproblems in which a functional form of the constraints is added to the objective function. The idea of this approach is that, in the limit, the solutions of the subproblems will converge to the solution of the original constrained problem. Penalty methods are classified into two groups depending on how the methods add the inequality constraints: (i) the *interior* point methods which generate a sequence of *feasible* points; and (ii) the *exterior* point methods which generate a sequence of *infeasible* points.

Penalty methods reformulate the constrained optimization problem into an unconstrained optimization problem. The unconstrained problem has two parts: the objective function of the original problem, and a penalty term. The unconstrained problem is expressed as

$$P(x, R) = f(x) + \Omega(R, g(x), h(x)) \tag{13.19}$$

In Eq. 13.19, $\Omega R, g(x), h(x)$ is the penalty term. It is a function of R and the constraints. In this function, R is called the penalty parameter. It can have more than one number. The penalty parameter and the constraint functions can take on different forms. The penalty parameter can be updated using different rules.

If the stationary point of $P(x, R)$ is infeasible, an exterior point method is being used. The updated parameter, R, forces the stationary point to be closer to the feasible region. In contrast, if the form of Ω forces the stationary point of the unconstrained function $P(x, R)$ to be feasible, an interior point method is being used. The interior point method is also called the barrier method, as the penalty term forms a barrier along the boundaries of the feasible region.

There are different choices of penalty forms. A commonly used penalty form for equality constraints is the parabolic penalty, which is given by

$$\Omega = R\{h_j(x)\}^2 \tag{13.20}$$

The parabolic penalty function is plotted in Fig. 13.2. The parabolic penalty term is 0 only at the point where $h(x) = 0$. As $h(x)$ moves farther away from 0, the value of Ω becomes larger. Thus, the parabolic penalty term discourages both positive and negative violations of $h(x)$.

Another useful penalty form is the logarithmic function. The logarithmic penalty form for inequality constraints is expressed as

$$\Omega = -Rlog[-g_i(x)] \tag{13.21}$$

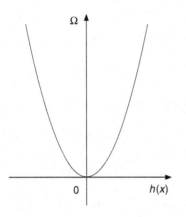

Figure 13.2. Parabolic Penalty

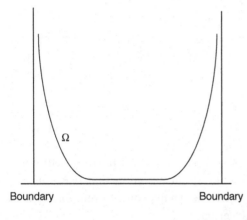

Figure 13.3. Generic Penalty Function for
Double-Sided Boundaries

The logarithmic penalty form is usually used for the interior point methods. The initial point is in the feasible region. As the point approaches the boundary of the feasible region and $-g_i(x)$ tends to 0, the penalty term, Ω, approaches positive infinity, and attempts to keep it from crossing the feasibility boundary. As the penalty parameter R approaches 0, the function $P(x, R)$ approaches the original objective function $f(x)$.

Another useful penalty function is the Inverse Penalty, expressed as

$$\Omega = R\frac{1}{g_i(x)} \tag{13.22}$$

In principle, if one is interested in a penalty function that keeps an objective function within two boundaries, a potential penalty function of the form illustrated in Fig. 13.3 could be devised.

It is important to note that some of these penalty functions are fraught with complications that must be attended to. In the case of the *log* function, when the argument should unintendedly become negative, we have a situation for which we must devise a recovery plan. In the case of the Inverse Penalty, when the argument should become too close to zero, the situation similarly becomes problematic. The published literature offers effective pertinent recovery mechanisms (see Ref. [4]).

The exterior point method and the interior method are illustrated using the following two examples.

Example: Solve the following optimization problem with an equality constraint using the parabolic penalty form.

$$\min_x f(x) = (x_1 - 3)^2 + (x_2 - 3)^2 \qquad (13.23)$$

subject to

$$h(x) = x_1 + x_2 - 4 = 0 \qquad (13.24)$$

The penalty function is formed using the parabolic penalty term.

$$P(x, R) = (x_1 - 3)^2 + (x_2 - 3)^2 + R(x_1 + x_2 - 4)^2 \qquad (13.25)$$

The stationary point of the unconstrained function $P(x, R)$ satisfies the following first-order conditions.

$$\frac{\partial P}{\partial x_1} = 2(x_1 - 3) + 2R(x_1 + x_2 - 4) = 0 \qquad (13.26)$$

$$\frac{\partial P}{\partial x_2} = 2(x_2 - 3) + 2R(x_1 + x_2 - 4) = 0 \qquad (13.27)$$

Then, the coordinates of the stationary point are given by

$$x_1 = x_2 = \frac{6 + 8R}{2 + 4R} \qquad (13.28)$$

As the value of R approaches positive infinity, the values of x_1 and x_2 become

$$x_1^* = x_2^* = \lim_{R \to \infty} \frac{6 + 8R}{2 + 4R} = 2 \qquad (13.29)$$

It can be verified that the minimum point, $[2, 2]^T$, satisfies the equality constraint. The numerical implementation of the penalty method begins with a small value of R. As R increases, the stationary point of the unconstrained optimization function approaches the actual optimal solution of the constrained problem. Table 13.1 provides the optimal results of the unconstrained optimization problem as R increases. In other implementations of the penalty method, R may start from a very large value and is decreased to zero in order to reach the constrained optimum.

Table 13.1. *Optimal Results for Different R Values*

R	x_1^*	x_2^*	f^*	$h(x)$
10	2.0476	2.0476	1.9048	0.0952
100	2.0050	2.0050	1.9900	0.0100
1,000	2.0005	2.0005	1.9990	0.0010
10,000	2.0000	2.0000	1.9999	0.0000
100,000	2.0000	2.0000	2.0000	0.0000

Example: Solve the constrained optimization problem with inequality constraints using the logarithmic penalty form.

$$\min_x f(x) = -2x_1 + x_2 + 5 \tag{13.30}$$

subject to

$$g_1(x) = x_1^2 - x_2 - 1 \leq 0 \tag{13.31}$$

$$g_2(x) = -x_1 \leq 0, \tag{13.32}$$

The penalty function is formed using the logarithmic penalty term,

$$P(x, R) = (-2x_1 + x_2 + 5)$$
$$+ R(-ln(-(x_1^2 - x_2 - 1))) + R(-ln(x_1)) \tag{13.33}$$

The stationary point of $P(x, R)$ should satisfy the following first-order conditions.

$$\frac{\partial P}{\partial x_1} = -2 + \frac{-2Rx_1}{x_1^2 - x_2 - 1} - \frac{R}{x_1} = 0 \tag{13.34}$$

$$\frac{\partial P}{\partial x_2} = 1 + \frac{R}{x_1^2 - x_2 - 1} = 0 \tag{13.35}$$

Then, the coordinates of the stationary point are given by

$$x_1 = \frac{1 + \sqrt{2R + 1}}{2} \tag{13.36}$$

$$x_2 = \frac{3R - 1 + \sqrt{2R + 1}}{2} \tag{13.37}$$

As the value of R approaches 0, the limits of x_1 and x_2 become

$$x_1^* = \lim_{R \to 0} \frac{1 + \sqrt{2R + 1}}{2} = 1 \tag{13.38}$$

$$x_2^* = \lim_{R \to 0} \frac{3R - 1 + \sqrt{2R + 1}}{2} = 0 \tag{13.39}$$

Table 13.2. *Optimal Results for Different R Values*

R	x_1^*	x_2^*	f^*	$g_1(x)$	$g_2(x)$
1	1.366	1.866	4.134	−1.000	−1.366
0.1	1.048	0.198	3.102	−0.100	−1.048
0.01	1.005	0.020	3.010	−0.010	−1.005
0.001	1.000	0.002	3.001	−0.001	−1.000
0.0001	1.000	0.000	3.000	0.000	−1.000

It can be verified that the minimum point, $(1,0)$, satisfies the inequality constraints.

The numerical implementation of the penalty method begins with a sufficiently large positive value of R. As R approaches zero, the stationary point of the unconstrained optimization function approaches the actual optimal solution of the constrained problem. Table 13.2 provides the optimal results of the unconstrained optimization problem as R approaches zero.

13.5 Karush-Kuhn-Tucker Conditions

For constrained nonlinear optimization, *the Lagrangian function* and *the Lagrange multiplier* are used to provide a strategy for finding and validating the optimum of a function subject to constraints. Consider the minimization of a nonlinear optimization problem subject to equality constraints.

$$\min_x \quad f(x) \tag{13.40}$$

subject to

$$h_j(x) = 0, j = 1, ..., n \tag{13.41}$$

The Lagrangian function converts the constrained problem into an unconstrained problem, as given by

$$L(x, v) = f(x) + \sum_{j=1}^{n} v_j h_j(x) \tag{13.42}$$

The unspecified constants, v_j, are the Lagrange multipliers. There are no sign restrictions on the values of v_j. Suppose the minimum for the unconstrained problem $L(x, v)$ is x^*, and x^* satisfies $h_j(x) = 0$. For all values of x that satisfy $h_j(x) = 0$, the minimum of $L(x, v)$ is the minimum of $f(x)$ subject to $h_j(x) = 0$. Then x^* minimizes the constrained optimization problem.

The following example shows us how to solve an optimization problem with equality constraints using the Lagrangian function.

Example: Minimize the following two variable function with a single equality constraint.

$$\min_{x} \quad f(x) = x_1^2 + x_2^2 + 5 \tag{13.43}$$

subject to

$$h_1(x_1, x_2) = x_1 + 2x_2 - 4 = 0 \tag{13.44}$$

The Lagrangian function of the problem is given by

$$L(x, v) = (x_1^2 + x_2^2 + 5) + v(x_1 + 2x_2 - 4) \tag{13.45}$$

The gradient of $L(x, v)$ with respect to x is equal to zero at the minimum.

$$\frac{\partial L}{\partial x_1} = 2x_1 + v = 0 \tag{13.46}$$

$$\frac{\partial L}{\partial x_2} = 2x_2 + 2v = 0 \tag{13.47}$$

The Hessian matrix of $L(x, v)$ with respect to x is given by

$$Hessian = \begin{bmatrix} 2 & 0 \\ 0 & 2 \end{bmatrix}$$

The Hessian is positive definite, implying that the optimal point is a minimum of the objective function.

From the above two functions of gradients and the equality constraint function, the point of the minimum is found to be $x_1^* = 0.8$ and $x_2^* = 1.6$. The Lagrange multiplier is $v = -1.6$. The minimum value of the objective function is 8.2.

The steps to solve constrained optimization problems with equality constraints using the Lagrangian function include

1. Construct the Lagrangian function $L(x, v)$ using the objective function and the equality constraints.
2. Solve $\nabla_x L(x, v) = 0$ and $h(x) = 0$.

The same method can be used to solve optimization problems with inequality constraints. The Lagrangian function in that case is constructed as

$$L(x, \lambda) = f(x) + \sum_{i=1}^{m} \lambda_i g_i(x) \tag{13.48}$$

However, there are sign restrictions on the Lagrange multipliers for inequality constraints. They should satisfy $\lambda_i \geq 0$. Additionally, the following functions should also be satisfied at the minimum.

$$\lambda_i^* g_i(x^*) = 0, \text{ for all } i = 1, ..., m \tag{13.49}$$

$$g_i(x^*) \leq 0, \text{ for all } i = 1, ..., m \tag{13.50}$$

A new concept known as *the active set* should be introduced before the discussion of this method. The active set at any feasible point x is comprised of the equality constraints and the inequality constraints for which $g_i(x) = 0$.

An inequality constraint, at the minimum point x^* in the feasible region, can be classified into the following two cases.

Case I: The inequality constraint is inactive at the minimum point x^*, which implies $g_i(x^*) < 0$. Since the minimum point x^* satisfies $\lambda_i^* g_i(x^*) = 0$, the Lagrange multiplier is $\lambda_i^* = 0$. Then $L(x^*, \lambda) = f(x^*)$.

Case II: The inequality constraint is active at the minimum point x^*, which implies $g_i(x^*) = 0$. Then $L(x^*, \lambda) = f(x^*)$.

At the minimum point x^*, $\nabla_x L(x^*, \lambda) = 0$.

The following example explains the above process.

Example: Minimize the following two-variable function with a single inequality constraint.

$$\min_x \quad f(x) = x_1 + x_2 \tag{13.51}$$

subject to

$$g(x_1, x_2) = x_1^2 + x_2^2 - 1 \leq 0, \tag{13.52}$$

The Lagrangian function is given by

$$L(x, \lambda) = (x_1 + x_2) + \lambda(x_1^2 + x_2^2 - 1) \tag{13.53}$$

The gradient of $L(x, \lambda)$ satisfies the following two equations:

$$\frac{\partial L}{\partial x_1} = 1 + 2\lambda x_1 = 0 \tag{13.54}$$

$$\frac{\partial L}{\partial x_2} = 1 + 2\lambda x_2 = 0 \tag{13.55}$$

The following two equations are satisfied:

$$\lambda(x_1^2 + x_2^2 - 1) = 0 \tag{13.56}$$

$$(x_1^2 + x_2^2 - 1) \leq 0 \tag{13.57}$$

$$\lambda \geq 0 \tag{13.58}$$

Since $\lambda = 0$ does not satisfy $\nabla_x L = 0$, λ is greater then zero. From Eq. 13.56, it is derived that $(x_1^2 + x_2^2 - 1) = 0$.

From Eq. 13.54 and 13.55, it is derived that $x_1 = x_2$.

Since $\lambda > 0$, it is derived from Eq. 13.54 and 13.55 that $x_1 < 0$ and $x_2 < 0$.

Solving $(x_1^2 + x_2^2 - 1) = 0$, the minimum point is found to be located at $(-\frac{\sqrt{2}}{2}, -\frac{\sqrt{2}}{2})$. The optimal value of the objective function is $-\sqrt{2}$.

The Lagrangian function for an optimization problem with multiple constraints is expressed as

$$L(x, \lambda, v) = f(x) + \sum_{i=1}^{n} \lambda_i g_i(x) + \sum_{j=1}^{m} v_j h_j(x) \tag{13.59}$$

At the feasible point x, if the gradients of the constraints in the active set are linearly independent, the *Linear Independence Constraint Qualification* (LICQ) holds.

Suppose there are two inequality constraints as given by

$$g_1(x) = x_1^2 + x_2^2 - 1 \le 0 \tag{13.60}$$

$$g_2(x) = 1 - x_1 \le 0 \tag{13.61}$$

In that case, the only feasible point is $(1, 0)$, as shown by Fig. 13.4. At $(1, 0)$, the gradients of the two constraints are given by

$$\nabla g_1(x) = [2x_1, 2x_2]^T = [2, 0]^T \tag{13.62}$$

$$\nabla g_2(x) = [-1, 0]^T \tag{13.63}$$

Equations 13.62 and 13.63 are not linearly independent. Therefore, the LICQ does not hold at that point. The Karush-Kuhn-Tucker (KKT) conditions are necessary for a solution to be a local minimum.

Theorem (Karush-Kuhn-Tucker Conditions)
Suppose that x^* is a local minimum solution of $f(x)$ subject to constraints $g_i(x) \le 0$ $(i = 1, ..., m)$ and $h_j(x) = 0$ $(j = 1, ..., n)$. The objective function, $f(x)$, and the constraints, $g_i(x)$ and $h_j(x)$, are continuously differentiable. The LICQ holds at x^*.

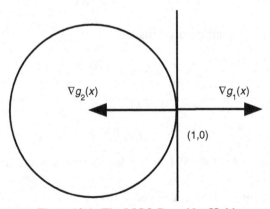

Figure 13.4. The LICQ Does Not Hold

Then there are Lagrange multipliers λ^* and ν^*, such that the following conditions are satisfied at (x^*, λ^*, ν^*).

$$\nabla_x f(x^*, \lambda^*, \nu^*) = 0 \tag{13.64}$$

$$\lambda_i^* \geq 0, (i = 1, ..., m) \tag{13.65}$$

$$\lambda_i^* g_i(x^*) = 0, (i = 1, ..., m) \tag{13.66}$$

$$\nu_j^* h_j(x^*) = 0, (j = 1, ..., n) \tag{13.67}$$

$$g_i(x^*) \leq 0, (i = 1, ..., m) \tag{13.68}$$

$$h_j(x^*) = 0, (j = 1, ..., n) \tag{13.69}$$

The following example is used to illustrate how to use the KKT conditions to solve an optimization problem.

Example: Minimize the following two variable function with two inequality constraints.

$$\min_x \qquad f(x) = x_1 + x_2 \tag{13.70}$$

subject to

$$g_1(x_1, x_2) = x_1^2 + x_2^2 - 1 \leq 0 \tag{13.71}$$

$$g_2(x_1, x_2) = -x_1 - \frac{1}{2} \leq 0 \tag{13.72}$$

This optimization problem can be solved using the graphical approach. The feasible region and the optimal point are plotted in Fig. 13.5.

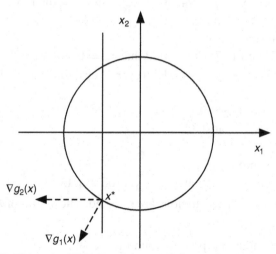

Figure 13.5. Using the KKT Conditions to Solve an Optimization Problem

The Lagrangian function is formulated as

$$L(x,\lambda) = x_1 + x_2 + \lambda_1(x_1^2 + x_2^2 - 1) + \lambda_2(-x_1 - \frac{1}{2}) \qquad (13.73)$$

This problem can be solved using the KKT conditions as follows.

$$\frac{\partial L}{\partial x_1} = 1 + 2\lambda_1 x_1 - \lambda_2 = 0 \qquad (13.74)$$

$$\frac{\partial L}{\partial x_2} = 1 + 2\lambda_1 x_2 = 0 \qquad (13.75)$$

$$\lambda_1 \geq 0 \qquad (13.76)$$

$$\lambda_2 \geq 0 \qquad (13.77)$$

$$\lambda_1(x_1^2 + x_2^2 - 1) = 0 \qquad (13.78)$$

$$\lambda_2(-x_1 - \frac{1}{2}) = 0 \qquad (13.79)$$

$$x_1^2 + x_2^2 - 1 \leq 0 \qquad (13.80)$$

$$-x_1 - \frac{1}{2} \leq 0 \qquad (13.81)$$

Based on Eqs. 13.74 — 13.81, the four possible cases for different values of λ_1 and λ_2 are discussed below.

Case I: $\lambda_1 = 0$ and $\lambda_2 = 0$.
If $\lambda_1 = 0$ and $\lambda_2 = 0$, Eq. 13.74 and 13.75 cannot be equal to 0.

Case II: $\lambda_1 > 0$ and $\lambda_2 = 0$.
From Eq. 13.74 and 13.75, it is found that $x_1 = x_2$ and they should be negative. From Eq. 13.78, $x_1 = x_2 = -\frac{\sqrt{2}}{2}$. However, they do not satisfy Eq. 13.81.

Case III: $\lambda_1 = 0$ and $\lambda_2 > 0$.
Equation 13.75 is not satisfied.

Case IV: $\lambda_1 > 0$ and $\lambda_2 > 0$.
From Eq. 13.79, it is found that $x_1 = -\frac{1}{2}$. From Eq. 13.75, $x_2 < 0$. From Eq. 13.78, $x_2 = -\frac{\sqrt{3}}{2}$. All the other equations are satisfied. The values of λ_1 and λ_2 are $\frac{\sqrt{3}}{3}$ and $(1 - \frac{\sqrt{3}}{3})$.
Therefore, the minimum point is found to be $(x_1, x_2) = (-\frac{1}{2}, -\frac{\sqrt{3}}{2})$, using the KKT conditions.

13.6 Sequential Linear Programming

In Chapter 11, we determined that certain algorithms can be used to efficiently and reliably solve linear programming problems. We note here that nonlinear optimization problems can be converted to approximate linear optimization problems. These converted problems are only partially equivalent to the original problem, and only in a certain neighborhood of the design space — near the operating point. Subsequently, in the neighborhood of the operating point, the problems can be solved using linear programming techniques.

The sequential linear programming method linearizes the objective function and constraints of an optimization problem, and expresses them as linear functions using Taylor series expansions. A Taylor series expansion at the point, x_k, is given by

$$f(x) = f(x_k) + \nabla f(x_k)(x - x_k) + O(\| x - x_k \|)^2 \qquad (13.82)$$

The higher order terms, $O(\| x - x_k \|)^2$, are ignored and only the linear term is retained. The linearization form of $f(x)$ at x_k is given by

$$\tilde{f}(x; x_k) = f(x_k) + \nabla f(x_k)(x - x_k) \qquad (13.83)$$

The most direct use of sequential linear programming is to replace a nonlinear problem with a complete linearization of the constituent functions at a sequential set of points that are intended to lead to the solution. This method can also be applied to a linearly constrained problem with a nonlinear objective function. A linearly constrained nonlinear programming problem has the following form.

$$\min_x \ f(x) \qquad (13.84)$$

subject to

$$Ax \le b \qquad (13.85)$$

$$x \ge 0 \qquad (13.86)$$

The objective function, $f(x)$, is linearized at a feasible point, x_k, and the problem is reformulated as follows.

$$\min_x \tilde{f}(x; x_k) \qquad (13.87)$$

subject to

$$Ax \le b \qquad (13.88)$$

$$x \ge 0 \qquad (13.89)$$

Assuming the feasible region is bounded, the above problem will possess an optimal solution, $\tilde{x}_k^*$, at a feasible corner point. The optimal solution, $\tilde{x}_k^*$, is not guaranteed to be improved over the current point, x_k. Since the feasible region is a polyhedron and since $\tilde{x}_k^*$ is a corner point of the feasible region, any point on the line between $\tilde{x}_k^*$

and x_k is feasible. Since $\tilde{f}(\tilde{x}_k^*; x_k) < f(x_k)$, the vector $(\tilde{x}_k^* - x_k)$ is a descent direction. A line search on this descent direction can lead to an improvement in $f(x)$. The formulated line search problem is stated as follows:

$$\min_{\alpha} f(x_k + \alpha(\tilde{x}_k^* - x_k)) \tag{13.90}$$

subject to

$$0 \le \alpha \le 1 \tag{13.91}$$

The line search will find a feasible point, x_{k+1}, that satisfies $f(x_{k+1}) < f(x_k)$. The new point, x_{k+1}, is used as a linearization point for the next linear approximation, and a new line search is then performed. After successive iterations of approximations and line searches, the process is expected to converge to the optimal point of the nonlinear problem. For nonlinear programming problems with linear constraints, the solution steps, using sequential linear programming, are:

1. Set an initial guess x_0. Specify the convergence tolerance $\epsilon > 0$.
2. Calculate the gradient of $f(x_k)$. If the gradient is $\|\nabla f_k\| \le \epsilon$, stop.
3. Approximate the original objective function at x_k using the Taylor expansion.
4. Solve the linearized problem to find the optimal solution, $\tilde{x}_k^*$.
5. Perform a line search for the original objective function between the points x_k and x_k^* to find an improved point x_{k+1}.
6. Set $k = k + 1$. Go to Step 2.

A general nonlinear optimization problem can involve nonlinear constraints, as defined by

$$\min_{x} \quad f(x) \tag{13.92}$$

subject to

$$g_i(x) \le 0, i = 1, ..., m \tag{13.93}$$

$$h_j(x) = 0, j = 1, ..., n \tag{13.94}$$

$$x_l \le x \le x_u \tag{13.95}$$

Using the Taylor expansions for the objective function and the constraints at the point x_k, the linearized approximation problem is constructed as

$$\min_{x} \quad f(x_k) + \nabla f(x_k)(x - x_k) \tag{13.96}$$

subject to

$$g_i(x_k) + \nabla g(x_k)(x - x_k) \le 0, i = 1, ..., m \tag{13.97}$$

$$h_j(x_k) + \nabla h(x_k)(x - x_k) = 0, j = 1, ..., n \tag{13.98}$$

$$x_l \le x \le x_u \tag{13.99}$$

Solving the above linear programming problem, a new point, x_{k+1}, is obtained in the feasible region of the linear constraints. A series of points can be generated through iterations. At each iteration, the solution to the previous linear approximate problem is used as the linearization point, and a new linear programming problem is constructed and solved. However, there is no assurance that the solution to the approximate linear programming problem lies within the feasible region of the original problem. In order to attain convergence to the true optimal solution of the nonlinear programming problem, at each iteration, an improvement in both the objective function and the constraint feasibility should be made. One way to achieve a satisfactory approximation of the linearization is to impose limits on the allowable increments in the variables, in order to keep the solution to the linear programming problem within a reasonably small neighborhood of the linearization point. The limits can be stated as

$$-\delta \leq x - x_k \leq \delta, \ \delta > 0. \tag{13.100}$$

The steps to solve a general nonlinear programming problem using the sequential linear programming method are:

1. Set an initial guess x_0. Specify the convergence tolerance $\epsilon > 0$.
2. Calculate the gradient of $f(x_k)$. If the gradient $\|\nabla f_k\| \leq \epsilon$, stop.
3. Approximate the original nonlinear functions at x_k using the Taylor expansion.
4. Impose increment limits. $-\delta \leq x - x_k \leq \delta, \delta > 0$.
5. Solve the linearized problem to find the optimal solution, x_{k+1}.
6. Set $k = k + 1$. Go to Step 2.

The optimization problem in the following example is a quadratic problem. It can be solved using the sequential linear programming method without imposing increment limits.

Example: Solve the following minimization problem.

$$\min_{x} f(x) = (x_1 - 1)^2 + (x_2 - 2)^2 \tag{13.101}$$

subject to

$$g(x) = x_1 - x_2^2 + 4x_2 - 5 \leq 0 \tag{13.102}$$

$$h(x) = x_1^2 - 2x_1 + x_2 - 3 = 0 \tag{13.103}$$

$$1 \leq x_1 \leq 4 \tag{13.104}$$

$$2.5 \leq x_2 \leq 4.5 \tag{13.105}$$

The feasible region lies on the curve $h_1(x) = 0$ between the point $(1, 4)$ determined by the linear bound $1 \leq x_1$ and the point $(2, 3)$ determined by the constraint $g_1(x) \leq 0$. The linearized approximation of the problem is constructed at the point $(2, 4)$ as shown below.

$$\min_x \tilde{f}(x) = 5 + 2(x_1 - 2) + 4(x_2 - 4) \tag{13.106}$$

subject to

$$\tilde{g}(x) = -3 + (x_1 - 2) - 4(x_2 - 4) \le 0 \tag{13.107}$$

$$\tilde{h}(x) = 1 + 2(x_1 - 2) + (x_2 - 4) = 0 \tag{13.108}$$

$$1 \le x_1 \le 4 \tag{13.109}$$

$$2.5 \le x_2 \le 4.5 \tag{13.110}$$

The solution to the approximate optimization problem is found to be $(1.8889, 3.2222)$. At this point, the optimization can be relinearized and solved. After several iterations, the minimum solution to the original problem is found to be $(2, 3)$. The estimated minimum value of the objective function is 2.

13.7 Sequential Quadratic Programming

Sequential quadratic programming (SQP) is a highly effective method to solve constrained optimization problems involving smooth nonlinear functions. This approach solves a series of quadratic subproblems.

A nonlinear optimization problem without inequality constraints is defined as

$$\min_x f(x) \tag{13.111}$$

subject to

$$h(x) = 0 \tag{13.112}$$

The Lagrangian function for the above problem is expressed as

$$L(x, v) = f(x) + v^T h(x) \tag{13.113}$$

The KKT conditions require that $\nabla L(x^*, v^*) = 0$ at the optimal point. The Newton method for unconstrained minimization can be expressed as

$$\begin{bmatrix} x_{k+1} \\ v_{k+1} \end{bmatrix} = \begin{bmatrix} x_k \\ v_k \end{bmatrix} + \begin{bmatrix} p_k \\ q_k \end{bmatrix} \tag{13.114}$$

In Equation 13.114, the step p_k and q_k are obtained as the solution to the following linear function.

$$\nabla^2 L(x_k, v_k) \begin{bmatrix} p_k \\ q_k \end{bmatrix} = -\nabla L(x_k, v_k) \tag{13.115}$$

Equation 13.115 can also be expressed as

$$\begin{bmatrix} \nabla_{xx}^2 L(x_k, v_k) & \nabla h(x_k) \\ \nabla h(x_k)^T & 0 \end{bmatrix} \begin{bmatrix} p_k \\ q_k \end{bmatrix} = \begin{bmatrix} -\nabla_x L(x_k, v_k) \\ -h(x_k) \end{bmatrix} \tag{13.116}$$

Equation 13.116 represents the first-order optimality conditions for the following optimization problem.

$$\min_{p} f(x_k) + p^T \nabla_x L(x_k, v_k) + \frac{1}{2} p^T \nabla_{xx}^2 L(x_k, v_k) p \qquad (13.117)$$

subject to

$$\nabla h(x_k)^T p + h(x_k) = 0 \qquad (13.118)$$

In Equation 13.114, q_k represents the Lagrange multiplier, v_k, in the above formulation.

At each iteration, a quadratic problem is solved to obtain $[p_k, q_k]^T$. These values are used to update $[x_k, v_k]$ using Eq. 13.114.

A nonlinear optimization problem with both equality and inequality constraints is defined as

$$\min_{x} \quad f(x) \qquad (13.119)$$

subject to

$$g(x) \leq 0 \qquad (13.120)$$

$$h(x) = 0 \qquad (13.121)$$

The above optimization problem can be reformulated as

$$\min_{p} p^T \nabla_x L(x_k, \lambda_k) + \frac{1}{2} p^T \nabla_{xx}^2 L(x_k, \lambda_k) p \qquad (13.122)$$

subject to

$$\nabla g(x_k)^T p + g(x_k) \leq 0 \qquad (13.123)$$

$$\nabla h(x_k)^T p + h(x_k) = 0 \qquad (13.124)$$

Example: Apply the SQP method to solve the following problem involving an equality constraint.

$$\min_{x} \quad f(x) = e^{-4x_1} + e^{3x_2} \qquad (13.125)$$

subject to

$$h(x) = x_1^2 + x_2^2 - 1 = 0 \qquad (13.126)$$

Considering the initial values of x and v as $X_0 = [1, -1]^T$ and $v_0 = 1$, we get

$$\nabla f(X_0) = \begin{bmatrix} -4e^{-4x_1} \\ 3e^{3x_2} \end{bmatrix} = \begin{bmatrix} -0.0732 \\ 0.1493 \end{bmatrix} \qquad (13.127)$$

$$\nabla^2 f(X_0) = \begin{bmatrix} 16e^{-4x_1} & 0 \\ 0 & 9e^{3x_2} \end{bmatrix} = \begin{bmatrix} 0.2931 & 0 \\ 0 & 0.4481 \end{bmatrix} \qquad (13.128)$$

$$h(X_0) = x_1^2 + x_2^2 - 1 = 1 \qquad (13.129)$$

$$\nabla h(X_0) = \begin{bmatrix} 2x_1 \\ 2x_2 \end{bmatrix} = \begin{bmatrix} 2 \\ -2 \end{bmatrix} \qquad (13.130)$$

$$\nabla^2 h(X_0) = \begin{bmatrix} 2 & 0 \\ 0 & 2 \end{bmatrix} \qquad (13.131)$$

$$\nabla_x L = \nabla f + v\nabla h = \begin{bmatrix} 1.9267 \\ -1.8506 \end{bmatrix} \qquad (13.132)$$

$$\nabla_{xx}^2 L = \nabla^2 f + v\nabla^2 h = \begin{bmatrix} 2.2931 & 0 \\ 0 & 2.4481 \end{bmatrix} \qquad (13.133)$$

Using the first-order optimality conditions, the solution to $[p_0, q_0]^T$ can be obtained from the following equation.

$$\begin{bmatrix} 2.2931 & 0 & 2 \\ 0 & 2.4481 & -2 \\ 2 & -2 & 0 \end{bmatrix} \begin{bmatrix} p_0^{(1)} \\ p_0^{(2)} \\ q_0 \end{bmatrix} = \begin{bmatrix} -1.9267 \\ 1.8506 \\ -1 \end{bmatrix} \qquad (13.134)$$

From Eq. 13.134, the values of steps p_0 and q_0 are given by

$$p_0 = [-0.2742, 0.2258]^T \qquad (13.135)$$

$$q_0 = 0.6490 \qquad (13.136)$$

The values of x and v can then be updated for the next iteration as

$$\begin{aligned} X_1 &= X_0 + p_0 \\ &= [1, -1]^T + [-0.2742, 0.2258]^T \\ &= [0.7258, -0.7742]^T \end{aligned} \qquad (13.137)$$

$$v_1 = v_0 + q_0 = 1 + 0.6490 = 1.6490 \qquad (13.138)$$

The above equations represent the process for the first iteration. The values of x and v are updated by the same procedure. The minimum point for this problem is found to be $[0.6633, -0.7483]^T$. The minimum function value is estimated as 0.1764.

13.8 Comparison of Computational Issues

In this chapter, the methods used to solve nonlinear programming problems with constraints were presented, including the elimination method, penalty methods, the sequential linear programming method, and the sequential quadratic programming method.

The elimination method is limited in its application. It can be useful for solving optimization problems involving equality constraints. It requires solving systems of equations, making it challenging to convert this method into numerical algorithms. This method is not practical for solving large scale problems.

The penalty methods solve a sequence of subproblems with updated penalty parameters. These methods require a large number of iterations. If the constraints values span several orders of magnitude, the penalty method may face scaling issues.

The sequential linear programming method is efficient for optimization problems with mild nonlinearities. This method is not well-suited for optimization problems involving highly nonlinear functions.

The sequential quadratic programming method is one of the most effective methods used to solve constrained nonlinear optimization problems. It can be leveraged to solve both small scale and large scale problems, as well as problems with significant nonlinearities. This method has been implemented in many optimization solvers and commercial software packages and has a wide applicability.

13.9 Summary

This Chapter delved further into the theory and application of nonlinear programming (NLP), with the description of the characteristics of constrained NLP, and introductions to the major approaches used for solving constrained NLP problems. The structure of constrained NLP problems is first presented, the general formulation of which includes inequality and equality constraints. Basic methods for solving constrained NLP, such as Elimination methods and Penalty methods, are then introduced. This was followed by the introduction of the Karush-Kuhn Tucker conditions, which are of paramount importance in NLP, both as a set of optimality criteria and as a means of solving simpler constrained NLP problems. Advanced methods for solving constrained NLP problems are also presented in this chapter, specifically including Sequential Linear Programming and Sequential Quadratic Programming. The chapter concluded with a comparative discussion of the computational capabilities and limitations of the different classes of constrained NLP methods.

13.10 Problems

Warm-up Problems

13.1   Consider the following optimization problem:

$$\max_{x} f(x) = x_1^2 - x_2 \tag{13.139}$$

subject to

$$x_1^2 - x_2^2 = 1 \tag{13.140}$$

1. Find all the points that satisfy the KKT condition using Lagrange multipliers.
2. Use the variable elimination method to solve the problem (eliminate x_1). Compare the solutions obtained using the variable elimination method with those obtained in Question 1.

13.2   Use the Lagrange multiplier method to solve the following problem:

$$\min_x f(x) = x_1^2 + x_2^2 + x_3^2 \tag{13.141}$$

subject to

$$x_1 + 2x_2 + 3x_3 = 7 \tag{13.142}$$

$$2x_1 + 2x_2 + x_3 = \frac{9}{2} \tag{13.143}$$

Intermediate Problems

13.3   Consider the following optimization problem:

$$\min_{H,D} f(H,D) = \pi DH + \frac{\pi D^2}{2} \tag{13.144}$$

subject to $$\tag{13.145}$$

$$g_1 \equiv 1{,}000\pi - \pi D^2 H \le 0 \tag{13.146}$$

$$g_2 \equiv 4.5 - D \le 0 \tag{13.147}$$

$$g_3 \equiv D - 12 \le 0 \tag{13.148}$$

$$g_4 \equiv 10 - H \le 0 \tag{13.149}$$

$$g_5 \equiv H - 18 \le 0 \tag{13.150}$$

1. Write down the Karush-Kuhn Tucker (KKT) necessary conditions for this problem
2. You are given that constraints g_1 and g_4 are active, and the other constraints are inactive. Given this information, simplify the KKT conditions found in No. 1.
3. Find all the possible KKT points using the simplified KKT conditions from No. 2.

13.4   Solve the constrained optimization problem below using the inverse penalty method.

$$\min_x f(x) = (x_1 - 1.5)^2 + (x_2 - 1.5)^2 \tag{13.151}$$

subject to
$$x_1 + x_2 - 2 \le 0 \tag{13.152}$$

1. Perform four unconstrained optimizations using the following values for the penalty parameter: $R = 1, 0.1, 0.01$, and 0.001.
2. For each R, prepare a table that provides the value of the penalty parameter, the values of the design variables, objective function and the constraint.
3. Can you guess where the constrained minimum might be?

13.5 Consider the following problem:

$$\min_{x} f(x) = (x_1 - 4)^2 + (x_2 - 4)^2 \tag{13.153}$$

subject to

$$5 - x_1 - x_2 \geq 0 \tag{13.154}$$

1. Solve the above optimization problem using the bracket operator penalty method.
2. Check your answer using `fminsearch`.
3. Plot contours of the objective function, and the path of the intermediate solutions as the optimal point is reached.

13.6 Let $f(x) = 2x^3 - 3x^2 - 12x + 1$. Answer the following questions:
   (a) Determine the minimum of $f(x)$ by hand. Report the values of the optimum x and the function value at this point.
   (b) This sub-problem is a simple demonstration of how constrained optimization is sometimes performed. Let us add another component to the above function. The function now takes the form $f(x) = 2x^3 - 3x^2 - 12x + 1 + R(x - 6)^2$, where R is a constant. Plot this new function for $R = 0, R = 1, R = 10, R = 100$, and $R = 1,000$ for $0 \leq x \leq 8$ on the same figure. Use different colors for each R. Keep your vertical axis limits between -100 and 800.
   (c) By looking at the plot in Part (b), can you tell what the minimum of the new function is for the different values of R? Indicate these minima on the plot, and compare them with the minimum of the original function in Part (a). What is the minimum of the new function if $R = \infty$?
   (d) Explain how this problem shows you one way to solve constrained optimization problems.

BIBLIOGRAPHY OF CHAPTER 13

[1] J. Johannes. *Introduction to the Theory of Nonlinear Optimization*. Springer, 3rd edition, 2007.
[2] I. Griva, S. G. Nash, and A. Sofer. *Linear and Nonlinear Optimization*. SIAM, 2nd edition, 2009.
[3] M. S. Bazaraa, H. D. Sherali, and C. M. Shetty. *Nonlinear Programming: Theory and Algorithms*. John Wiley and Sons, 3rd edition, 2013.
[4] A. Ravindran, G. V. Reklaitis, and K. M. Ragsdell. *Engineering Optimization: Methods and Applications*. John Wiley and Sons, 2006.

PART 5

MORE ADVANCED TOPICS IN OPTIMIZATION

This part introduces carefully chosen advanced topics that are appropriate for graduate students or undergraduate students who might pursue study in the design optimization field. These topics will generally be of great interest to industry practitioners as well. While these topics are advanced, a mathematically advanced presentation is not provided. Instead, an introductory presentation is provided, with an eye toward practical usefulness.

Specifically, the topics presented, with the chapter numbers, are given below:

14

Discrete Optimization

14.1 Overview

In most previous chapters, continuous optimization problems were considered where the design variables were assumed to be continuous; that is, design variables assume real values within given ranges. In many practical engineering problems, the acceptable values of the design variables do not form a continuous set. These problems are referred to as discrete optimization problems. For example, the number of rivets required in a riveted joint has to be an integer (such as 1, 2, 3). Another example is when the feasible region of the design variable is a set of given discrete numbers, such as {6.25, 6.95, 7.65}, which may be the available standardized sizes of nuts. The basics of discrete optimization were introduced in Chapter 9, where some pertinent elementary methods were presented. This chapter introduces more advanced approaches. The reader is advised to first review Chapter 9 as preparation for the current chapter.

This chapter is organized as follows. The next section (Sec. 14.2) provides the problem classes, examples, and definition (along with the notion of computational complexity of the solution algorithms). Section 14.3 discusses the basics of some popular techniques used to solve integer programming problems, with examples. The methods studied will be: the exhaustive search method (Sec. 14.3.1), the graphical method (Sec. 14.3.2), the relaxation method (Sec. 14.3.3), the branch and bound method (Sec. 14.3.4), and the cutting plane method (Sec. 14.3.5). Popular current software options (Sec. 14.3.7) are also discussed. The chapter concludes with a summary in Sec. 14.4.

14.2 Problem Classes, Examples and Definition

This section presents discrete optimization problem classes, problem examples, and problem definition. Computational complexity of the solution algorithms is also briefly addressed in connection with the discrete optimization problem definition.

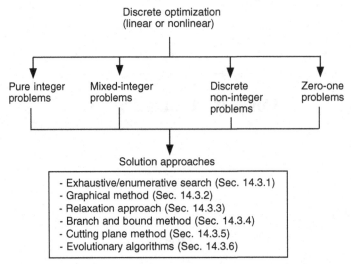

Figure 14.1. Discrete Optimization Overview

14.2.1 Problem Classes

Given the idea of discreteness of the feasible design variable set, there are several categories of optimization problems. As explained in Chapter 9, the following terms are commonly encountered within the umbrella of the discrete optimization literature; and define the main classes of discrete optimization problems (see Fig. 14.1).

1. **Pure integer programming problems:** When the design variables are allowed to assume only integer values. Example: Number of routes a traveling salesman can take (see Refs. [1, 2]).

2. **Mixed-integer programming problems:** When some design variables are allowed to assume only integer values, while others are allowed to assume continuous values.

3. **Discrete non-integer optimization problems:** When the design variables are allowed to assume only a given set of discrete, not necessarily integer, values. Example: Standardized sizes of machine components.

4. **Zero-one Programming problems:** When the design variables are allowed to assume values of either 0 or 1.

5. **Combinatorial optimization problems:** Where the set of possible feasible solutions is defined by a combinatorial set resulting from the given feasible discrete values of the design variables.

Next, we discuss some of the popular discrete optimization problems that you may encounter in the literature and in practice.

14.2.2 Popular Example Problems

In this subsection, descriptions of a few example problems in discrete optimization are introduced. The sample problems presented below span a wide range of fields, such as mathematics, transportation, and finance.

Knapsack problem: This is a classical example of a combinatorial optimization problem. A select set of items needs to be packed in a knapsack, or a bag. Each member of the set of items has a cost and a merit associated with it. How can the items be optimally chosen to be packed such that the cost is minimized, and the merit maximized?

Vehicle routing problem: The goal of this problem is to choose the route so as to minimize the *total distribution cost* of the goods to be delivered to a set of customers at different locations. The goods are assumed to be located in a central inventory.

Traveling salesman problem: A traveling salesman needs to plan a round-trip route. The salesman needs to visit a given number of cities, and each city exactly once. Each segment of the journey (from one city to another) has an associated cost. Which route yields the least expensive round-trip that visits each city exactly once?

Capital budgeting: This problem optimizes the allocation of capital budgets among different investment alternatives. Suppose a given amount of money in invested among four alternatives. Each investment alternative has a present value and a minimum required investment. How can the budgets be distributed among the alternatives so that the total present value is maximized?

The above discussion provides the reader with a basic idea of the various challenging discrete optimization problems commonly found in the literature. Some references in the area of discrete optimization contain a mathematically rigorous treatment of the subject, which is outside the scope of this introductory textbook. This chapter presents a basic treatment of the ideas involved, with illustrative examples.

14.2.3 Problem Definition and Computational Complexity

This section introduces a generic integer programming problem formulation. The notion of computational complexity of the solution algorithms is also briefly presented.

Problem Definition

A generic integer programming problem is given below.

$$\min_{x} \ f(x) \tag{14.1}$$

subject to

$$g(x) \leq 0 \tag{14.2}$$

$$h(x) = 0 \tag{14.3}$$

$$x \in Z \tag{14.4}$$

where x is the design variable vector; $f(x)$, $g(x)$, and $h(x)$ are the objective function, inequality constraints, and equality constraints, respectively; and Z is the set of given feasible integers. Note that for a generic *discrete optimization* problem, the set of

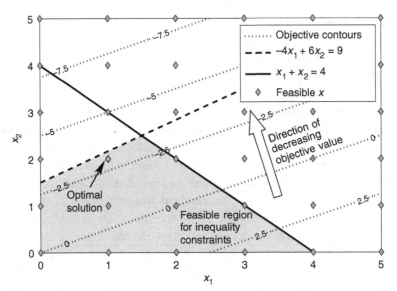

Figure 14.2. Graphical Solution for Example 1

integers Z in Eq. 14.4 will be replaced by the given set of *real numbers*, and are not necessarily integers.

Example: For illustration purposes, consider the linear integer programming problem shown below.

$$\min_x \; x_1 - 2x_2 \tag{14.5}$$

subject to

$$-4x_1 + 6x_2 \le 9 \tag{14.6}$$

$$x_1 + x_2 \le 4 \tag{14.7}$$

$$x_1, x_2 \ge 0 \tag{14.8}$$

$$x_1, x_2 \in Z \tag{14.9}$$

The feasible region of the above problem is shown in Fig. 14.2. The next section solves the above problem using some popular discrete optimization approaches.

Before studying the solution approaches, we briefly examine the computational complexity of the algorithms that can solve discrete optimization problems.

Computational Complexity

Discrete optimization problems are known to be computationally complex. The study of complexity is related to a branch of mathematics and computing known as *complexity theory*. One of the issues of interest in complexity theory is to quantify the performance of computational algorithms. The computation time of an algorithm is usually presented using the O (known as the Big O) notation.

Example: Say we are interested in finding the gradient of a function $f(x_1, x_2, ...x_n)$ using the finite difference method. The i-th component of the $n \times 1$ gradient vector can be computed as

$$\left. \frac{\partial f}{\partial x_i} \right|_{x=p} = \frac{f(p + \delta_p) - f(p)}{\delta_p} \tag{14.10}$$

where δ_p is the chosen step size. The above computation requires two function evaluations. To compute all of the n components of the gradient vector, we require $n + 1$ function evaluations. Therefore, this finite difference-based gradient computation algorithm has a complexity of the order of $O(n)$ for a problem size of n. Note that whether we use forward or backward difference (Chapter 7) to compute the gradient vector, we will have a complexity of the order of $O(n)$. An algorithm is said to be of *polynomial time* if its complexity is of the order of $O(n^k)$ for a problem size of n, where k is a finite non-negative integer. [3].

Another aspect of complexity theory is to categorize computational problems and algorithms into complexity classes. One important complexity class is known as the P class. The problems belonging to the P class are regarded as tractable, and easily solvable by algorithms in polynomial time. Unfortunately, several practical discrete optimization problems belong to the notoriously difficult complexity class of the so-called *NP-Complete* problems. A detailed presentation of complexity theory is beyond the scope of this chapter.

The methods employed to solve discrete optimization problems are studied next. Examples are provided where necessary.

14.3 Solution Approaches

An overview of discrete optimization solution approaches is provided in Fig. 14.1 (also see Sec. 14.2). Discrete problems can be linear (with linear constraints and linear objective function) or nonlinear problems (with nonlinear constraints and/or nonlinear objective function). From the formulation and solution approaches perspectives, discrete problems can be classified as *pure-integer* problems, *mixed-integer* problems, *discrete non-integer* problems, and *zero-one* problems. Nonlinear discrete problems are much more complicated to solve than linear discrete problems.

As reported in Fig. 14.1, although the broad solution approaches for various types of discrete problems are often the same, their specific implementation and complexity will vary. For example, variations of the *branch and bound* and *cutting plane* methods have been applied to linear, as well as nonlinear, discrete problems of several kinds. In this chapter, our goal is to introduce the basics of how these popular algorithms work. Linear integer programming problems will be considered for presentational simplicity. More rigorous treatment of the methods discussed in this section is available in [4, 5, 2].

Table 14.1. *Brute Force Method: Six Elements*

Element, i	1	2	3	4	5	6
Weight, w_i	1.5	6.4	2.0	3.2	5.7	4.3

Table 14.2. *Brute Force Method: Twenty Combinations*

Combination	Total weight	Combination	Total weight
1,2,3	9.9	2,3,4	11.6
1,2,4	11.1	2,3,5	14.1
1,2,5	13.6	2,3,6	12.7
1,2,6	12.2	2,4,5	15.3
1,3,4	**6.7**	2,4,6	13.9
1,3,5	9.2	2,5,6	16.4
1,3,6	7.8	3,4,5	10.9
1,4,5	10.4	3,4,6	9.5
1,4,6	9	3,5,6	12
1,5,6	11.5	4,5,6	13.2

14.3.1 Brute Force Method: Exhaustive Search

The most straightforward, and computationally expensive, approach to solving discrete problems is to perform an exhaustive search – where all the possible options are enumerated and evaluated. The optimal solution can then be readily selected from the enumerated solutions. This approach is a viable option only for small problems, as it can be computationally prohibitive for larger problems.

Example: Consider the following simple example. We have a set of 6 truss elements. The generic i-th element has a weight w_i (see Table 14.1). We need to choose three elements that yield a minimum total weight. This problem can be viewed as a combinatorial optimization problem, and can be stated as

$$\min_{x_i, x_j, x_k} \; w_i + w_j + w_k \tag{14.11}$$

where

$$x_i \neq x_j \neq x_k \tag{14.12}$$

$$x_i, x_j, x_k \in \{x_1, x_2, x_3, x_4, x_5, x_6\} \tag{14.13}$$

$$\{w_1, w_2, w_3, w_4, w_5, w_6\} = \{1.5, 6.4, 2.0, 3.2, 5.7, 4.3\} \tag{14.14}$$

There are $6C_3 = \frac{6!}{3!(6-3)!} = 20$ ways of choosing three elements from six possibilities. Table 14.2 lists the possible 20 combinations, along with the corresponding total weight.

The row shown in bold in Table 14.2 is the optimal solution. Note that as the number of possibilities increases, such exhaustive enumeration becomes computationally expensive.

The graphical method to solve discrete optimization problems is studied next.

14.3.2 Graphical Method

For simple problems where the objective function and the constraints can be visualized, the feasible region could be plotted to find the optimal discrete solution graphically. However, a graphical solution may not be a viable option for complex problems with a large number of design variables.

> **Example:** For the example problem previously discussed (Eqs. 14.5 through 14.9), the feasible region can be plotted as shown in Fig. 14.2. The direction of decreasing objective function value is shown by the block arrow. The grey shaded region represents the feasible region as defined by the inequality constraints alone, and the diamonds represent feasible integer solutions. Upon inspection of Fig. 14.2, the optimal solution is $x_1 = 1, x_2 = 2$, which is the integer solution within the feasible design space with the least objective function value.
> Once the feasible region is identified, one could also perform an exhaustive search, considering only the feasible design variable values, to find the optimum solution.

The next subsection discusses a simple and commonly used approach to solve discrete problems.

14.3.3 Relaxation Approach: Solve as Continuous Problem

In this method, the discrete formulation is *relaxed* by treating the discrete variables as if they were continuous. The optimization problem is then solved using the continuous optimization techniques learned earlier. The real-valued optimum design variables obtained are then rounded off to the nearest feasible discrete solution.

While this technique is used quite often due to its ease of implementation, the user is warned that the solution obtained can often be sub-optimal. In addition, the previously discussed rounding of the optimal solution can result in constraints violations at the approximate discrete solution.

> **Example:** Consider the linear discrete optimization problem presented in Eqs. 14.5 – 14.9. Now, solve it as a continuous optimization problem by ignoring the constraint in Eq. 14.9. The relaxed optimization problem then becomes

$$\min_{x} \; x_1 - 2x_2 \tag{14.15}$$

subject to

$$-4x_1 + 6x_2 \leq 9 \tag{14.16}$$

$$x_1 + x_2 \leq 4 \tag{14.17}$$

$$x_1, x_2 \geq 0 \tag{14.18}$$

Solving the above optimization problem using a continuous optimization algorithm, such as fmincon, yields the optimal solution as $x_1 = 1.5, x_2 = 2.5$. Rounding off to the nearest integer would yield an integer optimal solution of $x_1 = 2, x_2 = 3$. As seen in Fig. 14.2, the rounded solution lies in the infeasible region of the design space. If the continuous solution is instead rounded off to $x_1 = 1, x_2 = 2$, we obtain the correct solution. The relaxation approach for solving discrete problems must, therefore, be carefully employed.

14.3.4 Branch and Bound Method

The branch and bound method involves a systematic enumeration of the candidate solutions for a discrete optimization problem. We explain the basics of a branch and bound algorithm in a series of steps that can be used to solve a *linear integer programming problem* by hand. Software implementations of the algorithm could be much more involved.

A basic implementation of the branch and bound method consists of the following steps.

1. Formulate a relaxed continuous linear programming (LP) problem by ignoring the integer constraints – resulting in a continuous formulation. The resulting optimal solution may have some non-integer design variable values. If the resulting LP solution has only integer values, the obtained solution is indeed the optimal solution.

2. Notation: Say x_i^* contains a decimal part. Define the notation of $\lceil x \rceil$, also known as the ceiling function. This function returns the smallest integer value which is greater than or equal to x. For example, $\lceil 5.134 \rceil = 6$, $\lceil 10 \rceil = 10$, and $\lceil -8.6 \rceil = -8$. The second notation defined is the floor function, denoted by $\lfloor x \rfloor$, which returns the largest integer that is less than or equal to x. For example, $\lfloor 5.134 \rfloor = 5$, $\lfloor 10 \rfloor = 10$, and $\lfloor -8.6 \rfloor = -9$.

3. For those design variables with decimal parts in the solution, two subproblems that impose *bounds* on the design variable values are created. This process is known as *branching*. The following additional constraint is added to the first subproblem: $x_i \leq \lfloor x \rfloor$. The second subproblem is formulated by adding the constraint $x_i \geq \lceil x \rceil$. The subproblems are then solved as continuous problems. The solutions of the two subproblems are then examined for fractional parts, and the branching process is repeated.

4. For a given variable, the branching process is usually repeated until the relaxed continuous problem with the additional constraints yields either an integer solution or an infeasible solution. The branching process is repeated for all the variables that have fractional solutions.

5. Once the branching process is completed for all the variables, the best solution among the integer solutions from all branches is considered the optimal solution. Note that the above algorithm discussion applies only to linear integer programming problems.

Example: Consider the following example, and implement the branch and bound method.

$$\min_{x} \quad x_1 - 2x_2 \tag{14.19}$$

subject to

$$-4x_1 + 6x_2 \le 9 \tag{14.20}$$

$$x_1 + x_2 \le 4 \tag{14.21}$$

$$x_1, x_2 \ge 0 \tag{14.22}$$

$$x_1, x_2 \in Z \tag{14.23}$$

Figure 14.3 provides the subproblems and their solutions when the branching is performed on the variable x_1. The gray-shaded boxes are the integer solutions obtained during this branching. Note the branching process after the optimal solution $x_1^* = 0.75, x_2^* = 2, f^* = -3.25$. The two further possible branches are $x_1 \le 0$ and $x_1 \ge 1$. The first branch, $x_1 \le 0$, is not feasible, as it violates the first constraint, and is not solved further. The second branch, $x_1 \ge 1$, conflicts with the first branching, $x_1 \le 1$. Note that this conflict yields $x_1 = 1$ as the only possibility for the variable x_1, yielding the optimal integer solution, $x_1^* = 1, x_2^* = 2, f^* = -3$. Figure 14.4 reports the subproblems and the corresponding solutions when the branching is performed on x_2. The gray-shaded box is the integer solution obtained ($x_1^* = 1, x_2^* = 2, f^* = -3$), which is also the optimal integer solution for this problem.

14.3.5 Cutting Plane Method

The basic concept of the cutting plane method is to add inequalities, also known as cuts, to the existing integer problem. The purpose of adding the inequality is to cut off the non-integer solutions without eliminating integer points of the feasible region. The resulting formulation is then solved as a non-integer continuous problem, and is tested if the obtained solution is an integer. If not, a new cut is added and the procedure is repeated until an optimal integer solution is found. There are different approaches for finding effective inequalities or cuts that are added to the initial set of constraints. These effective inequalities are usually taken from pre-defined families, one of which is the *Gomory cut*, which will be studied in this chapter.

Consider a *linear integer programming* problem. A Gomory cut is a linear inequality constraint that does not exclude any feasible integer solutions from the integer problem. In this chapter, we will explain the cutting plane method by using the Simplex method that was presented in Chapter 11. The following is a generic discussion of Gomory's cutting plane method for linear integer programming problems.

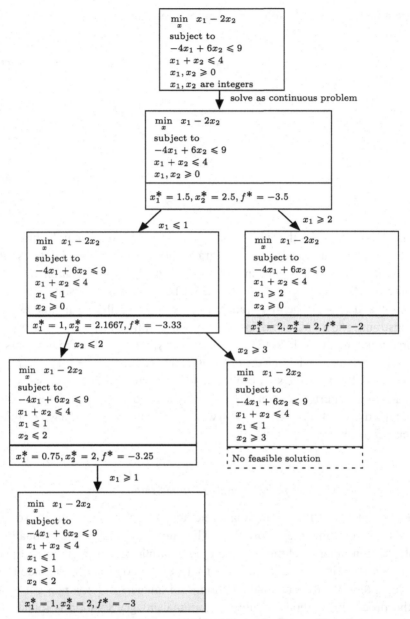

Figure 14.3. Branch and Bound Method Example: Integer Bounds on x_1

1. **Initial Simplex Tableau:**Begin by relasing the integer programming problem by ignoring the integer constraints on the design variables. The linear programming (LP) formulation can be written as

$$\min_x \quad c^T x \qquad (14.24)$$

subject to

$$Ax \leq b \qquad (14.25)$$

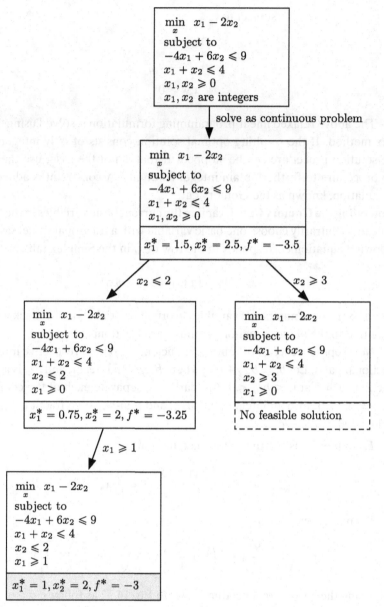

Figure 14.4. Branch and Bound Method Example: Integer Bounds on x_2

$$A_{eq}x = b_{eq} \tag{14.26}$$

$$x \geq 0 \tag{14.27}$$

Adding slack variables, denoted by the vector s, to the inequality constraints for the Simplex method, obtain the following formulation.

$$\min_{x} \ c^T x \tag{14.28}$$

subject to

$$Ax + s = b \tag{14.29}$$

$$A_{eq}x = b_{eq} \tag{14.30}$$

$$x, s \geq 0 \tag{14.31}$$

The above relaxed linear programming formulation is solved using the Simplex method. If the resulting optimal solution consists of only integer values, the solution procedure can be stopped. If that is not the case, use the following procedure to further obtain integer solutions. A constraint is added to the formulation, known as the Gomory cut.

2. **Generating the Gomory Cut:** Examine the current basic variables in the Simplex tableau. Arbitrarily choose one basic variable with a fractional value, say x_i. The following equation corresponds to the row of x_i in the Simplex tableau.

$$x_i = b_i - \{a_{i1}x_1 + \dots + a_{it}x_t\} \tag{14.32}$$

where t is the total number of variables: n original variables and m slack variables. Note that in the above equation, x_i and b_i are fractional.

Now separate each of the above coefficients into their respective integer and fractional parts. Let $b_i = b_{Z:i} + b_{f:i}$, where $b_{Z:i}$ is an integer, and $b_{f:i}$ is a positive fraction such that $0 \leq b_{f:i} \leq 1$. Similarly, we separate each of the coefficients, $a_{ij}, j = \{1, \dots, t\}$, into an integer part, $a_{Z:j}$, and non-negative fractional part, $a_{f:j}$, yielding $a_{ij} = a_{Z:j} + a_{f:j}$.

Example: Consider the following equation.

$$x_2 = \frac{53}{11} - \frac{2}{11}x_3 + \frac{4}{11}x_4 \tag{14.33}$$

It can be re-written as

$$x_2 = \left(4 + \frac{9}{11}\right) - \left[\left(0 + \frac{2}{11}\right)x_3 + \left(-1 + \frac{7}{11}\right)x_4\right] \tag{14.34}$$

Using the above nomenclature, rewrite Eq. 14.32 as follows.

$$x_i = b_{Z:i} + b_{f:i} - [\{a_{Z:1} + a_{f:1}\}x_1 + \dots + \{a_{Z:t} + a_{f:t}\}x_t] \tag{14.35}$$

The above equation can be further written as

$$b_{f:i} - [a_{f:1}x_1 + \dots + a_{f:t}x_t] = x_i - b_{Z:i} + [a_{Z:1}x_1 + \dots + a_{Z:t}x_t] \tag{14.36}$$

Note that the right-hand-side of the above equation is always an integer. Therefore, the left-hand-side should also be an integer.

$$b_{f:i} - [a_{f:1}x_1 + \dots + a_{f:t}x_t] \in Z \tag{14.37}$$

Since $a_{f:i}$ are non-negative fractions and $x \geq 0$, the quantity $[a_{f:1}x_1 + ... + a_{f:t}x_t]$ is non-negative. In addition, note that $0 \leq b_{f:i} \leq 1$. Therefore, the following equation holds.

$$b_{f:i} - [a_{f:1}x_1 + ... + a_{f:t}x_t] \leq b_{f:i} \leq 1 \tag{14.38}$$

From Eq. 14.37, note that the left-hand-side of the above equation should be an integer. This implies that $b_{f:i} - [a_{f:1}x_1 + ... + a_{f:t}x_t]$ should either be zero or a negative integer. This condition yields the *Gomory constraint*, given as

$$b_{f:i} - [a_{f:1}x_1 + ... + a_{f:t}x_t] \leq 0 \tag{14.39}$$

3. **Solving problem with the Gomory constraint:**
   Adding a slack variable to Eq. 14.39, we obtain

$$b_{f:i} - [a_{f:1}x_1 + ... + a_{f:t}x_t] + s_{t+1} = 0 \tag{14.40}$$

The above constraint is then added to the linear programming problem from STEP 1.. Note that the optimal solution obtained in STEP 1. (initial Simplex tableau) does not satisfy the Gomory constraint generated above. In this case, the basic Simplex method used in STEP 1. cannot be used to solve the new problem. Use the *dual Simplex method* (see Ref. [5]) for solving the new problem with the Gomory constraint.

4. **Repeat:** The process of generating the Gomory constraint and solving the new formulation with the dual Simplex method is repeated until integer solutions are obtained.

Note that the above-stated approach is a very simplified discussion of cutting plane methods. For a more rigorous mathematical discussion, see [2]. Some shortcomings of this method are: (1) As the number of variables grows, the size and the complexity of the problem also grows, since an additional slack variable and an additional constraint are added for each non-integer variable. (2) Implementing this method in a computer using decimal representation can result in rounding errors as the algorithm proceeds, and may result in a wrong integer optimal solution.

The cutting plane method is now illustrated with the help of the following example.

Cutting plane algorithm example:
The linear integer programming problem is given as follows.

$$\min_{x} \quad x_1 - 2x_2 \tag{14.41}$$

subject to

$$-4x_1 + 6x_2 \leq 9 \tag{14.42}$$

Table 14.3. *Initial Simplex Tableau Before Adding Cutting Planes*

	x_1	x_2	s_1	s_2	b	b_i/a_{ij}
R_1	-4	6	1	0	9	3/2
R_2	1	1	0	1	4	4
f	1	-2	0	0	f	

Table 14.4. *Final Simplex Tableau Before Adding Cutting Planes*

	x_1	x_2	s_1	s_2	b
R_1	0	1	1/10	2/5	2/5
R_2	1	0	−1/10	3/5	3/2
f	0	0	3/10	1/5	$f+2/7$

$$x_1 + x_2 \leq 4 \tag{14.43}$$

$$x_1, x_2 \geq 0 \tag{14.44}$$

$$x_1, x_2 \text{ are integers} \tag{14.45}$$

1. **Initial Simplex Tableau:** Temporarily ignore the integer constraints on x_1 and x_2. The standard linear programming formulation for the Simplex method can be written as follows.

$$\min_x \ x_1 - 2x_2 \tag{14.46}$$

subject to

$$-4x_1 + 6x_2 + s_1 = 9 \tag{14.47}$$

$$x_1 + x_2 + s_2 = 4 \tag{14.48}$$

$$x_1, x_2, s_1, s_2 \geq 0 \tag{14.49}$$

The initial Simplex tableau can be written as shown in Table 14.3. The final Simplex tableau that yields the continuous optimal solution is $x_1^* = \frac{3}{2}, x_2^* = \frac{5}{2}, f^* = -\frac{7}{2}$, and is shown in Table. 14.4.

2. **Generating the Gomory Cut:** If the above step yielded an integer solution, the integer problem would have been solved, and the solution procedure would end there. However, this is not the case in the example under consideration. A constraint is added using the theory of Gomory cut algorithm. From the Simplex tableau in Table 14.4, arbitrarily choose a basic variable that has a fractional value, say x_2 (from row R_1). The equation in R_1 can be written as

$$x_2 + \frac{1}{10}s_1 + \frac{2}{5}s_2 = \frac{5}{2} \tag{14.50}$$

Table 14.5. *Initial Simplex Tableau After Adding First*
Cutting Plane (Eq. 14.55)

	x_1	x_2	s_1	s_2	s_3	b
R_1	0	1	1/10	2/5	0	5/2
R_2	1	0	−1/10	3/5	0	3/2
R_3	**0**	**0**	**−1/10**	**-2/5**	**1**	**-1/2**
f	0	0	3/10	1/5	0	$f + 7/2$
$c_i/(-a_i)$	N/A	N/A	3	**1/2**	N/A	

or

$$x_2 = \frac{5}{2} - \frac{1}{10}s_1 - \frac{2}{5}s_2 \tag{14.51}$$

Now rewrite the above equation by separating the integer and fractional parts.

$$x_2 = 2 + \frac{1}{2} - (0 + \frac{1}{10})s_1 - (0 + \frac{2}{5})s_2 \tag{14.52}$$

$$\frac{1}{2} - \frac{1}{10}s_1 - \frac{2}{5}s_2 = x_2 - 2 \tag{14.53}$$

Using the earlier discussion and Eq. 14.39, the Gomory cut for the above case can be written as follows.

$$-\frac{1}{10}s_1 - \frac{2}{5}s_2 \leq -\frac{1}{2} \tag{14.54}$$

$$-\frac{1}{10}s_1 - \frac{2}{5}s_2 + s_3 = -\frac{1}{2} \tag{14.55}$$

3. **Solve problem with Gomory constraint (Iteration 1):**
   Including Eq. 14.55 in the Simplex tableau, the updated tableau can be written as shown in Table 14.5.
   The linear programming problem in Table 14.5 will now be solved using the dual Simplex method. Select R_3 from Table 14.5, as it has a negative b value. For each negative coefficient in R_3, find the corresponding cost coefficient (entries of the row f), c_i, and compute the following.

$$\min_{a_i < 0} \left(\frac{c_i}{-a_i} \right) \tag{14.56}$$

For example, the s_1 column can be evaluated as

$$\left(\frac{c_3}{-a_3} \right) = \frac{\frac{3}{10}}{-(\frac{-1}{10})} = 3 \tag{14.57}$$

Since the column for s_2 satisfied the above condition, we pivot on the element $-\frac{2}{5}$ in R_3 in Table 14.5. We then obtain the final Simplex tableau provided in

Table 14.6. *Final Simplex Tableau After Adding the First Cutting Plane*
(Eq. 14.55)

	x_1	x_2	s_1	s_2	s_3	b
R_1	0	1	0	0	1	2
R_2	1	0	-1/4	0	3/2	3/4
R_3	0	0	1/4	1	-5/2	5/4
f	0	0	1/4	0	1/2	$f + 13/4$

Table 14.7. *Initial Simplex Tableau After Adding Second Cutting Plane*
(Eq. 14.62)

	x_1	x_2	s_1	s_2	s_3	s_4	b
R_1	0	1	0	0	1	0	2
R_2	1	0	-1/4	0	3/2	0	3/4
R_3	0	0	1/4	1	-5/2	0	5/4
R_4	**0**	**0**	**-3/4**	**1**	**-1/2**	**1**	**-3/4**
f	0	0	1/4	0	1/2	0	$f + 13/4$
$c_i/(-a_i)$	N/A	N/A	**1/3**	N/A	1	N/A	N/A

Table 14.6. This is the final Simplex tableau as all the elements in the b vector
are positive.

The optimal values from Table 14.6 are $x_1^* = \frac{3}{4}, x_2^* = 2, f = -\frac{13}{4}$. Since the
optimal solution has fractional parts, another Gomory cutting plane will be
generated from Table 14.6.

4. **Solve problem with Gomory constraint (Iteration 2):**

   Choose x_1 as the variable for which a Gomory constraint is generated, using
   R_1. Using the discussion provided earlier, we separate the integer and the
   fractional parts of the coefficients, as shown below.

$$x_1 - \frac{1}{4}s_1 + \frac{3}{2}s_3 = \frac{3}{4} \tag{14.58}$$

$$x_1 + \left(-1 + \frac{3}{4}\right)s_1 + \left(1 + \frac{1}{2}\right)s_3 = \left(0 + \frac{3}{4}\right) \tag{14.59}$$

$$\frac{3}{4} - \frac{3}{4}s_1 - \frac{1}{2}s_3 = x_1 - s_1 + s_3 \tag{14.60}$$

$$\frac{3}{4} - \frac{3}{4}s_1 - \frac{1}{2}s_3 \leq 0 \tag{14.61}$$

Or, the Gomory constraint can be written as

$$-\frac{3}{4}s_1 - \frac{1}{2}s_3 + s_4 = -\frac{3}{4} \tag{14.62}$$

The initial Simplex tableau for the above problem is provided in Table 14.7.
Select row R_4 with the negative b value. Computing the $\frac{c_i}{-a_i}$ ratio for the

Table 14.8. *Final Simplex Tableau After Adding Second Cutting Plane (Eq. 14.62)*

	x_1	x_2	s_1	s_2	s_3	s_4	b
R_1	0	1	0	0	1	0	2
R_2	1	0	0	0	0	5/3	1
R_3	0	0	0	1	−8/3	1/3	1
R_4	0	0	1	0	2/3	−4/3	1
f	0	0	0	0	4/3	1/3	$f+3$

negative elements in the row leads to the choice of the s_1 column. Pivoting on $-\frac{3}{4}$ in row R_4 in Table 14.7, obtain the final Simplex tableau shown in Table 14.8, with optimal values of $x_1^* = 1, x_2^* = 2, f^* = -3$.

The above example concludes our discussion of the cutting plane method. We discuss other solution approaches next.

14.3.6 Evolutionary Algorithms

Evolutionary algorithms, and certain versions of swarm-based algorithms [6], are a popular choice for discrete optimization because of their ability to work directly with discrete search spaces. The application of evolutionary algorithms will be studied in Chapter 19.

14.3.7 Software Options for Discrete Optimization

This subsection presents some of the popular software tools available for solving discrete programming problems. Note that most of these software tools can solve more than just integer programming problems. For convenience, we repeat here the table listing software in Section 5.5, as Table 14.9, where software options are discussed. Here, we will focuss on the last column of Table 14.9, where discrete optimization software is presented.

1. The XPRESS suite of optimization algorithms is distributed by Dash Optimization [7]. The XPRESS MIP tool provides the capability to solve mixed-integer programming problems by using sophisticated implementations of the branch and bound and cutting plane algorithms. Further information about the XPRESS suite can be found at the following website.

   `www.fico.com/en/products/fico-xpress-optimization-suite`

2. CPLEX solvers, created by CPLEX Optimization, Inc., are designed to handle problems with mixed-integer programming features. The solver uses a combination of algorithms and heuristics [8]. Further information regarding the free CPLEX solver can be found at the following website.

   `openopt.org/cplex`

Table 14.9. *Broad Classification of Software for Optimization—with Discrete Case*

Stand-Alone (SO-SA)	Software for Optimization (SO) Within Design Framework (SO-WDF)	Within Analysis Package (SO-WAP)	Discrete Integer or Mixed
MATLAB Toolbox	iSIGHT	GENESIS	XPRESS
NEOS Server	PHX ModelCenter	NASTRAN	CPLEX
DOT-VisualDOC	modeFRONTIER	ABAQUS	Excel and Quattro
NAG	XPRESS	Altair	NEOS Server
NPSOL	LINDO/LINGO	ANSYS	MINLP
GRG2	GAMS	COMSOL	GAMS WORLD
LSSOL	Boss Quattro	MS Excel	
CPLEX		What'sBest!	
BTN		RISKOptimizer	
PhysPro		Busi. Spreadsh.	

3. Excel and Quattro Pro Solvers, developed by Frontline Systems[9], are available to solve small scale integer programming problems using Excel spreadsheets. The integer programming method employs a branch and bound technique, which uses a nonlinear programming tool known as GRG2 [10].

4. The NEOS Server is a website hosted by Argonne National Laboratory, and is free to use. Several state-of-the-art algorithms in optimization software, including those in integer programming, are available from the website www.neos-server.org/neos/solvers

5. The GAMS WORLD [11, 12] is a website hosted by the international GAMS community. It provides a large suite of source codes for solving mixed integer nonlinear programming (MINLP) problems, which are an important and complex class of discrete optimization problems. The MINLP solver codes can be found at the following website: www.gamsworld.org/minlp/solvers.htm

The above subsection concludes our discussion on the solution approaches of discrete optimization.

14.4 Summary

This chapter introduced important aspects of discrete optimization. The treatment of the subject in this chapter is fairly simple when compared to more detailed references. Simple and easy-to-implement solution approaches for discrete problems that can be readily implemented for engineering problems were introduced. Examples were provided to further illustrate how the algorithms work. A list of some popular software options was also provided, together with their pertinent characteristics.

14.5 Problems

14.1 Perform a literature survey and find at least two examples of discrete optimization problems that were not discussed in the chapter.

14.2 Solve the problem presented in Sec. 14.3.2 graphically. In addition, reproduce the details shown in Fig. 14.2. Clearly label your plot.

14.3 Solve the following integer problem graphically. Clearly label your plots.

$$\max_x \ 5x_1 + 8x_2 \tag{14.63}$$

subject to

$$x_1 + x_2 \leq 6 \tag{14.64}$$

$$5x_1 + 9x_2 \leq 45 \tag{14.65}$$

$$x_1, x_2 \geq 0 \tag{14.66}$$

$$x_1, x_2 \in Z \tag{14.67}$$

14.4 Solve the following integer problem graphically. Clearly label your plots.

$$\max_x \ 7x_1 + 9x_2 \tag{14.68}$$

subject to

$$-x_1 + 3x_2 \leq 6 \tag{14.69}$$

$$7x_1 + x_2 \leq 35 \tag{14.70}$$

$$x_1, x_2 \geq 0 \tag{14.71}$$

$$x_1, x_2 \in Z \tag{14.72}$$

14.5 Duplicate the results for the integer programming problem presented in Sec. 14.3.4.

14.6 Solve the integer problem shown in Problem 14.3 using the branch and bound method. Clearly write down the subproblem statement, and the corresponding optimal solutions.

14.7 Solve the integer problem shown in Problem 14.4 using the branch and bound method. Write down the detailed solutions.

14.8 Duplicate the results for the cutting plane method example discussed in Sec. 14.3.5.

14.9 Solve the integer problem shown in Problem 14.3 using Gomory's cutting plane method. Clearly write down the subproblem statement, and the corresponding optimal solutions.

14.10 Solve the integer problem shown in Problem 14.4 using Gomory's cutting plane method. Write down the detailed solutions.

BIBLIOGRAPHY OF CHAPTER 14

[1] J. K. Karlof. *Integer Programming: Theory and Practice*. CRC Press, 2005.

[2] D. S. Chen, R. G. Batson, and Y. Dang. *Applied Integer Programming: Modeling and Solution*. John Wiley and Sons, 2010.

[3] K. H. Rosen. *Discrete Mathematics and its Applications*. McGraw-Hill Science, 7th edition, 2012.

[4] M. M. Syslo, N. Deo, and J. S. Kowalik. *Discrete Optimization Algorithms: With Pascal programs*. Courier Dover Publications, 2006.

[5] S. S. Rao. *Engineering Optimization: Theory and Practice*. John Wiley and Sons, 4th edition, 2009.

[6] S. Chowdhury, W. Tong, A. Messac, and J. Zhang. A mixed-discrete particle swarm optimization algorithm with explicit diversity-preservation. *Structural and Multidisciplinary Optimization*, 47(3):367–388, 2013.

[7] Fair Issac Corp. Xpress optimization suite. `www.dashoptimization.com/home/index.html`.

[8] IBM. IBM ILOG CPLEX Optimization Studio. `www.ilog.com/products/cplex`.

[9] Frontline Systems, Inc. Frontline Solvers. `www.solver.com/exceluse.htm`.

[10] S. Smith and L. S. Lasdon. Solving sparse nonlinear programs using GRG. *INFORMS Journal on Computing*, 4(1):2–15, 1992.

[11] GAMS Development Corp. and GAMS Software GmbH. GAMS world. `www.gamsworld.org`.

[12] GAMS World. MINLP world. `www.gamsworld.org/minlp/index.htm`.

15

Modeling Complex Systems: Surrogate Modeling and Design Space Reduction

15.1 Overview

In this chapter, we introduce efficient mathematical approaches to model the behavior of complex systems in the context of optimizing the system. Specifically, complex systems often involve a large number of design parameters (to be tuned), and the optimization of these systems often demand large scale computational simulations or experiments to quantify the system behavior. The resulting high dimensionality of the design space, the prohibitive computation time (or expense), or the lack of mathematical models present important challenges to the quantitative optimization of these complex systems. This chapter introduces traditional and contemporary approaches, such as design variable linking and surrogate modeling, to address the modeling challenges encountered in solving complex optimization problems.

Section 15.2 discusses the generic challenges in complex optimization problems. The impact of problem dimension is addressed in Sec. 15.3, where design variable linking and design of experiments are introduced. Section 15.4 presents surrogate modeling, where the discussion includes: the process, the polynomial response surface methodology, the radial basis function method, the Kriging method, and the artificial neural network method. The chapter concludes with a summary provided in Section. 15.5.

15.2 Modeling Challenges in Complex Optimization Problems

Modeling is one of the primary activities in optimization. Leveraging computational tools and models, as opposed to purely depending on experiments, can be considered the way forward in the area of system optimization. The design of complex systems, such as aircrafts, cars, and smart-grid networks, are increasingly performed using simulation-based design and analysis tools such as Finite Element Analysis (FEA) and Computational Fluid Dynamics (CFD). The incorporation of these tools has dramatically transformed modern engineering design and optimization approaches.

These changes, however, are not free of challenges. For example, modern aerospace systems present significantly complex design requirements [1], where design objectives generally involve improving performance, reducing costs, and maximizing safety. Another example is the optimization of a vehicle structure to absorb crash energy, maintain adequate passenger space, and control crash deceleration pulse during collisions [2]. The collision process is an extremely complex multibody dynamic process.

The relationship between the objectives and the design variables in complex systems is generally not governed by functions of simple analytical form. The use of complex computational or simulation-based models to design these systems may demand unreasonably high computational costs. You will find that the techniques presented in this chapter are paramount to overcoming the computational barrier and efficiently exploring optimal designs for complex systems.

It is also important to remember that, as discussed in Chapter 4, inappropriate modeling can result in slow convergence, everlasting computation, or unrealistic solutions. In the context of modeling, practical optimization problems are faced with three major challenges:

1. Design-space dimensionality: Complex problems may involve a very large number of variables. For example, nearly 300 variables are involved in designing a vehicle powertrain system [3] and more than 1,000 design variables are involved in designing large scale wind farms or systems with multiple physical domains [4]. Models that are formulated to operate only on the entire (high-dimensional) variable space in these cases may pose challenging-to-prohibitive numerical burdens on the optimization process.

2. Computational expense: As we learned in the previous chapters, the behavior of complex physical systems are often modeled using computational simulations (*e.g.*, FEA or CFD), which are basically the computer-based implementation of complex (often multi-step) mathematical models. However, these simulations can take hours (even days) to run. Thus, their usage could become computationally/time-wise prohibitive in the context of optimization when simulation models have to be executed thousands (or tens of thousands) of times during optimization. For example, if a large eddy simulation [5] (which is a CFD model) is used to model the flow inside a wind farm, it may demand approximately 600 million CPU-hours for optimizing the configuration of a 25-turbine wind farm [6].

3. Lack of mathematical models: Alternatively, for certain systems or problems, there may not exist any mathematical models that define the underlying physics/behavior of the system (or the dynamics of the problem). This could be due to a lack of a clear understanding of the underlying physics or the relationship between the design parameters and the criteria functions. For example, without a physics-based model to represent the constitutive relationship of Ti-25V-15Cr-0.2 Si alloy during the hot deformation process, empirical models (based on experimental data) are used [7].

One class of approaches to address the first challenge (*i.e.*, design-space dimensionality) is design space reduction or design variable linking. Design variable linking is an approach that seeks to reduce the design dimensionality of a problem by developing relationships among the variables — thereby yielding models that quantify the system behavior in terms of a smaller (likely more tractable) set of variables than initially defined. These approaches are discussed in the following section. In addition, the following section introduces another aspect of model development, sampling (or design of experiments), which is also directly related to the dimensionality of the design space.

The other two challenges are generally addressed by a class of mathematical functions that have different names, based on the research community with which you are communicating (*e.g.*, mathematicians, material scientists, or aerospace engineers). The process of developing/using this class of mathematical functions is known by such names as metamodeling, surrogate modeling, response surface methodology, and function fitting. Essentially, they represent an approach to mathematical modeling that does not exploit the understanding of the underlying physics of the system behavior being modeled. An introductory discussion of this class of models, known as surrogate models, is provided in Sec. 15.4.

15.3 Impact of Problem Dimension

In this section, you will be introduced to approaches that can reduce the dimension of design optimization problems, which can be considered as an important aspect of modeling and problem formulation. Subsequently, we will explore other implications of the design space dimensionality. Specifically, we will learn how to design or plan a set of computational/physical experiments to analyze and model the behavior of the system (that is being optimized); and how this *design of experiments* is closely related to problem dimensionality (Ref. [8]).

15.3.1 Design Variable Linking

In certain optimization problems, when the number of design variables is large, it is possible to reduce the dimensionality through design variable linking [9]. Equality constraints are used to define this linking. The value of one design variable can be expressed in terms of one or more design variables. The behavior of symmetric components can be identical under certain circumstances, which may offer opportunities to reduce the number of independent design variables. This reduction of the number of design variables is generally intended to reduce the computational cost of optimization.

Assume that an optimization problem is defined by n-dimentional design variable vector x, where n is deemed large and computationally undesirable. It may be possible to redefine the optimization problem in terms of a new m-dimensional design variable vector $\tilde{x}$, where $m < n$, through a linear transformation defined by the constant matrix T. This transformation can be expressed as

$$x_{n \times 1} = T_{n \times m} \, \tilde{x}_{m \times 1}$$

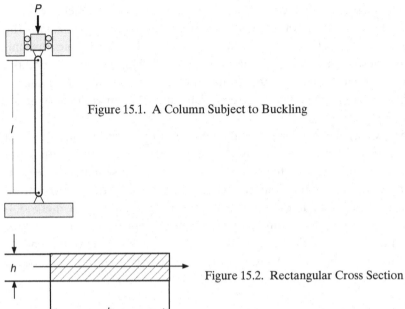

Figure 15.1. A Column Subject to Buckling

Figure 15.2. Rectangular Cross Section

For example, the reduction from 3 to 2 design variables can take the form

$$x_{3 \times 1} = \begin{bmatrix} 1 & 0 \\ 1 & 0 \\ 0 & 1 \end{bmatrix} \tilde{x}_{2 \times 1} \tag{15.1}$$

Lets look at an example where *design variable linking* is used.

Example: Buckling is a common failure mode when a structure is subjected to high compressive stresses. A vertical aluminum column is pinned at both ends, as shown in Fig. 15.1. The height of the column is $l = 0.5m$. Its cross section is depicted in Fig. 15.2. The value of h is a quarter of b. The column is subjected to a force, $P = 35.5N$, on its top. The Young's modulus of aluminum is 69 GPa at room temperature (21° C). Only buckling failure is considered in this optimization problem, where the objective is to minimize the volume of the column.

The area moment of inertia of a rectangle is given by

$$I_0 = \frac{bh^3}{12} \tag{15.2}$$

The critical force that the column can accept is given by

$$P_{cr} = \frac{\pi^2 E I_0}{(Kl)^2} \tag{15.3}$$

where P is the critical force, E is the modulus of elasticity, I_0 is the area moment of inertia, l is the unsupported length of the column, and K is the column effective length factor. For a column where both ends are pinned, the factor K is equal to 1. The optimization problem is formulated as

$$\min f(h, b) = lhb \tag{15.4}$$

subject to

$$h - 0.25b = 0 \tag{15.5}$$

$$P - P_{cr} \leq 0 \tag{15.6}$$

$$h, b \geq 0 \tag{15.7}$$

The above formulation of the optimization problem involves two design variables. Using design variable linking [9], we can reduce the number of design variables to one. Constraint (15.5) is used to eliminate h from the formulation. Therefore, the optimization problem is reduced to

$$\min f(h, b) = 0.25lb^2 \tag{15.8}$$

subject to

$$P - P_{cr} \leq 0 \tag{15.9}$$

$$h, b \geq 0 \tag{15.10}$$

The optimal solution to the above optimization problem is $b = 0.01m$ and $h = 0.0025m$. The volume is $1.25 \times 10^{-5}m^3$. Thus, in this problem, design variable linking [9] is used to reduce the number of design variables from two to one, thereby simplifying the optimization problem and likely saving computational costs.

15.3.2 Design of Experiments

Design of experiments (DoE) techniques were originally developed to study (and empirically model) the behavior of systems through physical experiments (*e.g.*, in experimental chemistry [10]). *DoE can be defined as the design of a controlled information gathering exercise for a system or phenomena where variation is known to be present.* DoE techniques have existed in some form since the late 1700s [11], and is considered a discipline with very broad applications across the different natural and social sciences, and engineering.

The primary objective of DoE is to determine multiple combinations of the controlled parameters (or conditions) at which the experiments will be conducted. In mathematical terms, each combination of controlled parameters can be considered a sample. As such, DoE can also be perceived as a process of generating parameter/condition samples to conduct controlled experiments. Although, traditionally controlled experiments referred only to physical experiments, in modern times, it

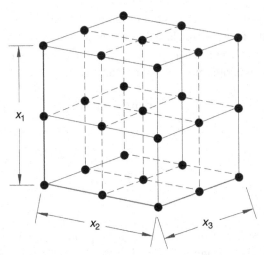

Figure 15.3. A 3-Level and 3-Dimensional Full Factorial Design (27 Points)

would include both physical and computational experiments (*i.e.*, simulation-based experiments).

Once experiments have been performed for each planned sample condition, the data acquired can be used to (i) investigate a theoretical hypothesis, (ii) analyze the system behavior (*e.g.*, sensitivity analysis), and/or (iii) develop empirical or surrogate models to represent the relationship between different system parameters. In this chapter, we will primarily focus on the third application of DoE, where the objective is to understand and model the relationship between "*the input parameters (that comprised the planned sample set)*," and "*the output parameters of interest (whose values were acquired through the experiment)*."

DoE techniques have a large influence on the accuracy of the surrogate model developed thereof. To develop effective surrogate models, it is necessary to acquire adequate information about the underlying system. Assuming no prior knowledge of the system behavior, the typical DoE strategy is to generate a distribution of sample points throughout the design space in an uniform fashion. There are several techniques available to distribute the sample points, to provide an adequate coverage of the design space without any particular variable bias. These techniques include Factorial design [12], Central Composite designs [13], and Latin Hypercybe design [14], and Sobol sequence [15].

The most straightforward approach to uniform sampling is the *factorial design* method. In this technique, the range of each design variable is divided into different levels between the upper and lower limits of a design space. In a *full factorial design*, sample points are located at all the combination of the different levels of all the design variables. Figure 15.3 illustrates a three level factorial design in a three-variable space. In high-dimensional problems, the full factorial design approach may be cost/time prohibitive. For example, in a 10 dimension problem, even a 2-level full factorial design would require as many as $2^{10} = 1,024$ sample points. If we assume that each experiment takes only 1 hr, 42 days would be required to run through the entire sample set.

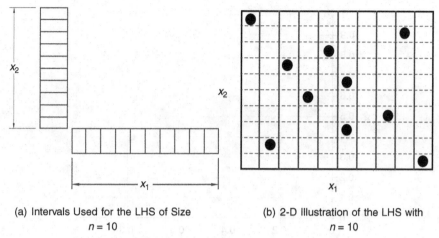

(a) Intervals Used for the LHS of Size
$n = 10$

(b) 2-D Illustration of the LHS with
$n = 10$

Figure 15.4. Latin Hypercube Sampling (LHS) with 10 Points

In these situations of excessive resources requirment, only a fraction of the sample points can be used for conducting experiments. Designs that use a fraction of the full factorial design sample points are called *fractional factorial design*. Central composite design (CCD) combines full or fractional factorial designs with additional points to allow the fitting of full quadratic polynomial models. The MATLAB functions available to perform full factorial design and fractional factorial design are `fullfract` and `fracfact`, respectively. A simple two-level factorial design can be performed using the function `ff2n` as follows.

```
% To generate a 2-level factorial design with 3 factors
dFF2 = ff2n(3)
dFF2 =
    0    0    0
    0    0    1
    0    1    0
    0    1    1
    1    0    0
    1    0    1
    1    1    0
    1    1    1
```

In a Latin Hypercube design (LHD) or Latin Hypercube Sampling (LHS), the range of each design variable is divided into n non-overlapping intervals with equal probability. A sample point is then located randomly on each interval of every design variable. Consider a simple example where we wish to generate a LHD of size $n = 10$ for two design variables. The ten intervals of the variable x_1 and x_2 in that case are presented in Fig. 15.4(a). The next step is to randomly select specific values for x_1 and x_2 in each of their ten intervals as shown in Fig. 15.4(b).

MATLAB provides a set of in-built functions to perform different types of Latin Hypercube Sampling (LHS). The most straight-forward implementation of LHS can

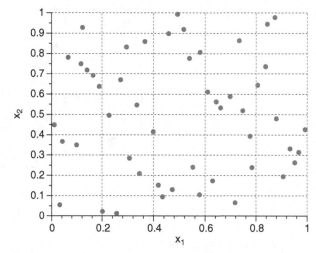

Figure 15.5. A LHS for a 2-Variable Problem, Generated Using MATLAB

be performed using lhsdesign. Direct implementation of this function generates a sample set in the range of $[0,1]$ for each variable. A two-variable implementation is shown below.

```
% To generate and plot a LHS set for two variables
x = lhsdesign(50,2);
plot(x(:,1),x(:,2))
```

The plot of the sample set generated by using the MATLAB LHS function is provided in Fig. 15.5. It is noted from Fig. 15.5 that the LHS provides uniform yet non-deterministic coverage of the multi-dimensional design space. Note that using such methods as LHS for *sample generation* is also helpful when creating the initial population of candidate designs in evolutionary algorithms, as we will see in a later chapter in this book.

In the next section, we show that DoE or effective sampling is the first step in the development of surrogate models [16]. This is because surrogate models are trained by using (i) the input data generated by the DoE and (ii) the output data generated by the experiments conducted under that DoE. Thus, the greater the number of samples or the more dense the coverage of the design space, the greater is the expected accuracy of the surrogate models trained with these samples.

Unfortunately, using a generously large number of samples contradicts the very purpose of developing surrogate models, which is to avoid the unreasonable cost of running too many expensive computational/physical experiments. To develop accurate surrogate models, performing a comprehensive set of simulations or experiments is desirable, but often unreasonable. We learned from the discussion of the different DoE methods, the size of the sample set is closely related to the dimension of the problem. Richard Bellman coined the term "curse of dimensionalit" to describe the rapid increase of the sample size resulting from an increase in the number of variables [17]. The DoE techniques that you learned in this section are uniquely helpful in planning effective sample sets for the surrogate modeling of systems of different design dimensions.

15.4 Surrogate Modeling

The need to quantify the economic and engineering performance of complex systems often demands highly complex and computationally/time-wise expensive simulations and experiments. The direct use of these computational simulations or experiments in optimization could be anywhere from challenging to prohibitive. Surrogate models are one of the most popular methodologies to deal with this issue. They provide a significantly less expensive and often more tractable alternative toward model-based optimization.

Surrogate modeling is concerned with the construction of purely mathematical models to estimate system performance or, in other words, to define relationships between specific system inputs and outputs. Over the past couple of decades, function estimation methods and approximation-based optimization have progressed remarkably. Surrogate models are being extensively used in the analysis and optimization of complex systems or in the solution of complex problems. Surrogate modeling techniques have been used for a variety of applications from multidisciplinary design optimization to the reduction of analysis time and the improvement of the tractability of complex analysis codes [1, 18]. Figure 15.6 illustrates the diverse applicability of surrogate modeling.

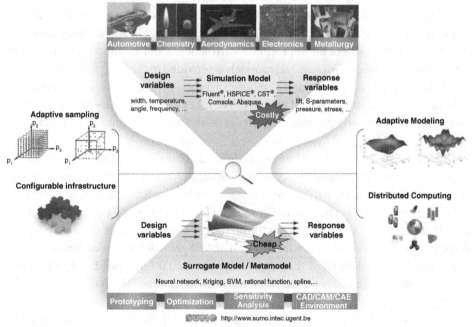

Figure 15.6. Applications of Surrogate Modeling (Graphics Courtesy SUMO)

The general surrogate modeling problem can be stated as follows: "Given a set of data points $x^i \in R^m, i = 1, \cdots, n_p$, and the corresponding function values, $f(x^i)$, obtain a global approximation function, $\tilde{f}(x)$, that adequately represents the original/actual functional relationship over a given design domain." In this section, you are provided an introduction to the overall surrogate modeling process, and a brief description of the major surrogate modeling methods.

15.4.1 Surrogate Modeling Process

The process of surrogate modeling generally involves three stages: (i) design of experiments (DoE); (ii) construction or training of the surrogate model; and (iii) model validation.

We have already learned about DoE and sampling in the previous section. However, it is important to realize that in some cases, it might not be practically feasible to control the DoE. This is especially true in practice, when the source of experimental data is significantly different from where the surrogate model is being constructed. For example, data from the literature, historical data from previously reported experiments or measurements, or data from commercial sources may need to be used to construct a surrogate model. In these cases, the user constructing the model does not have any direct control over the data generation. For example, Zhang et al. [19] used data from the National Renewable Energy Laboratory [20] to construct a wind farm cost model, where the levelized cost of wind farms was represented as a function of turbine features and reported labor costs.

The sample points used to construct the surrogate models are generally called *training points*, while the construction of the surrogate models is often called model training. Once the *training points* have been defined (through DoE or other sources), the next step is to select an appropriate surrogate model or functional form. An introductions to four major surrogate models is provided in the following subsections. These models are: (i) polynomial response surfaces, (ii) radial basis functions, (iii) Kriging, and (iv) artificial neural network. We take this opportunity to note that the quantification of surrogate modeling errors is an important current research area (see Refs. [21, 22]).

Once the surrogate model has been constructed, the final step is to evaluate the performance or expected accuracy of the surrogate model. The two most popular measures of model error are the *root mean squared error* (RMSE) and the *maximum absolute error* (MAE). The *root mean squared error* (RMSE) is a global error measure that provides an understanding of the model's accuracy over the entire design domain; while the *maximum absolute error* (MAE) provides an understanding of the maximum local deviations of the model from the actual function. The most prominent approaches used to estimate these error measures are [1]: (i) split sample, (ii) cross-validation, and (iii) bootstrapping.

In the split sample strategy, the sample data is divided into training and test points. The former is used to construct the surrogate; while the latter is used to test the performance of the surrogate. The cross-validation technique operates through the following five steps:

1. Splits the sample points randomly into q (approximately) equal subsets;
2. Removes each of these subsets in turn (one at a time);
3. Trains an intermediate surrogate model to the remaining $q - 1$ subsets;
4. Computes the error of the intermediate surrogate using the omitted subset; and
5. Once each one of the q subsets has been used as the omitted subset, the q sets of errors evaluated therein are generally aggregated to yield a global error measure.

The bootstrapping approach generates m subsamples from the sample points. Each subsample is a random sample with replacements from the full sample. Different variants of the bootstrapping approach can be used for (i) model identification, and (ii) identifying confidence intervals for surrogate models [1].

The four major surrogate models are discussed next.

15.4.2 Polynomial Response Surface Methodology

The Polynomial Response Surface (PRS) methodology is motivated by the Taylor series expansion [23]. A Taylor series generally requires an infinite number of terms to obtain the exact value of the real function. The approximation takes the form of a polynomial. The number of terms included in a PRS depends on the desired accuracy. We may seek a zero-th order (constant), a first order (linear), a second order (quadratic), or an even higher order approximation. The approximation is accurate in the neighborhood of a chosen point, and becomes progressively inaccurate as we move away from that point. Depending on the form of a real function, the availability of training data, and users' accuracy requirement, PRS may be selected as a linear, second-order, or higher-order polynomial of the vector of the design variables. Low-order polynomials, defined in a small region of the variable space, are generally used in practice when PRS is the model of choice.

The Kth-order PRS of a single variable x has the following form.

$$\tilde{f}(x) = a_0 + \sum_{k=1}^{K} a_k x^k \tag{15.11}$$

where $a_0, a_1, \ldots, a_K$ are arbitrary coefficients to be determined by training.

In n-dimensional space, variable x has n components: x_j, where $j = 1, \cdots, n$. The linear PRS of n-dimensional variable x has the following form.

$$\tilde{f}(x) = a_0 + \sum_{j=1}^{n} a_j x_j \tag{15.12}$$

where the generic a_j are arbitrary coefficients to be determined by training.

The most popular PRS is the the 2nd-order PRS, or the Quadratic Response Surface (QRS). The QRS of an n-dimensional variable x has the following form.

$$\tilde{f}(x) = a_0 + \sum_{j=1}^{n} a_j x_j + \sum_{j=1}^{n} \sum_{i=j}^{n} a_{ji} x_j x_i \tag{15.13}$$

where the generic a_j and a_{ji} are arbitrary coefficients to be determined by training.

The generalized kth-order PRS of an n-dimensional variable x can be represented as

$$\tilde{f}(x) = a_0 + \sum_{j=1}^{n} a_j x_j + \sum_{j_1=1}^{n} \sum_{j_2=j_1}^{n} a_{j_1 j_2} x_{(j_1)} x_{(j_2)} + \cdots$$

$$+ \sum_{j_1=1}^{n} \sum_{j_2=j_1}^{n} \cdots \sum_{j_k=j_{k-1}}^{n} a_{j_1 j_2 \cdots j_k} x_{(j_1)} x_{(j_2)} \cdots x_{(j_k)} \tag{15.14}$$

where the generic a_j to $a_{j_1 j_2 \cdots j_k}$ are arbitrary coefficients to be determined by training.

The PRS methodology is frequently used in regression analyses. In regression analyses, the number of training points is generally greater than that of the unknown coefficients, a_i. Only lower order PRS are used in practice. Thus, the resulting PRS does not necessarily pass through the training sample data (*i.e.*, does not necessarily have a zero error at the training points). This is why PRS and other regression functions are often called approximation functions. One of the approaches used to evaluate the unknown coefficients (a) is the least squares method. The least squares method solves a regression as an optimization problem. The overall solution minimizes the sum of the squares of the errors between the real function values and the corresponding estimated PRS values, where the summation is taken over all the training points. The following example illustrates how to estimate the parameters of a QRS for a one-dimensional problem using the least squares method.

Example: The following four training points and their corresponding real function values are given as: $f(x_1 = 2) = 4.2$, $f(x_2 = 7.1) = 8.7$, $f(x_3 = 4) = 14.8$, and $f(x_4 = 5) = 11.1$. Using the least squares method, fit a quadratic response surface (QRS), $\tilde{f}(x)$ of the single variable x. Find the maximum value of the fitted function, which is the approximated maximum value of the real function. The second-order response surface of one variable is given as

$$\tilde{f}(x) = a_0 + a_1 x + a_2 x^2 \tag{15.15}$$

The parameters, a_0, a_1, and a_2, need to be determined using the least squares method. In order to solve the least squares method, the following unconstrained optimization problem is formulated.

$$\min_{a_0, a_1, a_2} h(a_0, a_1, a_2) = \sum_{j=1}^{4} \left(\tilde{f}(x_j) - f(x_j) \right)^2 \tag{15.16}$$

The solution to the optimization problem is $a_0 = -12.10$, $a_1 = 10.06$, and $a_2 = -1.05$. The mean squared error is estimated to be 2.38.

To find the maximum value of the fitted function, we need to solve the following optimization problem.

$$\max_x \tilde{f}(x) = a_0 + a_1 x + a_2 x^2 \qquad (15.17)$$

The optimal point is $x = 4.78$, and the corresponding maximum function value is 11.93.

The process to fit PRS for practical optimization problems is the same as described in the example. First, the least squares method or other methods are used to fit an appropriate PRS based to the training data. The accuracy of the PRS is generally given by the RMSE error.

For a one-dimensional problem, the in-built MATLAB function, polyfit, can also be used to fit a PRS of a desired order. The process of fitting a QRS for the above example using polyfit is shown below.

```
% To fit a 1-D QRS to a set of four training points
x = [2 3.5 4 5];
y = [4.2 7.1 14.8 11.1];
P = polyfit(x,y,2)
P =
-1.0533 10.0627 -12.1013
```

Note that the coefficients of the PRS, as estimated and displayed by polyfit, appear from the highest order term to the lowest order term (*i.e.*, from a_2 to a_0). It is also observed that the QRS trained using polyfit is similar to that trained using the least squares method.

15.4.3 Radial Basis Function Method

The idea of using *Radial Basis Functions* (RBFs) as approximation functions was first proposed by Hardy [24] in 1971, where he used multiquadric RBFs to fit irregular topographical data. Since then, RBFs has been used for numerous applications that require a global representation of multidimensional scattered data [25, 26, 27].

Radial Basis Function (RBF) expresses surrogate models as linear combinations of a particular type of basis function ($\psi(r)$), where each constituent basis function is defined in terms of the Euclidean distance (r) between a training point and the point of evaluation. The Euclidean distance (r) can be expressed as

$$r = \|x - x_i\| \qquad (15.18)$$

where x_i is the ith training point, and x is the point of evaluation.

The commonly used non-parametric basis functions are:

1. Linear: $\psi(r) = r$,
2. Cubic: $\psi(r) = r^3$, and
3. Thin plate spline: $\psi(r) = r^2 \ln r$.

The commonly used parametric basis functions are:

1. Gaussian: $\psi(r) = e^{-r^2/(2\delta^2)}$,
2. Multiquadric: $\psi(r) = (r^2 + \delta^2)^{1/2}$, and
3. Inverse multiquadric: $\psi(r) = (r^2 + \delta^2)^{-1/2}$.

The RBF model is then expressed as a linear combination of the basis functions across all the training points, $x_i \in R^n, i = 1, \cdots, m$, as given by

$$\tilde{h}(x) = \sum_{i=1}^{m} w_i \psi(\|x - x_i\|) \tag{15.19}$$

where w_i are the generic weights of the basis functions (to be determined by training). The standard process of estimating the weights (w_i) is described below.

The weights, w_i, are evaluated using all the training points x_i and their corresponding function values $f(x_i)$. In the evaluation of w_i, Ψ is used to represent the matrix of the basis function values at the training points, as given by

$$\Psi = \begin{pmatrix} \psi(\|x_1 - x_1\|) & \psi(\|x_1 - x_2\|) & \cdots & \psi(\|x_1 - x_m\|) \\ \psi(\|x_2 - x_1\|) & \psi(\|x_2 - x_2\|) & \cdots & \psi(\|x_2 - x_m\|) \\ \vdots & \vdots & \ddots & \vdots \\ \psi(\|x_m - x_1\|) & \psi(\|x_m - x_2\|) & \cdots & \psi(\|x_m - x_m\|) \end{pmatrix} \tag{15.20}$$

The vector of the weights is W.

$$W = \begin{pmatrix} w_1 \\ w_2 \\ \vdots \\ w_m \end{pmatrix} \tag{15.21}$$

The vector Y has the function values at the training points ($f(x_i)$).

$$Y = \begin{pmatrix} f(x_1) \\ f(x_2) \\ \vdots \\ f(x_m) \end{pmatrix} \tag{15.22}$$

The weights w_i are then evaluated by solving the following matrix equation.

$$\Psi W = Y \tag{15.23}$$

It is important to keep in mind that RBFs are essentially interpolating functions (*i.e.*, a trained RBF model will pass through all the training points). This property enables RBFs to represent highly nonlinear data, which is otherwise often challenging to accomplish using low order PRSs.

The training date for the example in Sec. 15.4.2 is used to illustrate the process of constructing a RBF model.

Example: The training points and their corresponding real function values are $f(x_1 = 2) = 4.2$, $f(x_2 = 3.5) = 7.1$, $f(x_3 = 4) = 14.8$, and $f(x_4 = 5) = 11.1$. Fit a RBF model $\tilde{f}(x)$ of a single variable x. Find the maximum value of the fitted function in the range [2,5], which will give the approximated maximum value of the real function.

In this example, the multiquadric basis function, $\psi(r) = (r^2 + \delta^2)^{1/2}$, is used. The parameter, δ, is set to 0.9..

The matrix Ψ of the basis function values at training points, and the vectors Y and W are given below.

$$\Psi = \begin{pmatrix} \psi(\|x_1 - x_1\|) & \psi(\|x_1 - x_2\|) & \psi(\|x_1 - x_3\|) & \psi(\|x_1 - x_4\|) \\ \psi(\|x_2 - x_1\|) & \psi(\|x_2 - x_2\|) & \psi(\|x_2 - x_3\|) & \psi(\|x_2 - x_4\|) \\ \psi(\|x_3 - x_1\|) & \psi(\|x_3 - x_2\|) & \psi(\|x_3 - x_3\|) & \psi(\|x_3 - x_4\|) \\ \psi(\|x_4 - x_1\|) & \psi(\|x_4 - x_2\|) & \psi(\|x_4 - x_3\|) & \psi(\|x_4 - x_4\|) \end{pmatrix} \tag{15.24}$$

$$= \begin{pmatrix} (0^2 + 0.9^2)^{1/2} & (1.5^2 + 0.9^2)^{1/2} & (2.0^2 + 0.9^2)^{1/2} & (3.0^2 + 0.9^2)^{1/2} \\ (1.5^2 + 0.9^2)^{1/2} & (0^2 + 0.01^2)^{1/2} & (0.5^2 + 0.9^2)^{1/2} & (1.5^2 + 0.9^2)^{1/2} \\ (2.0^2 + 0.9^2)^{1/2} & (0.5^2 + 0.9^2)^{1/2} & (0^2 + 0.9^2)^{1/2} & (1.0^2 + 0.9^2)^{1/2} \\ (3.0^2 + 0.9^2)^{1/2} & (1.5^2 + 0.9^2)^{1/2} & (1.0^2 + 0.9^2)^{1/2} & (0^2 + 0.9^2)^{1/2} \end{pmatrix} \tag{15.25}$$

$$Y = \begin{pmatrix} 4.2 \\ 7.1 \\ 14.8 \\ 11.1 \end{pmatrix} \tag{15.26}$$

and

$$W = \begin{pmatrix} w_1 \\ w_2 \\ w_3 \\ w_4 \end{pmatrix} \tag{15.27}$$

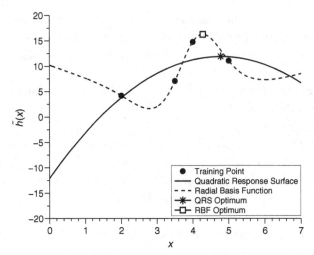

Figure 15.7. Fitted Function

By solving the following matrix equation for the weights

$$\Psi W = Y \tag{15.28}$$

we obtain $w_1 = -4.947$, $w_2 = 66.437$, $w_3 = -82.098$, and $w_4 = 23.144$.
To find the maximum of the RBF function, we solve the following optimization problem.

$$\max_{x} \tilde{h}(x) = \sum_{l=1}^{4} w_l \psi(\|x - x_i\|) \tag{15.29}$$

The optimal point is found to be $x = 4.28$, and the corresponding maximum function value is 16.29.

Figure 15.7 shows the plot of the fitted function in the range [0,7]. The fitted function is unimodal in the range [2,5]. However, it is multimodal when we consider the entire range from 0 to 7. As an interpolation method, RBFs are highly capable of representing such multimodal data. However, on the negative side, as an interpolating function, the accuracy of RBFs is highly unreliable beyond the range of the training data.

Although the maximum function values obtained using the PRS model and the RBF model are similar, they are meaningfully different. This observation indicates that the choice of surrogate models can have an important impact on the optimal solutions obtained.

There exists an advanced version of RBF, called the Extended Radial Basis Functions or E-RBF [28]. The E-RBF is essentially a combination of radial and non-radial basis functions, where *non-radial basis functions* (N-RBFs) are defined in terms of the individual coordinates of generic points x, relative to a given data point x^i, in each dimension separately. Further description of the E-RBF surrogate model can be found in the paper by Mullur and Messac [28].

15.4.4 Kriging Method

Another popular surrogate model is Kriging, which is most commonly (but not restricted to be) used as an interpolating model. Kriging [29, 30] is an approach to approximate irregular data. The Kriging approximation function consists of two components: (i) a global trend function, and (ii) a deviation function representing the departure from the trend function. The trend function is generally a polynomial (*e.g.*, constant, linear, or quadratic). The general form of the Kriging surrogate model is given by [31]:

$$\tilde{f}(x) = G(x) + Z(x) \tag{15.30}$$

where $\tilde{f}(x)$ is the unknown function of interest, $G(x)$ is the known approximation (usually polynomial) function, and $Z(x)$ is the realization of a stochastic process with the mean equal to zero, and a nonzero covariance. The $i, j - th$ element of the covariance matrix of $Z(x)$ is given as

$$COV[Z(x^i), Z(x^j)] = \sigma_z^2 R_{ij} \tag{15.31}$$

where R_{ij} is the correlation function between the ith and the jth data points, and σ_z^2 is the process variance. Further description of the Kriging model can be found in [31].

15.4.5 Artificial Neural Networks (ANN)

A more recent and increasingly popular choice of surrogate model is the Artificial Neural Network (ANN). A neural network generally contains an input layer, one or more hidden layers, and an output layer (Ref. [32]). Figure 15.8 shows a typical three-layer feedforward neural network. An ANN is developed by defining the following three types of parameters:

1. The interconnection pattern between different layers of neurons;
2. The learning process for updating the weights of the interconnections; and
3. The activation function that converts a neuron's weighted input to its output activation.

   One of the drawbacks associated with neural networks for function approximation is the fairly large number of parameters that need to be prescribed by the user, thereby demanding adequate user experience in implementing ANN. These prescribed parameters include the number of neurons, the number of layers, the type of activation function, and the optimization algorithm used to train the network. In addition, the training process generally needs to be supervised in order to avoid "over-fitting" [33]. MATLAB provides a dedicated toolbox for ANN, which can be started using the command `nnstart` (Ref. [34]).

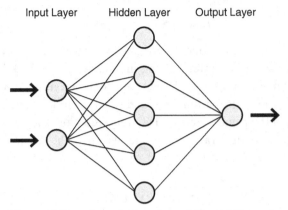

Figure 15.8. A Generic Topology of Neural Networks

15.5 Summary

This chapter provided a brief overview of the modeling issues encountered in the process of optimizing complex and/or large dimensional systems. The chapter introduced methods for reducing the dimension of the design space in the course of modeling (*e.g.*, design variable linking), and methods for addressing empirical modeling and analysis in the design space (*e.g.*, design of experiments). This was followed by the development of a special class of mathematical models that provide an approximation of the system behavior, by leveraging an affordable and carefully designed set of expensive experiments (simulation-based or physical experiments). These models are known as surrogate models, of which four major types were presented. They are used to provide quick/computationally benign approximations of complex system behavior for the purpose of optimizing the system. The methodologies and modeling approaches provided in this chapter are crucial in overcoming the barrier of computational expense and complexity in designing current and next generation complex systems, such as aircraft, ground vehicles, and wind turbines.

15.6 Problems

15.1   The 9 training points are (2,3), (4,3), (6,3), (2,5), (4,5), (6,5), (2,7), (4,7), and (6,7). Their corresponding function values are 12.12, 5.97, 11.98, 8.04, 2.18, 7.97, 12.03, 5.97, and 12. Use the training data to fit a second-order polynomial response surface of two variables. Find the minimum value of the fitted function.
The second-order response surface has the following form.

$$\tilde{f}(x) = a_0 + a_1 x^{(1)} + a_2 x^{(2)} + a_{11}(x^{(1)})^2 + a_{22}(x^{(2)})^2 + a_{12}x^{(1)}x^{(2)}. \quad (15.32)$$

15.2   For Problem 15.1, fit RBF functions using the multiquadric basis $\psi(r) = (r^2 + \delta^2)^{1/2}$. The value of δ can be set to 0.01. Find the minimum point in the

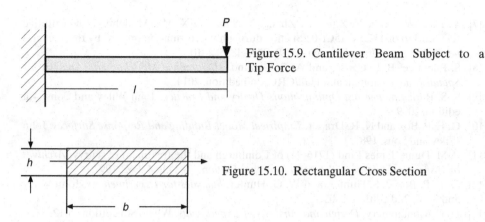

Figure 15.9. Cantilever Beam Subject to a Tip Force

Figure 15.10. Rectangular Cross Section

region spanned by the training points. Plot the figure of the fitted function in a region slightly larger than that surrounded by the training points.

15.3 The 9 training points are (2,3), (4,3), (6,3), (2,5), (4,5), (6,5), (2,7), (4,7), and (6,7). Their corresponding function values are -12.12, -5.97, -11.98, -8.04, -2.18, -7.97, -12.03, -5.97, and -12. Fit RBF functions using the multiquadric basis $\psi(r) = (r^2 + \delta^2)^{1/2}$. The value of δ can be set to 0.01. Find the maximum point in the region spanned by the training points. Plot the figure of the fitted function in a region slightly larger than that surrounded by the training points.

15.4 The aluminum cantilever beam shown in Fig. 15.9 is subjected to a tip force, $P = 35.5N$. The length of the column is $l = 0.5m$. Its cross section is shown in Fig. 15.10. The Young's modulus of aluminum is 69 GPa at room temperature ($21°$ C). Minimize the volume of the beam. Reduce the minimization problem using design variable linking.

BIBLIOGRAPHY OF CHAPTER 15

[1] N. V. Queipo, R. T. Haftka, W. Shyy, T. Goel, R. Vaidyanathan, and P. K. Tucker. Surrogate-based analysis and optimization. *Progress in Aerospace Sciences*, 41(1):1–28, January 2005.

[2] R. N. Cadete, J. P. Dias, and M. S. Pereira. Optimization in vehicle crashworthiness design using surrogate models. In *6th World Congresses of Structural and Multidisciplinary Optimization*, Rio de Janeiro, Brazil, June 2005.

[3] H. S. Chung and J. J. Alonso. Using gradients to construct coKriging approximation models for high-dimensional design optimization problems. In *40th AIAA Aerospace Sciences Meeting and Exhibit*, Reno, NV, USA, January 2002.

[4] O. Sigmund. Design of multiphysics actuators using topology optimization – Part I: One-Material structures. *Computer Methods in Applied Mechanics and Engineering*, 190(49-50):6577–6604, 2001.

[5] J. Annoni, P. Seiler, K. Johnson, P. J. Fleming, and P. Gebraad. Evaluating wake models for wind farm control. In *American Control Conference*, Portland, OR, USA, June 2014.

[6] S. Chowdhury, J. Zhang, A. Messac, and L. Castillo. Optimizing the arrangement and the selection of turbines for wind farms subject to varying wind conditions. *Renewable Energy*, 52:273–282, 2013.

[7] Y. Han, W. Zeng, Y. Zhao, X. Zhang, Y. Sun, and X. Ma. Modeling of constitutive relationship of Ti-25V-15Cr-0.2Si alloy during hot deformation process by fuzzy-neural network. *Materials and Design*, 31(9):4380–4385, 2010.

[8] S. Banerjee, B. P. Carlin, and A. E. Gelfand. *Hierarchical Modeling and Analysis for Spatial Data*. Chapman and Hall/CRC, 2nd edition, 2014.

[9] S. S. Rao. *Engineering Optimization: Theory and Practice*. John Wiley and Sons, 4th edition, 2009.

[10] G. E. P. Box and N. R. Draper. *Empirical Model-Building and Response Surfaces*. John Wiley and Sons, 1987.

[11] P. M. Dunn. James Lind (1716-94) of Edinburgh and the treatment of scurvy. *Archives of Disease in Childhood-Fetal and Neonatal Edition*, 76(1):F64–F65, 1997.

[12] G. E. P. Box, J. S. Hunter, and W. G. Hunter. *Statistics for Experimenters*. John Wiley and Sons, 2nd edition, 2005.

[13] D. C. Montgomery. *Design and Analysis of Experiments*. Wiley, 8th edition, 2012.

[14] M. D. McKay, R. J. Beckman, and W. J. Conover. Comparison of three methods for selecting values of input variables in the analysis of output from a computer code. *Technometrics*, 21(2):239–245, 1979.

[15] T. J. Santner, B. J. Williams, and W. I. Notz. *The design and analysis of computer experiments*. Springer, 2003.

[16] A. I. J. Forrester, A. Sóbester, and A. J. Keane. *Engineering Design via Surrogate Modelling: A Practical Guide*. John Wiley adn Sons, 2008.

[17] R. E. Bellman. *Adaptive Control Processes: A Guided Tour*. Princeton University Press, 1961.

[18] G. Wang and S. Shan. Review of metamodeling techniques in support of engineering design optimization. *Journal of Mechanical Design*, 129(4):370–380, 2007.

[19] J. Zhang, S. Chowdhury, A. Messac, and L. Castillo. A response surface-based cost model for wind farm design. *Energy Policy*, 42:538–550, 2012.

[20] M. Goldberg. *Jobs and Economic Development Impact (JEDI) Model*. National Renewable Energy Laboratory, Golden, Colorado, US, October 2009.

[21] J. Zhang, S. Chowdhury, A. Mehmani, and A. Messac. Characterizing uncertainty attributable to surrogate models. *Journal of Mechanical Design*, 136(3):031004, 2014.

[22] A. Mehmani, S. Chowdhury, J. Zhang, W. Tong, and A. Messac. Quantifying regional error in surrogates by modeling its relationship with sample density. In *54th AIAA/ASME/ASCE/AHS/ASC Structures, Structural Dynamics and Materials Conference*, Boston, MA, USA, 2013.

[23] A. Ravindran, G. V. Reklaitis, and K. M. Ragsdell. *Engineering Optimization: Methods and Applications*. John Wiley and Sons, 2006.

[24] R. L. Hardy. Multiquadric equations of topography and other irregular surfaces. *Journal of Geophysical Research*, 76(8):1905–1915, 1971.

[25] R. Jin, W. Chen, and T. W. Simpson. Comparative studies of metamodelling techniques under multiple modelling criteria. *Structural and Multidisciplinary Optimization*, 23(1):1–13, 2001.

[26] J. B. Cherrie, R. K. Beatson, and G. N. Newsam. Fast evaluation of radial basis functions: Methods for generalized multiquadrics in r^n. *SIAM Journal of Scientific Computing*, 23(5):1549–1571, 2002.

[27] M. F. Hussain, R. R. Barton, and S. B. Joshi. Metamodeling: Radial basis functions, versus polynomials. *European Journal of Operational Research*, 138(1):142–154, 2002.

[28] A. A. Mullur and A. Messac. Extended radial basis functions: More flexible and effective metamodeling. *AIAA Journal*, 43(6):1306–1315, 2005.

[29] A. A. Giunta and L.T. Watson. A comparison of approximation modeling techniques: Polynomial versus interpolating models. In *7th AIAA/USAF/NASA/ISSMO Symposium on Multidisciplinary Analysis and Optimization*, St. Louis, MO, USA, September 1998.

[30] S. Sakata, F. Ashida, and M. Zako. Structural optimization using Kriging approximation. *Computer Methods in Applied Mechanics and Engineering*, 192(7-8):923–939, 2003.

[31] N. Cressie and C. K. Wikle. *Statistics for Spatio-Temporal Data*. John Wiley and Sons, 2011.

[32] J. Han and M. Kamber. *Data Mining: Concepts and Techniques*. Morgan Kaufmann, 3rd edition, 2011.

[33] A. A. Mullur. *A New Response Surface Paradigm for Computationally Efficient Multiobjective Optimization*. PhD thesis, Mechanical Engineering, Rensselaer Polytechnic Institute, Troy, NY, December 2005.

[34] The MathWorks, Inc. MATLAB neural network toolbox. `www.mathworks.com/products/neural-network/`, 2014.

16

Design Optimization Under Uncertainty

16.1 Overview

The previous chapters presented *deterministic* optimization methods. These do not implicitly take into account the inherent uncertainties typically present in the design process, in the system being designed, or in the models describing the system behavior. Uncertainties emanates from myriad sources. These include imperfect manufacturing processes and material properties, fluctuating loading conditions, over-simplified engineering models, or uncertain operating environment. All of these may have a direct impact on the system performance in its use or in the market place. To obtain a *reliable* and *robust* design (these terms are defined later), these uncertainties must be considered as part of the design process.

In the past, empirical safety factors were often used to guard against engineering design failure [1]. Safety factors resulted in overly conservative designs, increasing the probability that businesses may lose their competitive edge in terms of cost and performance. More recently, design optimization methods under uncertainty have gained broad recognition. These approaches explicitly consider uncertainties of various forms, and search for designs that are insensitive to variations or uncertainty to a significant extent.

This chapter introduces the concept of design optimization under uncertainty, and discusses the pertinent popular approaches available. Section 16.2 defines a generic example that is used throughout the book to facilitate the presentation of the material. The next Section (Sec. 16.3) defines five generic components/STEPS involved in design under uncertainty. This is followed by the presentation of these STEPS in the following five Sections: (1) uncertainty types identification (Sec. 16.4), (2) uncertainty quantification (Sec. 16.5), (3) uncertainty propagation (Sec. 16.6), (4) formulation of optimization under uncertainty (Sec. 16.7), and (5) results analysis (Sec. 16.8). Section 16.9 briefly discussed other popular methods. A chapter summary is provided in Sec 16.10.

16.2 Chapter Example

This section defines a generic example that it used throughout the chapter to illustrate the various concepts involved in design under uncertainty. Consider the optimization of a two bar truss shown in Fig. 16.1. The objective is to identify the optimal position of node P from the left, b, and the optimal cross sectional areas for the bars, a_1 and a_2, such that the squared displacement at the node P is minimized. The loads W_1 and W_2 are applied as shown in Fig. 16.1. The deterministic optimization problem statement for the truss example is given as follows.

$$\min_{a_1,a_2,b,\theta,\beta} \quad J = u_1^2 + u_2^2 \tag{16.1}$$

subject to

$$g_i \equiv \frac{S_i}{S_{\max}} - 1 \leq 0 \tag{16.2}$$

$$h_1 \equiv \left\{\theta - \arctan\left(\frac{L}{b}\right)\right\} = 0 \tag{16.3}$$

$$h_2 \equiv \left\{\beta - \arctan\left(\frac{L}{2L - b}\right)\right\} = 0 \tag{16.4}$$

$$h_3 \equiv \left\{\left(\frac{a_1 L}{\sin \theta} + \frac{a_2 L}{\sin \beta}\right)/4,000 - 1\right\} = 0 \tag{16.5}$$

$$a_{i\min} \leq a_i \leq a_{i\max} \qquad i = (1,2) \tag{16.6}$$

$$b_{\min} \leq b \leq b_{\max} \tag{16.7}$$

The variables u_1 and u_2 are the horizontal and vertical deflections at node P, respectively; S_i is the normal stress induced in each bar; $S_{\max}$ is the maximum allowable stress; θ is the angle between the horizontal and Bar 1 (defined by h_1); and β is the

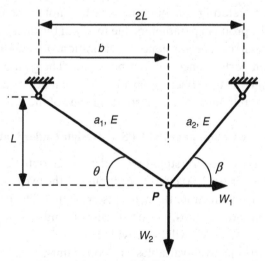

Figure 16.1. Two-Bar Truss Example

angle between the horizontal and Bar 2 (defined by h_2). The volume of the truss is required to be equal to 4,000 in^3 (defined by h_3). The above constraints are presented in a normalized form. The fixed parameters for this problem are the Young's Modulus, $E = 29 \times 10^3$ ksi; the truss height, $L = 60$ ft; the maximum allowable stress, $S_{max} = 350$ ksi; and the loads, $W_1 = 100$ kips, and $W_2 = 1,000$ kips. Additionally, $a_{imin} = 0.8$ in^2, and $a_{imax} = 3.0$ in^2 for $i = (1,2)$; $b_{min} = 30$ ft; and $b_{max} = 90$ ft. The expressions for the stresses are given as

$$S_1 = \frac{W_1 \sin\beta + W_2 \cos\beta}{a_1(\cos\theta\sin\beta + \sin\theta\cos\beta)} \tag{16.8}$$

$$S_2 = \frac{-W_1 \sin\theta + W_2 \cos\theta}{a_2(\cos\theta\sin\beta + \sin\theta\cos\beta)} \tag{16.9}$$

The deterministic optimum of this problem (using an initial guess $[(a_{1min} + a_{1max})/2, (a_{2min}+a_{2max})/2, (b_{min}+b_{max})/2, 0.1, 0.1]$) is: $J = 187.5$ in^2, $a_1 = 2.55$, $a_2 = 1.5$, $b = 3600$ in.

Discussion: The above example is used to illustrate why the consideration of uncertainty is important.

The deterministic problem presented does not take into account the various uncertainties that are present in the problem. For example, in the cross sectional areas of the bars, the loads applied are subject to random variations. The material properties, such as the Young's modulus, and the maximum allowable stress could be uncertain. The engineering model we have assumed for the analysis may also be over-simplified.

Given these uncertainties, optimizing the truss in a deterministic fashion as shown above may result in design failure due to a possible stress constraint violation (Eq. 16.2) under uncertainty. The design might not be *reliable*. In addition, the deterministically optimized design may not consistently perform as intended under uncertainty (the objective function is viewed as a measure of the performance of the design). Since deterministic optimization does not consider uncertainties, a small variation in the problem parameters due to any of the uncertainties mentioned above may lead to significant performance deterioration, or even catastrophic failure. The resulting design performance will not be *robust*. The goal of design optimization under uncertainty is to explicitly consider the influence of uncertainty, and to improve the design's reliability and robustness in view of these uncertainties.

16.3 Generic Components/STEPS of Design Under Uncertainty

This chapter presents five generic steps involved in design optimization under uncertainty, as descrived in Fig. 16.2. At every step discussed, the two bar truss example will be used to quantitatively illustrate the concepts presented. The popular approaches used for each of the above steps will be discussed, and pertinent references for theoretical and practical details will be provided.

Given the design objectives, variables, and constraints, a typical design optimization problem under uncertainty involves the following five steps.

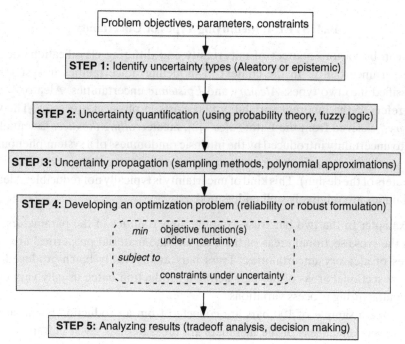

Figure 16.2. Overview of Design Optimization Under Uncertainty

1. **STEP 1: Identifying Types of Uncertainty:** What types of uncertainties exist in the problem (*e.g.*, uncertainties in parameters, in design variables, or in the model itself)? We will study this issue in Sec. 16.4 with examples.

2. **STEP 2: Uncertainty Quantification:** How do we model the uncertainty mathematically (*e.g.*, using probability theory)? How can the uncertainty be defined in terms of a set of parameters (*e.g.*, mean and standard deviation)? This step is discussed in more detail in Sec. 16.5 with examples.

3. **STEP 3: Uncertainty Propagation:** How do we propagate the uncertainty in the models (*e.g.*, given the uncertainty parameters in the cross sectional areas in the truss problem, how do we calculate the uncertainty parameters of the stress)? Sec. 16.6 covers this topic in detail with examples.

4. **STEP 4: Development of an Optimization Formulation:** How do we incorporate the uncertainty into the optimization problem? We will study this issue in Sec. 16.7 with examples.

5. **STEP 5: Analyzing Results:** How do we interpret the results? What are the tradeoffs involved? Sec. 16.8 presents more details concerning this topic with examples.

The reader should keep in mind that each of the above steps are evolving areas of active research. What follows is an introductory presentation of the popular techniques available in the literature, illustrated with the two bar truss example.

16.4 STEP 1: Identifying Types of Uncertainty

There can be various sources of uncertainty and differing classifications of these sources of uncertainty. In the context of modeling, uncertainties in systems can be classified into two types: *Aleatory* and *Epistemic* uncertainties. Aleatory uncertainty refers to the inherent variability that exists in physical processes. The word *"aleatory,"* derived from the Latin *"alea"* [2], means *rolling the dice*. Essentially, it refers to uncertainty introduced by the intrinsic randomness of a system/phenomena. It is sometimes also referred to as parametric uncertainty (*i.e.*, uncertainty in the parameters of the design). This kind of uncertainty is typically not reducible. Aleatory uncertainties are relatively well understood.

> **Example:** In the two bar truss example, uncertainties in the parameters, such as the cross sectional areas of the bars and the material properties, are examples of aleatory uncertainties. Truss bars are typically batch-produced. The cross sectional areas of the bars across a production batch usually vary due to manufacturing process variations.
>
> Assume a sample of 100 bars are collected from a production batch, and the cross sectional areas of the members are measured, as given in Table 16.1. In deterministic optimization, the tolerances in the areas are not considered. In most design optimization methods under uncertainty, a statistical measure of the variation in the data (Table 16.1) is used in the optimization (*e.g.*, mean and standard deviation). The data in Table 16.1 is also provided in the book website (`www.cambridge.org/Messac`).
>
> The mean of the observed set of values in Table 16.1 is $0.995\,\mathrm{in}^2$, and the standard deviation is $0.096\,\mathrm{in}^2$. The MATLAB commands `mean` and `std` can be used to find the mean and standard deviation of a data set, as shown below.

```
% Using the data given in Table 1.1, define the
% variable "set" as a vector of 100 elements.
m = mean(set);
s = std(set);
```

> As we will see later, the mean and standard deviation of uncertain quantities are used in optimization under uncertainty.

Epistemic uncertainty [3] arises because of the following related factors: (1) lack of knowledge in the quantity, environment, and/or physical process being modeled, (2) insufficient data, (3) over-simplification of complex coupled physical phenomena, and (4) lack of knowledge of the possible failure modes of the design. The word *"epistemic,"* derived from the Greek *"episteme,"* means knowledge [2]. Thus, it refers to uncertainty introduced by the lack of knowledge of a system/phenomena. Epistemic uncertainty is also commonly referred to as modeling uncertainty, and is usually more difficult to model than aleatory uncertainty. This kind of uncertainty can be reduced by developing a better understanding of the involved phenomena, (*e.g.*, by conducting more experiments).

Table 16.1. *Sample Set of One Hundred Cross-Sectional Areas for a Batch-Produced Truss Bar (in^2)*

0.943	0.966	0.903	0.978	1.023	1.142	1.115	1.121	0.893	0.959
1.229	0.857	1.066	1.167	1.088	1.116	0.947	0.906	0.948	1.110
0.879	0.929	1.133	1.072	0.940	0.988	0.966	0.816	0.867	1.192
1.026	1.015	1.063	0.751	0.819	1.033	1.142	1.148	0.921	1.024
0.917	0.956	0.935	1.071	0.937	1.094	1.006	1.071	0.958	1.054
0.998	0.960	0.921	0.969	0.955	0.931	0.971	0.820	0.871	1.050
1.037	1.060	1.060	0.853	1.028	0.978	0.954	1.072	0.959	0.880
1.096	1.099	0.997	1.074	0.925	1.136	1.017	0.919	0.849	1.047
1.038	1.028	1.161	1.013	0.910	1.053	0.998	1.029	1.159	0.929
0.849	0.875	0.966	0.951	1.011	0.947	0.989	1.045	1.035	0.802

Example: In the two bar truss problem, epistemic uncertainty could arise from the fact that we have included only compressive failure mode for the bars. We ignored other possible failure modes, (*e.g.*, buckling). In other words, we have over-simplified the failure analysis for the problem. This is typically viewed as a modeling or epistemic uncertainty. In a complex design problem, such as aircraft design, the designer might not even be fully aware of the modes of failure that have been ignored, which might adversely impact the reliability and robustness. For the truss problem, the modeling uncertainty in the stresses in the bars, S_i {i=1,2}, can be represented by considering a multiplicative term, S_i^*, and an additive term, $\tilde{S}_i$, as shown below.

$$S_1 = \left[\frac{W_1 \sin \beta + W_2 \cos \beta}{a_1(\cos \theta \sin \beta + \sin \theta \cos \beta)} \right] S_1^* + \tilde{S}_1 \qquad (16.10)$$

$$S_2 = \left[\frac{-W_1 \sin \theta + W_2 \cos \theta}{a_2(\cos \theta \sin \beta + \sin \theta \cos \beta)} \right] S_2^* + \tilde{S}_2 \qquad (16.11)$$

The additive and multiplicative terms could be functions of the design variables with problem-specific definitions. The additive term, $\tilde{S}_i$, could be the higher order term of an expansion, such as a Taylor series. In keeping with the scope of this introductory chapter, we will henceforth use $\tilde{S}_i = 0$ and $S_i^* = 1$.

Engineering design often involves both types of uncertainties. As a result, it is sometimes challenging to classify a source of uncertainty (at the modeling stage) as exclusively aleatory or epistemic. In recent years, interesting models and methodologies have been developed to address systems with mixed aleatory-epistemic uncertainties (simultaneous presence of both uncertainties) [4, 5, 6]. The report by Eldred and Swiler [5] provides a helpful review of these methods and the corresponding benchmark results. Once the uncertainties in a problem have been identified, the next step is to quantify those uncertainties.

16.5 STEP 2: Uncertainty Quantification

Thus far, we have not defined how to model uncertainty in a mathematical form. The question of interest in this section is: how to parameterize uncertainty? That is, how to describe uncertainty using a set of parameters? These uncertainty parameters will become part of the optimization problem in various forms. Quantifying aleatory uncertainties will be discussed. A detailed discission of epistemic uncertainties is beyond the scope of this introductory chapter (see Ref. [7]).

This section presents uncertainty quantification methods that are based on the availability of data for a particular problem. What is meant by "data"? If a quantity of interest is uncertain or random, data refers to a set of possible measured outcomes or samples of the quantity. This data may be available from a manufacturer's catalog or recorded information from past history. Given that data may or may not always be available, uncertainty quantification in two different cases will be studied, as follows.

16.5.1 Sufficient Data Available: Probability Theory

To mathematically represent a random quantity, probability theory can be used when sufficient data is available. Basic knowledge of probability theory by the reader is assumed [8] in this chapter.

In probability theory, a *random variable* is associated with the outcome of an uncertain event. For example, the cross sectional areas of the truss, a_1 and a_2 (in Fig. 16.1), are random variables, since they are batch-produced and have associated tolerances. In this chapter, random variables are denoted by upper case letters, A_1 and A_2. A probability density function (PDF), denoted by $f_X(x)$, is used to compute the probability that a continuous random variable X lies between two limits x_1 and x_2, as given below.

$$P(x_1 < X < x_2) = \int_{x_1}^{x_2} f_X(x)dx \qquad (16.12)$$

There are several standard PDF types that are used in engineering applications, such as Uniform, Normal, Weibull, and Lognormal distributions [8]. Normal distributions are often used for design optimization problems. The two quantities, mean and standard deviation, define the bell-shape of the normal distribution depicted in Fig. 10.4. The shaded area in Fig. 10.4 represents the probability that a random variable lies within one standard deviation of the mean value.

When sufficient data is available, techniques such as histogram or probability plots may be used to explore the underlying distribution [9]. While more than one standard distribution may "fit" a data set, the underlying physical quantity may suggest a particular distribution. For example, normal or lognormal distributions are used to represent physical dimensions and material properties. Weibull distribution is commonly used in reliability problems to model quantities such as time to failure. Statistical tests, such as the Chi-square test or Kolmogorov-Smirnov test, can be used to determine the goodness of a PDF fit [9].

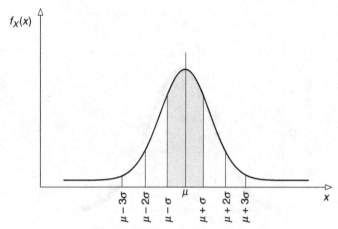

Figure 16.3. Normal Distribution

Once a PDF is fit to a data set, the uncertainty parameters that are of interest are usually the first two moments of the PDF, namely, mean and standard deviation.

Example: In the two bar truss problem, the cross sectional areas could be modeled as normal random variables. Material properties are usually given as lognormal variables, since most of them are positive quantities.

For the given data set of cross sectional areas (Table 16.1) for the two bar truss problem, find the underlying distribution using the two MATLAB commands: histfit and normplot. The histfit command plots the histogram of the given set, while normplot command plots the normal probability plot for the given data.

```
% Using the data given in Table 16.1, define the
% variable "set" as a vector of 100 elements.
histfit(set);
figure(2)              % Open a new figure
normplot(set);
```

Figure 16.4 presents the pertinent plots for the data set given in Table 16.1. Note that a data set is considered normally distributed if its normal probability plot is approximately linear, as is the case in Fig. 16.4. Therefore, the cross sectional area can be assumed to be normally distributed. The shape of the histogram further validates the normality assumption.

Once the assumed distribution of the data is tested and deemed satisfactory, the uncertainty parameters of interest are the mean and the standard deviation of the PDF. For the given data for the cross sectional areas, the mean and the standard deviation are 0.9947 and 0.0961 in^2, respectively.

Thus far, we discussed the case where the data set for the cross sectional areas was provided to us. How can we quantify uncertainty when sufficient data is not available? That is the next topic of our discussion.

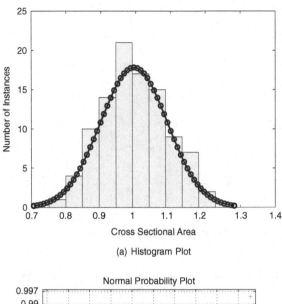

(a) Histogram Plot

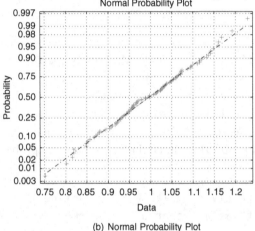

(b) Normal Probability Plot

Figure 16.4. Finding the Underlying Distribution of the Given Cross-Sectional Area Data Set

16.5.2 Insufficient Data: Non-Probabilistic Methods

As previously discussed, probability theory for uncertainty quantification can be used when sufficient data is available. However, the data sufficiency requirement is not always satisfied (*e.g.*, during early conceptual design when sufficient data is not available). Gunawan and Papalambros [10] propose the notion that insufficient data has become a major bottle-neck in engineering analysis involving uncertainty. For these cases, the use of evidence theory [11, 12, 13], possibility theory [14, 15, 16, 17, 18], Bayes theory [19, 20, 21], and imprecise probabilities is an emerging trend [22, 17, 23]. Evidence theory uses fuzzy measures called plausibility and belief to measure the likelihood of events. These fuzzy measures are the upper and lower bounds on the probability of an event. Plausibility and belief measures are, respectively, associated with evidence theory and the classical probability theory. We note that there is some controversy associated with these non-probabilistic approaches.

In recent years, other methods have also evolved to address insufficient data scenarios. A combination of Evidence Theory and the Bayesian approach was suggested by Zhou and Mourelatos [24] to deal with insufficient data. Wang et al. [25] presented a new paradigm in system reliability prediction that allowed the use of evolving, insufficient, and subjective data sets. To deal with these data sets, a combination of probability encoding methods and Bayesian updating mechanism was used.

How to quantify the basic uncertainties (*i.e.*, uncertainties in design variables and design parameters) has been discussed. The next step is to compute the uncertainty parameters of the functions of random variables (*i.e.*, constraints and objectives in the optimization problem).

16.6 STEP 3: Uncertainty Propagation

The goal of uncertainty propagation is to compute the uncertainty characteristics of a function of random variables (known as a *random function*), given the uncertainty characteristics of the variables (known as *input variables*) present in those functions (see Ref. [26]). The uncertainty characteristics of interest could be the moments of the function, or the probability of failure of the function.

The random function of interest may be a given linear or nonlinear function of the constituent random variables, or may be a black-box function with no explicit/provided functional form. In some cases, where the random function is given, it may be possible to analytically compute the moments of a function. When this is not possible, other methods to propagate the input uncertainties must be employed. This section introduces some popular uncertainty propagation methods in the literature. From an optimization perspective, uncertainty propagation is an important step that is required for objective and constraint function formulation.

This section introduces four popular approaches for uncertainty propagation: (i) sampling methods, (ii) analytical methods, such as the First Order Reliability Method (FORM) and the Second Order Reliability Method (SORM), (iii) polynomial approximation using Taylor series, and (iv) advanced methods, such as stochastic expansion. Illustrative examples are provided for discussion.

> **Example:** Let's revisit the two bar truss problem. Once the uncertainties in the variables have been quantified, (*e.g.*, cross sectional areas), we then need to calculate the uncertainty parameters of the quantities that are functions of the areas (*e.g.*, compressive stresses, S_1 and S_2). These estimated parameters are needed for the optimization process.

16.6.1 Sampling Methods

Sampling methods are used to generate a set of sample points for the input variables as per their uncertainty distributions. At each generated sample, the values of the random functions are computed, and a set of sample points of the function values are subsequently generated. This generated function sample set can then be used

to compute statistics of interest for the random function. An illustrative example
is provided shortly after the important practical issues are discussed. More details
regarding this topic of sampling methods in optimization can be found in [27].

The first issue of importance in sampling methods is how to distribute the sample
set for the input random variables. Should they be uniformly distributed (uniform
sampling), randomly distributed (Monte Carlo sampling), or should we concentrate
the samples in a desired region of importance (importance sampling)? Other popular sampling techniques include stratified sampling and Latin Hypercube sampling.
Monte Carlo and Latin Hypercube sampling are very commonly used in the optimization community. The selection of a sampling scheme is dependent on the level
of computational resources available, the acceptable error in the estimated parameter, and also on the nature of the data. The focus will be on Monte Carlo sampling
techniques.

The second issue of importance in sampling methods is *how many input samples are needed* (*e.g.*, is 10,000 samples sufficient? or is 10^6 sufficient?) This number
depends on the accuracy required in the quantity being estimated. How are these
sampling considerations important in the present context of optimization under
uncertainty?

For this case, the goal of the sampling method is to estimate failure probabilities
and/or moments of the random function. The failure probability can be estimated
from a Monte Carlo simulation as follows. Generate N input random samples, and
compute the corresponding constraint values. Then, identify those N_f instances out
of N that violate the constraint feasibility. The probability of failure can then be
estimated as $\frac{N_f}{N}$.

The failure probability in engineering problems of interest may be as low as
10^{-6}. To observe at least one failure in a Monte Carlo simulation for such a case, the
sample set should have at least 10^6 simulations. The number of samples N should
be chosen to be at least one order of magnitude higher than 10^6. Several problems
at the end of the chapter are provided to illustrate the issue of the number of input
samples.

The advantage of sampling methods is that (i) they are relatively accurate, and (ii)
the pertinent (sampling) errors are usually quantifiable. However, sampling methods, especially when used in optimization, can be computationally expensive. Many
mathematical computational software packages have built-in functions to generate
random numbers using standard distributions. As shown in the following example,
MATLAB has built-in functions to generate normal random variables.

Example: Assume for the truss problem that the cross sectional area, A_1, of the
truss is normally distributed with a mean value of 1.6 in and a standard deviation
of 0.05 in. Use the MATLAB normrnd function to generate 10,000 instances of
the variable A_1.

```
% To generate a vector of 10,000 X 1
% normal random variables
A1 = normrnd(1.6,0.05,10000,1)
```

At each generated instance, compute the value of S_1 (using Eq. 16.8). Assume for this particular case that b and L are deterministic. In addition, assume that $\theta = 30°, \beta = 60°$, yielding $S_1 = \frac{\sqrt{3}W_1 + W_2}{2A_1}$. Using this expression, obtain a set of 10,000 stress values, that is used to compute the uncertainty parameters of the stress. The generated data set can be used to evaluate the probability of failure of the constraint $S_1 < S_{max}$.

From the results of MATLAB, the probability of failure ranges from approximately 0.92 to 0.94. If the program is run multiple times, the resulting estimated probability of failure may change from run to run because of the random nature of the simulation. The amount of variation in the failure probability value depends on the number of simulation samples considered. As the number of simulation samples increases, the variation in the failure probability from run to run decreases. This point is further illustrated through the problems at the end of the chapter. The high failure probability obtained above is a function of the input design variable uncertainties. The uncertainty-based optimization helps find input uncertainty parameters that yield acceptable failure probabilities in a systematic manner.

16.6.2 First-Order and Second-Order Reliability Methods (FORM and SORM)

The First Order Reliability Method (FORM) and the Second Order Reliability Method (SORM) are more popularly used as major components of the Reliability-based Design Optimization (RBDO) architecture (see Rozvany and Maute in Ref. [28]). However, they can also be leveraged to estimate uncertainty propagation. Zaman et al. [29] reported that if the uncertainty described by intervals can be converted to a probabilistic format, well established probabilistic methods of uncertainty propagation, such as the Monte Carlo methods [30] and the optimization-based methods (FORM and SORM), can be readily used. The application of probabilistic methods for uncertainty propagation will avoid the computational expense of interval analysis, as it allows for the treatment of aleatory and epistemic uncertainties together without nesting [29]. Du [31] also proposed a unified uncertainty analysis method based on the FORM, where aleatory uncertainty and epistemic uncertainty were modeled using probability theory and evidence theory, respectively. Further discussion of the FORM and the SORM can be found in Sec. 16.7.1.

16.6.3 Polynomial Approximation Using Taylor Series

Using this approach, the mean and standard deviation of the random function are approximated by first building a polynomial approximation of the function, followed by finding the moments of this polynomial approximation. A linear or quadratic polynomial approximation of the random function is constructed using a Taylor series expansion about a point of interest, usually the mean value vector.

Unlike sampling methods, this approach is not commonly used to estimate probability of failures because of its limited accuracy, especially for non-normal cases. It is more commonly used in robust design optimization formulations. The mean and the variance of a function of random variables, $g(X)$, can be approximated using the first-order Taylor series [9], given as

$$\mu_g = g(\mu_X) \tag{16.13}$$

$$\text{Var}[g] = \sum_{i=1}^{n_x} \left[\frac{\partial g}{\partial X_i} \bigg|_{X=\mu_X} \sigma_{X_i} \right]^2$$

$$+ \sum_{i=1}^{n_x} \sum_{j=1, i \neq j}^{n_x} \frac{\partial g}{\partial X_i} \bigg|_{X=\mu_X} \frac{\partial g}{\partial X_j} \bigg|_{X=\mu_X} \text{Cov}(X_i, X_j) \tag{16.14}$$

where n_x is the number of elements in the input random vector X, μ_X and σ_X denote the vectors of the mean and standard deviation, respectively, of X; and $\text{Cov}(X_i, X_j)$ denotes the covariance between the variables X_i and X_j, with $i, j = \{1, ..., n_x\}$. Covariance is a measure of correlation between two sets of random variables [8].

The standard deviation of g can then be computed as $\sigma_g = \sqrt{\text{Var}[g]}$. If the design variables are assumed to be statistically independent, the covariance between them is zero, and the variance expression given in Eq. 16.14 reduces to

$$\text{Var}[g] = \sum_{i=1}^{n_x} \left[\frac{\partial g}{\partial X_i} \bigg|_{X=\mu_X} \sigma_{X_i} \right]^2 \tag{16.15}$$

Using the above method to estimate moments is quite simple, as it uses a linear approximation of the random function to approximate its moments. In the case of black box functions, where it is not possible to compute the analytical partial derivatives in the above equations, finite differences may be used instead. A careful choice of step size for the finite differences is necessary. This issue is further explored in Chapter 7.

Example: Consider the expression for the stress, $S_1 = \frac{\sqrt{3}W_1 + W_2}{2A_1}$, from the previous subsection. Assuming that A_1 is the input random variable with a mean and a standard deviation of $\mu_{A_1} = 1$ in^2 and $\sigma_{A_1} = 0.1$ in^2, respectively; the mean and standard deviation of S_1 can be estimated as

$$\mu_{S_1} = \frac{\sqrt{3}W_1 + W_2}{2\mu_{A_1}} = 586.6 \text{ ksi} \tag{16.16}$$

$$\sigma_{S_1} = \left[\frac{\sqrt{3}W_1 + W_2}{2\mu_{A_1}^2} \right] \sigma_{A_1} = 58.6 \text{ ksi} \tag{16.17}$$

16.6.4 Advanced Methods Overview

The previous subsection reviewed the Taylor series method that approximates the moments of a random function by taking the moments of a polynomial approximation. The First-order Taylor series approximation is commonly used, since it is simple to implement and requires only gradient information. However, the method can yield erroneous results in some cases [32], and must be applied carefully. A first-order Taylor series approximation builds a linear approximation of the random function to compute its approximate moments. The approximation is valid only in the neighborhood of the linearization point, and may not be suitable for highly nonlinear functions.

One could also use a second order Taylor series approximation of the mean and the variance of a random function, which may yield higher accuracy when compared to the first order approximation at the cost of high computational expense.

Another method to compute moments is to approximate the distribution with respect to which the moments are computed [33]. Use of advanced methods, such as polynomial chaos [34] for uncertainty propagation, is also popular.

16.6.5 An Important Note on Uncertainty: Analysis vs. Optimization

Thus far, we have discussed the details of uncertainty quantification and propagation. These two steps together are commonly referred to as *uncertainty analysis* or, in some cases, *probabilistic analysis* or *stochastic analysis*. Note that uncertainty analysis alone is a complex and computationally expensive task, and may be the end goal of some engineering studies. These studies are performed to estimate the reliability or performance robustness of a system, and may involve complex physical phenomena modeled by computationally expensive simulation tools.

Design optimization problems under uncertainty can be viewed as one level above uncertainty analysis, where each iteration of optimization usually requires uncertainty analysis of the objective and constraint functions. Optimization adds another layer of complexity to these problems.

Why then optimize these problems? As was discussed with the truss example previously, each set of input design variables yields a particular failure probability of the design. Ideally, the design will have the least possible constraint violation. Given a set of constraints in the problem, optimization helps automate the search for the least constraint violation for all constraints, while simultaneously minimizing the objectives. Without optimization, uncertainty analysis alone will merely allow us to *explore* the system's probabilistic behavior, but without the means to favorably impact it.

16.7 STEP 4: Embedding Uncertainty into an Optimization Framework

This step considers how the optimization problem under uncertainty is posed mathematically, and how the concepts studied thus far can be used in an optimization

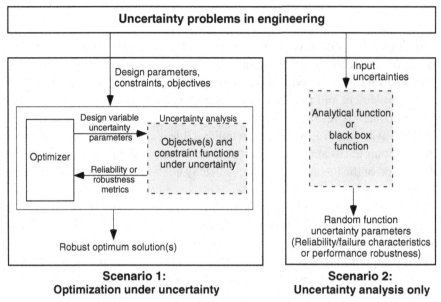

Figure 16.5. Uncertainty Problems in Engineering Design

framework. Probabilistic methods, such as robust design optimization (RDO) and reliability based design optimization (RBDO) techniques, are popular approaches that include the effects of uncertainty in an optimization framework. First, a brief overview of the pertinent issues is provided; illustrative examples are given later.

There are two primary challenges that designers face in optimization problems under uncertainty. The first challenge is to ensure that the design does not fail under uncertainty. This involves controlling the constraints' failure probabilities by repeating the constraint uncertainty analysis for different values of the design variables, as illustrated in Fig. 16.5. One of the primary differences between RBDO and RDO techniques is the approach used for uncertainty analysis for constraints.

The second challenge in optimization problems under uncertainty is to maintain robust performance of the system design. Consider the design of an air conditioning system that is required to maintain an ambient temperature of 27° C in a room. Several conditions are uncertain in this design (*e.g.*, the outside temperature). In spite of the uncertainty, the design of the air conditioning system must have the ability to maintain the desired ambient temperature in the room (*i.e.*, the design should remain *optimal*). Moreover, it is required that the temperature of the room not significantly fluctuate under uncertainty. In other words, the performance of the design must be *robust*. A robust design is one that is minimally sensitive to uncertain conditions. The second important difference between the RBDO and RDO approaches is that RDO methods typically consider *robustness* metrics as one of the objectives to be minimized, while RBDO approaches typically do not.

The above two challenges (maintaining feasibility and robustness) are generally conflicting. Because of the computational complexity involved, most current research in this area tends to focus either (1) on rigorous constraint formulations (RBDO approaches) alone, with simplified problem architecture, or (2) on a rigorous problem architecture alone, with a simplified constraint structure.

Example: The two bar truss problem is used to explore the above discussion in concrete terms. What is the goal of the optimization problem under uncertainty? Recall that in the deterministic formulation, the goal was to minimize the squared nodal displacement. In the uncertainty problem, this quantity is now random and has its own PDF, which is not a normal distribution. We use the uncertainty propagation tools previously discussed to determine the approximate mean and standard deviations of the squared displacement. More pertinent quantitative details are discussed in Sec. 16.7.3.

The objective of the robust problem can then be posed in terms of minimizing the mean of the squared displacement (*optimality*), or the standard deviation of the squared displacement (*robustness*), or both. In addition, the constraints must be suitably posed. The constraint can be posed in terms of the probability of constraint violation (RBDO approach), or by shifting the constraint boundary by a suitable amount (RDO approach).

We examine the RBDO methods first, followed by the RDO method.

16.7.1 Reliability-Based Design Optimization (RBDO)

The development in this field is primarily derived from the structural reliability community [9]. In this field of research, significant progress has been made over the last decade. This section provides a high level summary of the ideas involved in this area of research. Interested readers are directed to the references for details.

The RBDO approach emphasizes high reliability of the design by ensuring a desired probability of constraint satisfaction. The mean of a desired performance metric is usually used as the objective function for RBDO problems. A general formulation for the constraint $g(X) \leq 0$ in RBDO approach can be given as

$$P\{g(X) \leq 0\} \geq R \qquad (16.18)$$

where R is the desired reliability of the constraint. The probability of failure in Eq. 16.18 is given by the following integral.

$$P\{g(X) \leq 0\} = \int \ldots \int_{g(X) \leq 0} f_X(x_1, x_2, \ldots x_n) dx_1 dx_2 \ldots dx_n \qquad (16.19)$$

where $f_X(x_1, x_2, \ldots x_n)$ is the joint probability density function of the n random variables $\{X_1, \ldots X_n\}$. In practice, the joint probability density function of the design variables is almost impossible to obtain. Even if it can be obtained, evaluating the multiple integral in Eq. 16.19 is difficult. Analytical approximations of the integral that yield the probability of failure are typically used: the First-Order Reliability Method (FORM) and the Second-Order Reliability Method (SORM) [9].

The inequality constraint $g(X) \leq 0$ is usually called the limit state equation. The limit state equation can be a linear or a nonlinear function of the design variables. The FORM method can be used when the limit state equation is a linear function of uncorrelated normal random variables, or is represented as a linear approximation of equivalent normal variables. The SORM method estimates the probability of failure using a second order approximation of the limit state.

The concept of the Most-Probable Point (MPP) is used to approximate the multiple integral [35]. The MPP is usually a point in the design space that is at the minimum distance from the origin (closest to failure), after performing requisite coordinate transformations. In the case of nonlinear constraints, the computation of the MPP distance from the origin is an optimization problem. The overall RBDO problem is a nested optimization problem, where the computation of the probability of failure at each iteration itself is an optimization problem. Several computationally efficient techniques, such as sequential methods [36], and hybrid methods [37], have been developed for efficient implementation of the nested ("double loop") RBDO formulations.

It is important to note that there exists multiple optimization architectures within RBDO (*e.g.*, "double loop" and "single loop" approaches). The more traditional approach is the nested or the "double loop" approach as discussed above. In this approach, each iteration step of the design optimization process requires a loop of iterations of the reliability analysis. Two popular "double loop" approaches are the *reliability index approach* (RIA) [38] and the *performance measure approach* (PMA) [39, 37]. These approaches apply FORM to perform the reliability analysis, which requires an inner nonlinear constrained optimization loop. When the constraints are active, the two approaches yield similar results. However, in the literature, PMA has been reported to be more efficient and stable than RIA [39, 37]. When the reliability analysis (comprising the inner loop) searches for the MPP, the overall "double loop" computation can become prohibitive, particularly if the concerned function evaluation is computationally expensive [40, 41, 42].

In order to reduce the computational burden of RBDO, several approximate RBDO approaches have been developed that decouple the double loop problem. These approaches are popularly known as "single loop" approaches. A list of "single loop" approaches is summarized in the paper by Nguyen et al. [43]. One of the major "single loop" approaches uses the Karush Kuhn Tucker (KKT) optimality condition to approximate the solution of the inner-loop optimization [44]. The inner-loop is replaced by a deterministic constraint, which transforms a double-loop RBDO problem into an equivalent single-loop optimization problem. More advanced single loop approaches have been proposed in recent years [45, 43].

16.7.2 Use of Approximation Methods Under Uncertainty

Incorporating reliability evaluations within an optimization framework can be computationally prohibitive. This problem is compounded in the presence of computationally expensive limit states, where each evaluation of the function could take hours or possibly days. To alleviate this computational burden, a popular approach is to use approximation methods, such as response surfaces and Kriging, in optimization problems under uncertainty. The approximation could be done at an optimization level, at the uncertainty propagation level, or both. More details can be found in the reference [46].

> **Example:** The two bar truss problem that has been considered thus far has a simple failure surface, which is the normal stress equation. Instead, if the stresses were computed using a finite element analysis, the computational expense would significantly increase. At each iteration of the optimization problem, an uncertainty propagation for the finite element code would be performed; which, in itself, is a computationally expensive task. Approximation methods can significantly reduce computational costs.

16.7.3 Robust Design Optimization (RDO)

As previously mentioned, the focus in RDO methods is to minimize the effects of uncertainty on product performance [47, 48, 49, 50, 51, 52, 53, 54, 55]. While the constraint handling in RDO problems is typically simpler than that in RBDO problems, RDO has been used in problems in a multidisciplinary and multiobjective setting. The challenges in these problems and the pertinent formulation methods are discussed in this section. An illustrative example is provided after the theoretical developments are discussed.

To reduce the computational burden associated with probabilistic constraints, a more simplistic approach known as the moment matching approach is widely used in RDO [56, 48]. The moment matching approach employs moments of the constraints and objectives in the optimization framework, unlike the computationally expensive probabilistic formulation in the RBDO approach. For inequality constraints, a worst-case formulation is usually used, where the constraints are shifted so that the worst case uncertainty still results in a feasible design [48] (see Fig. 16.6).

Equality constraints need careful consideration when formulated under uncertainty because of the strictness associated in their feasibility. An equality constraint can be classified into two types in a robust problem[57, 58]: (1) Type I: those that are always exactly satisfied (*e.g.*, laws of nature, such as static and dynamic equilibrium), and (2) Type II: those that are not always exactly satisfied (*e.g.*, designer-imposed dimensional constraints, such as Eq. 16.5).

Classification of constraints into the above two types depends on the nature of the design variables present in them, and/or the designer preferences [58]. Some of the existing equality constraint formulations are [58]: (1) TYPE I constraints: elimination of the constraint by substituting for a dependent variable, or satisfy the constraint at its mean value, and (2) TYPE II constraints: satisfy the constraint as

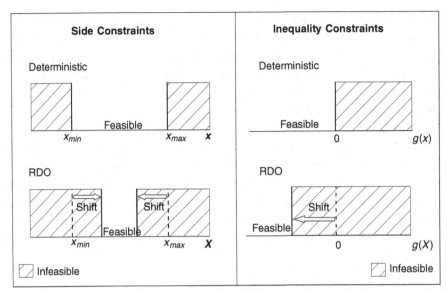

Figure 16.6. Inequality Constraints: Deterministic vs. Robust

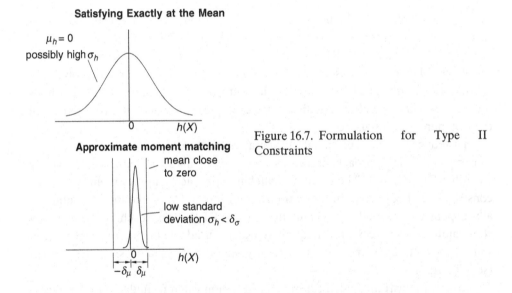

Figure 16.7. Formulation for Type II Constraints

closely as possible using the approximate moment matching method [58], or satisfy exactly at its mean value (see Fig. 16.7). An illustrative example is presented shortly.

The objectives of the RDO problem are usually to optimize the mean of the objective function, to minimize the standard deviation of the objective function, or both. The standard deviation of the objective function can be estimated using a first order Taylor series expansion. The RDO mathematical formulation is presented using the two bar truss example.

Example: Recall the deterministic truss formulation presented earlier. Using the discussion provided in this section regarding inequality and equality constraints, objective function, and design variables, the following RDO formulation can be obtained (explained in detail next).

$$\min_{\mu_{A_1}, \mu_{A_2}, \mu_B} \{\mu_J, \sigma_J\} \tag{16.20}$$

subject to

$$g_i + 3\sigma_{g_i} \leq 0 \tag{16.21}$$

$$\mu_{h_3} = 0 \tag{16.22}$$

$$A_{i\min} + 3\sigma_{A_i} \leq A_i \leq A_{i\max} - 3\sigma_{A_i}, \quad i = (1,2) \tag{16.23}$$

$$B_{\min} + 3\sigma_B \leq B \leq B_{\max} - 3\sigma_B \tag{16.24}$$

1. **Design variables:** Assume that a_1, a_2, and b are uncertain normal random variables. The corresponding random design variables are denoted by A_1, A_2, and B. Assume $\sigma_{a_i} = 0.005$ in and $\sigma_b = 6$ in.

2. **Inequality constraints:** The inequality constraints, S_1 and S_2, and the design variable bounds are shifted by three respective standard deviations. This makes the constraints more conservative by shrinking the feasible design space, as illustrated in Fig. 16.6. The mean and the standard deviation of the inequality constraints can be computed using the uncertainty propagation methods discussed in Sec. 16.6.

3. **Equality constraints:** Constraints h_1 and h_2 of the deterministic problem are connectivity constraints at the node P, which must be satisfied. These are TYPE I constraints. Eliminate the constraints h_1 and h_2 by substituting for θ and β. Constraint h_3 is a dimensional constraint, which restricts the structural volume of the truss to be equal to $4,000\,\text{in}^3$. This is a TYPE II constraint. We could either satisfy the constraint exactly at its mean value, or use the approximate moment matching method. We chose the first option because of its simplicity.

4. **Objectives:** Consider minimization of the mean and the standard deviation of the squared nodal displacement, μ_J and σ_J, which can be calculated using a first order Taylor series expansion.

5. **Solutions:** Note that the deterministic single objective problem (Eq. 16.1) has now become multiobjective under uncertainty (Eq. 16.20). Solve this multiobjective problem and obtain its Pareto set (see Fig. 16.8(a)). The above formulation presents an interesting tradeoff situation, since a Pareto set of robust solutions is now available to choose from – as opposed to a single deterministic solution. Each of the Pareto solutions represents a different tradeoff between the μ_J and σ_J objectives.

6. **Observation:** Obtaining the partial derivatives for the above expressions for the first order Taylor series approximation can be tedious. The MATLAB

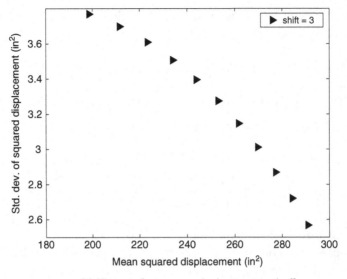

(a) Mean performance and robustness tradeoff

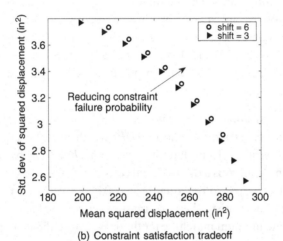

(b) Constraint satisfaction tradeoff

Figure 16.8. Two-Bar Truss Results

Symbolic Toolbox was used in this case to ease the burden. In addition, note that the objectives are of different magnitudes, and scaling may be needed.

Next is the final step of the uncertainty-based optimizations as shown in Fig. 16.2. Once we have the solutions, how do we interpret them?

16.8 STEP 5: How to Analyze the Results

This section presents a series of issues that are important in analyzing the results. These include: (1)mean performance and robustness tradeoff, (2) deterministic vs. robust solution, (3) constraint tradeoffs, (4) final design choice, and (5) multiobjective problems under uncertainty: decision making problem.

16.8.1 Mean Performance and Robustness Trade-off

The mean performance and the variation in performance are usually conflicting, and a tradeoff decision must be made. Several researchers have studied how to model this tradeoff [47, 48] using multiobjective formulations. Constraint satisfaction also affects these tradeoff decisions.

Example: Figure 16.8(a) presents the Pareto frontier for the two bar truss RDO problem, which reflects the tradeoff between the mean performance and the robustness objectives. The Pareto frontier provided in Fig. 16.8(a) has been generated using the normalized normal constraint method [59], which is described in detail in Chapter 17. Use of the weighted sum method for this problem cannot generate the complete representation of the Pareto frontier.

Next, we compare the robust solutions with the deterministic solution.

16.8.2 Deterministic vs. Robust Solutions

The deterministic optimization does not consider uncertainty explicitly, and not a good choice for the final design when uncertainty is important. If the deterministic design was chosen as the final design, a small change in the design variable values will likely push the constraints into the infeasible region, or yield a higher standard deviation for the objective function. Instead, uncertainty-based methods take the uncertainty into account to minimize constraint violations.

Example: Assume that the deterministic optimum is used for the truss design. Now assume that the design variables are uncertain, with $\sigma_{a_i} = 0.005$ and $\sigma_b = 6$ in. Assume the deterministic optimum to be the mean design variable vector, and use the above standard deviation values for the design variables. Compute the constraint violation for the S_1 constraint using a Monte Carlo simulation. Since the deterministic solution did not take uncertainty into account, the failure probability for the S_1 constraint is 0.53, which is a constraint violation.

When the violation of the S_1 constraint *about the robust optima* is computed, the failure probabilities range from 0.0008 to 0.0025. This is one of the primary advantages of optimization under uncertainty. Better constraint feasibility is obtained by accounting for the uncertainties beforehand.

Now examine how the robustness of the objective function compares for the deterministic vs. RDO cases. As above, use the deterministic optimum for the mean design variable values and the above prescribed standard deviation values. If the robustness metric for this scenario, σ_J, is computed, a value of $\sigma_J = 3.8$ in^2 will be obtained. This implies that if the truss is designed deterministically, and there are uncertainties in the design variables as given above, there will be important changes in the deflection of the node P. However, if the robust formulation is used, the output σ_J is explicitly minimized, thereby ensuring a robust design. Note that the least σ_J for the RDO formulation along the Pareto frontier is approximately 2.5 in^2, which is more desirable than the deterministic scenario (see Fig. 16.8(a)).

The next issue of interest is the tradeoffs that arise between constraint satisfaction and objective minimization.

16.8.3 Constraint Trade-offs

The RDO formulation usually formulates the inequality constraints by shifting them (*i.e.*, making them stricter than does the deterministic formulation). While doing so can reduce the constraint violations under uncertainty, it also usually makes the RDO mean objective worse than the deterministic objective. The higher the desired constraint satisfaction, the worse the RDO mean objective. Similarly, for equality constraints, the constraint satisfaction under uncertainty can be increased only if a deteriorated mean performance is acceptable. The approximate moment matching method is generally more suitable for exploring equality constraint tradeoffs. Engineering problems display this kind of tradeoff between constraint satisfaction and objective minimization.

> **Example:** For the two bar truss RDO formulation, note that as you increase the inequality constraint shift from three to six standard deviations, you will observe that the estimated failure probability for the S_1 constraint is reduced to zero along the Pareto frontier. However, this improved failure probability also results in the worsening of the μ_J and/or σ_J, as shown by the hollow circles in Fig. 16.8(b). As the shift increases, the Pareto frontier shifts into the northeast region of the design space. This is an interesting tradeoff in optimization problems under uncertainty.

16.8.4 Final Design Choice

The various tradeoff studies and trends regarding constraint satisfaction and robustness metrics have been presented. Now that a set of candidate solutions have been generated, how do we choose one desirable design? This final choice entails some subjectivity, and is usually made after an extensive design space exploration has been conducted. Visualization techniques can help understand the tradeoffs graphically [60, 61].

16.8.5 Multiobjective Problems Under Uncertainty: Decision-Making Problem

The problem formulations discussed thus far considered only a single deterministic objective function. Design problems are often multiobjective in nature, and can be challenging to model in the robust domain. Managing the tradeoffs due to constraint satisfaction, robustness, and mean performance objectives of a single objective problem can be a complicated task by itself. If there are multiple design objectives, the tradeoff scenario can be very challenging. Interested readers in this topic are referred to [47, 57].

16.9 Other Popular Methods

This section introduces two other popular approaches used for robust design and optimization under uncertainty: (1) The Taguchi Method, and (2) Stochastic Programming.

16.9.1 Taguchi's Robust Design Methods

Taguchi defines robustness as "the state where the product or process design is minimally sensitive to factors causing variability, at the lowest cost" [62]. Taguchi's product design approach consists of three stages: system design, parameter design, and tolerance design. System design is the conceptual design stage. Parameter design, also known as the robust design, enhances the robustness of the design by identifying factors that reduce the design's sensitivity to noise. Tolerance design is the phase where appropriate tolerances for parameter values are specified. Taguchi methods use metrics, such as signal-to-noise ratio and quality loss functions, to perform parameter design. However, within the basic Taguchi method, constraints are not incorporated. Further enhancements of the Taguchi method can be found in these references [63, 64].

16.9.2 Stochastic Programming

Stochastic programming and similar techniques are used extensively in the fields of mathematics, finance, and artificial intelligence. Many ideas and issues that were studied in this chapter provide the building blocks for these approaches as well. Interested readers are pointed to the following references in stochastic programming [65, 66, 67, 68].

16.10 Summary

This chapter provided an introductory presentation of the issues involved in design optimization under uncertainty. This area is an active field of research, and this chapter provides a summary of the popular methods available, with references to more detailed information for interested readers. Optimization under uncertainty is a process that can be defined as including five main steps as defined in Fig. 16.2. We explained the basics of each of the five steps involved, and provided examples to illustrate the discussion. Using a truss example, optimization problems under uncertainty were presented as decision making problems, where a tradeoff must be made among multiple conflicting issues of interest.

16.11 Problems

Intermediate Problems

16.1   Derive the equations for the stresses S_1 and S_2 for the truss problem shown in Eqs. 16.8 and 16.9.

16.2   Using MATLAB, duplicate the deterministic two bar truss results presented in Sec. 16.2.

16.3   Duplicate the results found in Fig. 16.4 using the `normplot` and `normrnd` commands in MATLAB.

16.4   Consider the results of Sec. 16.6.1.
   (1) Duplicate the results of the example discussion of Sec. 16.6.1.
   (2) Now let the number of samples be 10. Estimate the failure probability for this case. Run your program 10 different times and observe the results. Do all the 10 runs match? Explain.
   (3) Increase the number of samples to 1,000, 10,000, and 1,000,000. For each case, run your program 10 times and record the probability of failure values. As the number of samples increase, report your observations regarding the failure probability values of the multiple runs.

16.5   Explore the `histfit` command in MATLAB. Understand what the X-axis and Y-axis values of the plot represent.
   (1) For each case of number of samples given in Problem 16.4, plot the histogram of the stress values and label your plot. Observe the change in the shape and position of the histogram as you increase the number of samples. Explain your observations.
   (2) Where does the maximum allowable stress, S_{max}, lie on the plot? Show the failure region on the plot.
   (3) Comment as to why the failure probability in this case appears sensible using the histogram plot and the S_{max} value.

16.6   Repeat Parts (2) through (3) of Problem 16.4 with a mean A_1 of 1.8 in, and the standard deviation given in Sec. 16.6.1. How do the values of the failure probabilities change when compared to Problem 16.4?

16.7   Our objective is to minimize the failure probability obtained in Sec. 16.6.1. Setup an optimization problem that uses the mean of A_1 as a design variable. Assume that the standard deviation of A_1 is as prescribed in Sec. 16.6.1. Solve the optimization problem to find the value of the mean of A_1 that yields the least failure probability for the S_1 constraint. (Hint: Each iteration of your objective function requires one Monte Carlo simulation. Use a sample size of 100,000.)

16.8   Repeat Problem 16.5 using the assumptions in Problem 16.6.

(1) Observe the change in the shape and position of the histogram when compared to those obtained in Problem 16.5. What do you observe? (Hint: Pay special attention to the tails of the histogram distribution, and compare it with the maximum allowable stress value.)

(2) Identify the failure regions on the plot. Justify the failure probability values for this case using the histogram plot.

16.9 Consider the stress S_2 given in Eq. 16.9. Assume that A_2 is normally distributed with a mean of 1 in, and a standard deviation of 0.1 in. Assume that $\theta = 60°, \beta = 30°$. Derive the expression for S_2 based on the data given above. Repeat Questions (2) through (3) of Problem 16.4.

16.10 In the previous problem, the failure probability of S_2 was very high. What can be done to reduce its value? What are the parameters you can change to reduce the failure probability? Discuss how optimization can help in this context.

16.11 Setup an optimization problem that uses the mean of A_2 as a design variable. Assume that the standard deviation of A_2 is as prescribed in Problem 16.9. Solve the optimization problem to find the value of the mean of A_2 that yields the least failure probability for the S_2 constraint. (Hint: Each iteration of your objective function requires one Monte Carlo simulation. Use a sample size of 100,000.)

16.12 Consider the equation $h = \frac{A_1 L}{\sin \theta} + \frac{A_2 L}{\sin \beta}$, which is part of the equality constraint of the truss problem. This expression denotes the structural volume of the truss. Assume that A_1 and A_2 are normal random variables with mean values of 2.4 and 1.5 in, and standard deviations of 0.1 and 0.1, respectively. Assume that $b = L$. Using a Monte Carlo simulation, estimate the mean and standard deviation of the structural volume.

16.13 Using the `histfit` command, plot the histogram of the structural volume from the previous problem.

(1) In the constraint shown in Eq. 16.5, observe that the structural volume is restricted to be 4,000 in^3. Show this value on your histogram plot. Comment on the failure region of the constraint.

(2) How does the failure region of the equality constraint differ from that of the inequality constraint studied in the earlier examples?

(3) What can we possibly change in the given data to reduce the failure region for the equality constraint? How can optimization help in this context?

16.14 Consider a linear function $P = 2X_1 + 3X_2 + 4X_3$, where X_1, X_2, and X_3 have means of 0.1, 0.4, and 1, respectively; and standard deviations of 0.01, 0.01, and 0.01, respectively. Estimate the moments of P using a first order Taylor series approximation.

16.15 Consider a nonlinear function, $f = 5X_1^2 - X_2 X_3 + \cos(X_4)$, where X_1, X_2, X_3, X_4 have means of 1, 1, 1, 0, and standard deviations of 0.1, 0.1,

0.05, 0.1, respectively. Estimate the moments of f using first order Taylor series approximation.

16.16 Duplicate the results discussed in Sec. 16.6.3 for the truss problem. Derive the expressions shown in Eqs. 16.16 and 16.17.

16.17 We are interested in finding the estimates of the mean and standard deviations of the stress S_2 of the truss problem. In Eq. 16.9, assume that A_2 is normally distributed with a mean of 1 in, and a standard deviation of 0.1 in. Assume that $\theta = 60°, \beta = 30°$. Estimate the mean and standard deviation of S_2 using a first order Taylor series expansion.

16.18 Consider the equation $h = \frac{A_1 L}{\sin \theta} + \frac{A_2 L}{\sin \beta}$, which is part of the equality constraint of the truss problem. This expression denotes the structural volume of the truss. Assume that A_1 and A_2 are normal random variables with mean values of 2.4 and 1.5 in, and standard deviations of 0.1 and 0.1, respectively. Assume that $b = L$. Find the mean and standard deviation of h using a first order Taylor series approximation.

16.19 Read the following paper and prepare a 2-page summary of its key contributions. The paper reference is: Messac, A., and Ismail-Yahaya, A., "Multiobjective Robust Design Using Physical Programming," *Structural and Multidisciplinary Optimization Journal*, of the International Society of Structural and Multidisciplinary Optimization (ISSMO), Springer, Vol. 23, No. 5, 2002, pp. 357-371.

16.20 Implement the RDO formulation for the two-bar truss problem in MATLAB shown in Eqs. 16.20 through 16.24. Using this MATLAB code, generate the results shown in Sec. 16.8.2 for the deterministic case.

16.21 Duplicate the deterministic case results shown in Sec. 16.8.3.

16.22 Duplicate the deterministic case results shown in Sec. 16.8.1.

BIBLIOGRAPHY OF CHAPTER 16

[1] J. N. Siddall. *Probabilistic Engineering Design: Principles and Applications*. Mechanical Engineering Series. CRC Press, 1983.

[2] A. D. Kiureghian and O. Ditlevsen. Aleatory or epistemic? Does it matter? *Structural Safety*, 31(2):105–112, 2009.

[3] H. Agarwal, J. E. Renaud, E. L. Preston, and D. Padmanabhan. Uncertainty quantification using evidence theory in multidisciplinary design optimization. *Reliability Engineering and System Safety*, 85(1-3):281–294, 2004.

[4] R. W. Youngblood, V. A. Mousseau, D. L. Kelly, and T. N. Dinh. Risk-informed safety margin characterization (rismc): Integrated treatment of aleatory and epistemic uncertainty in safetyanalysis. In *The 8th International Topical Meeting on Nuclear Thermal-Hydraulics, Operation and Safety (NUTHOS-8)*, Shanghai, China, October 2010.

[5] M. S. Eldred and L. P. Swiler. Efficient algorithms for mixed aleatory-epistemic uncertainty quantification with application to radiation-hardened electronics – Part I: Algorithms and benchmark results. Technical Report SAND2009-5805, Sandia National Laboratory, Albuquerque, NM and Livermore, CA, September 2009.

[6] A. Urbina, S. Mahadevan, and T. L. Paez. An approximate epistemic uncertainty analysis approach in the presence of epistemic and aleatory uncertainties. *Reliability Engineering and System Safety*, 77(3):229–238, 2002.

[7] R. C. Smith. *Uncertainty Quantification: Theory, Implementation, and Applications*. SIAM, 2013.

[8] S. M. Ross. *A First Course in Probability*. Prentice Hall, Upper Saddle River, NJ, 5 edition, 1998.

[9] A. Haldar and S. Mahadevan. *Probability, Reliability, and Statistical Methods in Engineering Design*. John Wiley and Sons, Inc, New York, 2000.

[10] S. Gunawan and P. Y. Papalambros. A Bayesian approach to reliability-based optimization with incomplete information. *ASME Journal of Mechanical Design*, 128(4):909–918, 2006.

[11] K. Sentz and S. Ferson. Combination of evidence in Dempster-Shafer theory. Technical Report SAND2002-0835, Sandia National Laboratories, Setauket, New York, April 2002.

[12] H. R. Bae, R. V. Grandhi, and R. A. Canfield. Uncertainty quantification of structural response using evidence theory. *AIAA Journal*, 41(10):2062–2068, 2003.

[13] J. C. Helton, J. D. Johnson, W. L. Oberkampf, and C. J. Sallaberry. Sensitivity analysis in conjunction with evidence theory representations of epistemic uncertainty. *Reliability Engineering and System Safety*, 91(10-11):1414–1434, 2006.

[14] L. V. Utkin and S. V. Gurov. A general formal approach for fuzzy reliability analysis in the possibility context. *Fuzzy Sets Systems*, 83:203–213, 1996.

[15] X. G. Bai and S. Asgarpoor. Fuzzy-based approaches to substation reliability evaluation. *Electric Power Systems Research*, 69(2-3):197–204, 2004.

[16] L. Du, K. K. Choi, B. D. Youn, and D. Gorsich. Possibility-based design optimization method for design problems with both statistical and fuzzy input data. *ASME Journal of Mechanical Design*, 128(4):928–935, 2006.

[17] J. Zhou and Z. P. Mourelatos. A sequential algorithm for possibility-based design optimization. *ASME Journal of Mechanical Design*, 130(1), 2008.

[18] B. D. Youn, K. K. Choi, and D. Gorsich. Integration of possibility-based optimization to robust design for epistemic uncertainty. *ASME Journal of Mechanical Design*, 129(4):876–882, 2008.

[19] F. P. N. Coolen and M. J. Newby. Bayesian reliability analysis with imprecise prior probabilities. *Reliability Engineering and System Safety*, 43(1):75–85, 1994.

[20] H. Z. Huang, M. J. Zuo, and Z. Sun. Bayesian reliability analysis for fuzzy lifetime data. *Fuzzy Sets Systems*, 157(12):1674–1686, 2006.

[21] B. D. Youn and P. Wang. Bayesian reliability-based design optimization using eigenvector dimension reduction method. *Structural Multidisciplinary Optimization*, 36(2):107–123, 2008.

[22] Z. P. Mourelatos and J. Zhou. A design optimization method using evidence theory. *Journal of Mechanical Design*, 128(4):901–908, 2006.

[23] J. M. Aughenbaugh and C. J. J. Paredis. The value of using imprecise probabilities in engineering design. *ASME Journal of Mechanical Design*, 128(4):969–979, 2006.

[24] J. Zhou and Z. P. Mourelatos. Design under uncertainty using a combination of evidence theory and a Bayesian approach. *SAE International Journal of Materials and Manufacturing*, 1(1):122–135, April 2009.

[25] P. Wang, B. D. Youn, Z. Xi., and A. Kloess. Bayesian reliability analysis with evolving, insufficient, and subjective data sets. *ASME Journal of Mechanical Design*, 131(11):111008–11, 2009.

[26] M. Grigoriu. *Stochastic Systems: Uncertainty Quantification and Propagation*. Springer, 2012.

[27] T. W. Simpson, D. Lin, and W. Chen. Sampling strategies for computer experiments: Design and analysis. *International Journal of Reliability and Application*, 2(3), 2001.

[28] G. I. Rozvany and K. Maute. Analytical and numerical solutions for a reliability-based benchmark example. *Structural and Multidisciplinary Optimization*, 43(6):745–753, 2011.

[29] K. Zaman, M. McDonald, and S. Mahadevan. Probabilistic framework for uncertainty propagation with both probabilistic and interval variables. *ASME Journal of Mechanical Design*, 133(2):021010–14, February 2011.

[30] C. P. Robert and G. Casella. *Introducing Monte Carlo Methods*. Springer-Verlag, 2010.

[31] X. Du. Unified uncertainty analysis by the first order reliability method. *ASME Journal of Mechanical Design*, 130(9):091401, 2008.

[32] G. W. Oehlert. A note on the Delta method. *The American Statistician*, 46(1):27–29, February 1992.

[33] R. N. Bhattacharya and R. R. Rao. *Normal Approximation and Asymptotic Expansions*. SIAM, 2010.

[34] D. Xiu and G. E. Karniadakis. Modeling uncertainty in flow simulations via generalized polynomial chaos. *Journal of Computational Physics*, 187(1):137–167, 2003.

[35] A. M. Hasofer and N. C. Lind. Exact and invariant second-moment code format. *Journal of Engineering Mechanics Division*, 100(1):111–121, 1974.

[36] X. Du and W. Chen. Sequential optimization and reliability assessment method for efficient probabilistic design. *ASME Journal of Mechanical Design*, 126(2):225–233, 2004.

[37] B. D. Youn, K. K. Choi, and Y. H. Park. Hybrid analysis method for reliability-based design optimization. *ASME Journal of Mechanical Design*, 125(2):221–232, 2003.

[38] I. Enevoldsen and J. D. Sorensen. Reliability-based optimization in structural engineering. *Structural Safety*, 15(3):169–196, 1994.

[39] J. Tu., K. K. Choi, and Y. H. Park. A new study on reliability-based design optimization. *ASME Journal of Mechanical Design*, 121(4):557–564, 1999.

[40] B. D. Youn, K. K. Choi, and L. Du. Enriched performance measure approach (+pma) for relibaility-based design optimization. *AIAA Journal*, 43(4):874–884, 2005.

[41] R. Yang and L. Gu. Experience with approximate reliability-based optimization methods II: An exhaust system problem. *Structural and Multidisciplinary Optimization*, 29(6):488–497, 2005.

[42] S. Shan and G. Wang. Reliable design space and complete single-loop reliability-based design optimization. *Reliability Engineering and System Safety*, 93(8):1218–1230, 2008.

[43] T. H. Nguyen, J. Song, and G. H. Paulino. Single-loop system reliability-based design optimization using matrix-based system reliability method: Theory and applications. *ASME Journal of Mechanical Design*, 132(1):011005–11, 2010.

[44] J. Liang, Z. Meorelatos, and J. Tu. A single-loop method for reliability-based design optimisation. *International Journal of Product Development*, 5(1-2):76–92, 2008.

[45] M. Ba-Abbad, E. Nikolaidis, and R. Kapania. New approach for system reliability-based design optimization. *AIAA Journal*, 44(5):1087–1096, 2006.

[46] R. Jin, X. Du, and W. Chen. The use of metamodeling techniques for optimization under uncertainty. *Structural and Multidisciplinary Optimization*, 25(2):99–116, 2003.

[47] A. Messac and A. Ismail-Yahaya. Multiobjective robust design using Physical Programming. *Structural and Multidisciplinary Optimization*, 23(5):357–371, 2002.

[48] X. Du and W. Chen. Towards a better understanding of modeling feasibility robustness in engineering design. *ASME Journal of Mechanical Design*, 122:385–394, December 2000.

[49] X. Gu, J. E. Renaud, S. M. Batill, R. M. Brach, and A. Budhiraja. Worst case propagated uncertainty of multidisciplinary systems in robust optimization. *Structural Optimization*, 20(3):190–213, 2000.

[50] S. Gunawan and S. Azarm. Multi-objective robust optimization using a sensitivity region concept. *Structural and Multidisciplinary Optimization*, 29(1):50–60, 2005.

[51] W. Chen, R. Garimella, and N. Michelena. Robust design for improved vehicle handling under a range of maneuver conditions. *Engineering Optimization*, 33(3):303–326, 2001.

[52] C. D. McAllister and T. W. Simpson. Multidisciplinary robust design optimization of an internal combustion engine. *ASME Journal of Mechanical Design*, 125(1):124–130, 2003.

[53] J. K. Allen, C. Seepersad, H. Choi, and F. Mistree. Robust design for multiscale and multidisciplinary applications. *ASME Journal of Mechanical Design*, 128(4):832–843, 2006.

[54] H. Liu, W. Chen, M. Kokkolaras, P. Y. Papalambros, and H. M. Kim. Probabilistic analytical target cascading: A moment matching formulation for multilevel optimization under uncertainty. *ASME Journal of Mechanical Design*, 128(4):991–1000, 2006.

[55] B. Dodson, P. Hammett, and R. Klerx. *Probabilistic Design for Optimization and Robustness for Engineers*. John Wiley and Sons, 2014.

[56] A. Parkinson, C. Sorensen, and N. Pourhassan. A general approach for robust optimal design. *ASME Journal of Mechanical Design*, 115(1):74–80, 1993.

[57] S. Rangavajhala, A. A. Mullur, and A. Messac. Equality constraints in multiobjective robust design optimization: Decision making problem. *Journal of Optimization Theory and Applications*, 140(2):315–337, 2009.

[58] S. Rangavajhala, A. A. Mullur, and A. Messac. The challenge of equality constraints in robust design optimization: Examination and new approach. *Structural and Multidisciplinary Optimization*, 34(5):381–401, 2007.

[59] A. Messac, A. Ismail-Yahaya, and C. A. Mattson. The normalized normal constraint method for generating the Pareto Frontier. *Structural and Multidisciplinary Optimization*, 25(2):86–98, 2003.

[60] G. Stump, T. W. Simpson, M. Yukish, and L. Bennett. Multidimensional visualization and its application to a design byshopping paradigm. In *9th AIAA/ISSMO Symposium on Multidisciplinary Analysis and Optimization Conference*, Atlanta, GA, USA, September 2002.

[61] S. Rangavajhala, A. A. Mullur, and A. Messac. Uncertainty visualization inmultiobjective robust design optimization. In *2nd AIAA Multidisciplinary Design Optimization Specialist Conference*, Newport, RI, USA, May 2006.

[62] G. Taguchi, Subir Chowdhury, and S. Taguchi. *Robust Engineering: Learn How to Boost Quality While Reducing Costs and Time to Market*. McGraw-Hill, 1999.

[63] K. N. Otto and E. K. Antonsson. Extensions to the Taguchi method of product design. *ASME Design Theory and Methodology*, 115(1):5–13, 1993.

[64] K. L. Tsui. Robust design optimizaton for multiple characteristic problems. *International Journal of Production Research*, 37(2):433–445, 1999.

[65] J. R. Birge and F. Louveaux. *Introduction to Stochastic Programming*. Springer, 2011.

[66] I. M. Stancu-Minasian. *Stochastic Programming with Multiple Objective Functions*. Editura Academiei, 1984.

[67] D. Bertsimas and M. Sim. The price of robustness. *Operations Research*, 52(1):35–53, 2004.

[68] J. Mulvey, R. J. Vanderbei, and S. Zenios. Robust optimization of large scale systems. *Operations Research*, 43(2):264–281, 1995.

17

Methods for Pareto Frontier
Generation/Representation

17.1 Overview

Multiobjective optimization plays an important role in engineering design, management, and decision making in general. In Chapter 6, the formulation of multiobjective optimization problems and the concept of Pareto frontier were introduced. Several approaches to obtain Pareto solutions, including the weighted sum method, compromise programming, goal programming, and physical programming, were introduced in Chapter 6.

In this chapter, new approaches used to obtain Pareto solutions are presented. These approaches can generate an evenly spaced set of Pareto points in the design space, with the objective of capturing the entire Pareto frontier. This chapter is structured as follows. Section 17.2 presents the requisite mathematical preliminaries. Section 17.3 presents the Normal Boundary Intersection (NBI) method. In Sec. 17.4, the normalized Normal Constraint (NC) method is introduced. Section 17.5 develops the Pareto filter concept, which eliminates dominated solutions from the generated set of points. Numerical examples are provided in Sec. 17.6, while the summary of this chapter is presented in Sec. 17.7.

17.2 Mathematical Preliminaries

This section provides the requisite mathematical preliminaries. A generic formulation of the multiobjective optimization problem is defined as follows. This problem is named **Problem P1**.

$$\min_{x}\{\mu_1(x),\ \mu_2(x),\ \cdots,\mu_n(x)\},\ (n \geq 2) \tag{17.1}$$

subject to

$$g_j(x) \leq 0,\ (1 \leq j \leq r) \tag{17.2}$$

$$h_k(x) = 0,\ (1 \leq k \leq s) \tag{17.3}$$

$$x_{li} \leq x_i \leq x_{ui}\ (1 \leq i \leq n_x) \tag{17.4}$$

The vector x denotes the design variables and μ_i denotes the i_{th} generic design objective. Equation 17.2 and 17.3 denote the inequality and equality constraints, respectively, while Eq. 17.4 denotes the side constraint. The above problem does not generally yield a unique solution.

Before the approaches used to obtain Pareto frontiers are introduced, several definitions need to be introduced.

Anchor point (μ^{i*}): The generic i^{th} anchor point (or end point of the Pareto frontier) is obtained when the generic i^{th} objective is minimized independently. When the global optimum is not unique, the one non-dominated is chosen. It is also called an optimum vertex.

Utopia line: The line joining two anchor points in bi-objective cases is called the Utopia line.

Utopia hyperplane: The plane that comprises all the anchor points in the multiobjective case is called the Utopia hyperplane.

Utopia point: It is defined by a vector where it's generic i^{th} components is the value of the design objective minimum obtained by minimizing only the corresponding i^{th} design objective.

The word Utopia is used here to indicate that the plane contains the n optimum vertices, components of which form the Utopia point. We note that since the Utopia point is generally unattainable, it is not part of the Utopia plane. If the optimization problem has n objectives, it involves n anchor points. The ith anchor points can be obtained by solving **Problem PUi**, which is defined as follows:

$$\min_{x} \mu_i(x), \ (1 \leq i \leq n) \tag{17.5}$$

subject to

$$g_j(x) \leq 0, \ (1 \leq j \leq r) \tag{17.6}$$

$$h_k(x) = 0, \ (1 \leq k \leq s) \tag{17.7}$$

$$x_{li} \leq x_i \leq x_{ui} \ (1 \leq i \leq n_x) \tag{17.8}$$

We now define the following quantities, which result from solving the above PUi problem:

x^{i*}: Optimal decision vector ($x^{i*} \in R^{n_x}$);

μ_i^*: Generic i^{th} optimal objective - specifically, $\mu_i^* = \mu_i(x^{i*})(\mu_i^* \in R)$;

μ^u: Utopia Point $\mu^u = [\mu_1^*, \mu_2^*, ..., \mu_n^*]^T (\mu^u \in R^n)$;

μ^{i*}: i^{th} Anchor Point $\mu^{i*} = [\mu_1(x^{i*}), \mu_2(x^{i*}), ..., \mu_n(x^{i*})]^T, (\mu^{i*} \in R^n)$;

P^u: Utopia Plane. Hyperplane in n-dimensions that comprises the n anchor points, $\mu^{i*}(i = 1, ..., n)$.

We provide a Pareto frontier of a bi-objective optimization problem in Fig. 17.1. Anchor points and the Utopia point are illustrated in the figure. Please note that the design space is not normalized.

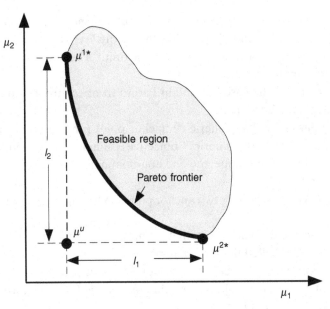

Figure 17.1. General Design Objective Space for a Bi-Objective Case

17.3 Normal Boundary Intersection Method

This section introduces the Normal Boundary Intersection (NBI) Method [1, 2] for finding several Pareto optimal points for a general nonlinear multiobjective optimization problem. This method is successful in producing an evenly distributed set of points in the Pareto frontier, and it method can be applied to optimization problems with more than two objectives.

In Section 17.2, the Utopia point μ^u and the anchor point μ^{i*} are defined. We define the pay-off matrix, Φ, as follows: Φ is an $n \times n$ function whose i^{th} column is $\mu^{i*} - \mu^u$. The pay-off matrix has the following form:

$$\Phi = \begin{pmatrix} 0 & \mu_1(x^{2*}) - \mu_1(x^{1*}) & \cdots & \mu_1(x^{n*}) - \mu_1(x^{1*}) \\ \mu_2(x^{1*}) - \mu_2(x^{2*}) & 0 & \cdots & \mu_2(x^{n*}) - \mu_2(x^{2*}) \\ \vdots & \vdots & \ddots & \vdots \\ \mu_n(x^{1*}) - \mu_n(x^{n*}) & \mu_n(x^{2*}) - \mu_n(x^{n*}) & \cdots & 0 \end{pmatrix} \quad (17.9)$$

The pay-off matrix translates anchor points in the design objective space. For a bi-objective design space, Fig. 17.2 shows the following: (i) in the translated coordinate system of the design objective space, the Utopia point is at the origin, (ii) the two anchor points lie on the axes lines, and (iii) the coordinates of the two anchor points are determined by their distances from the Utopia point. For optimization problems with more than two design objectives, in the translated coordinate system of the design objective space, the Utopia point is at the origin. One coordinate of each anchor point is 0, and the other coordinates are determined by the distance between the anchor point and the Utopia point.

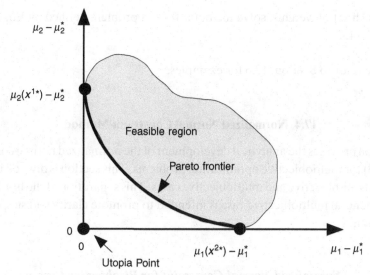

Figure 17.2. Translated Design Objective Space for a Bi-Objective Case

Using the pay-off matrix, the convex combinations of $\mu^{i*} - \mu^u$ are $\{\Phi\beta : \beta \in R^n, \sum_{i=1}^{n} \beta_i = 1, \beta_i \geq 0\}$. For a bi-objective optimization problem, $\Phi\beta$ are the points on its Utopia line. For an optimization problem with more than two objectives, $\Phi\beta$ are the points on its Utopia hyperplane.

In the translated design objective space, let $\hat{n}$ denote the unit normal to the points on the Utopia line (bi-objective) or the Utopia hyperplane (more than two objectives) toward the origin; then, $\Phi\beta + t\hat{n}, t \in R$ represents the set of points on that normal. The point of intersection of the normal and the Pareto frontier is the solution to the following subproblem:

$$\max_{x,t} t \tag{17.10}$$

subject to

$$\Phi\beta + t\hat{n} = [\mu_1(x), \mu_2(x), ..., \mu_n(x)]^T \tag{17.11}$$

$$g_j(x) \leq 0, \ (1 \leq j \leq r) \tag{17.12}$$

$$h_k(x) = 0, \ (1 \leq k \leq s) \tag{17.13}$$

$$x_{li} \leq x_i \leq x_{ui} \ (1 \leq i \leq n_x) \tag{17.14}$$

The steps of the NBI method are:

1. Evaluate all the anchor points, μ^{1*} to μ^{n*}, by solving **Problem PUi** ($i = 1, 2, ..., n$).
2. Generate the pay-off matrix using Eq. 17.9.
3. Generate the weights, β, so that $\Phi\beta$ samples the Utopia line for a bi-objective optimization problem, or samples the Utopia hyperplane for an optimization problem with more than two objectives.

4. For each set of weights, solve the optimization problem defined by Eq. 17.10 to Eq. 17.14.

Please refer to Section. 17.6 for examples.

17.4 Normalized Normal Constraint Method

This section provides the analytical development of the normalized normal constraint method [3] for multiobjective optimization problems. This section is divided into two subsections: bi-objective and multiobjective cases. This separation of the bi-objective from the general multiobjective case is intended to promote clarity and simplicity of presentation.

Normalized Normal Constraint for Bi-objective Case

Figure 17.3 represents the normalized Pareto frontier in the normalized design space. In the normalized objective space, all the anchor points are one unit away from the Utopia point, and the Utopia point is at the origin. A bar over a variable implies that it is normalized.

We begin with a graphical perspective of the normalized normal constraint method. Figure 17.1 illustrates the non-normalized design space and the Pareto frontier of a generic bi-objective problem. Figure 17.3 represents the normalized Pareto frontier in the normalized design space. In the normalized objective space, all the anchor points are one unit away from the Utopia point, and the Utopia point is at the origin by definition.

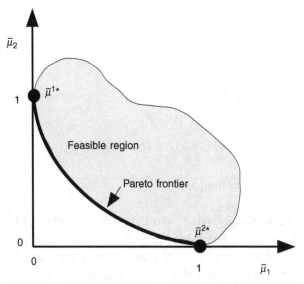

Figure 17.3. Normalized Design Objective Space for a Bi-Objective Case

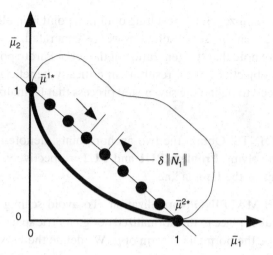

Figure 17.4. A Set of Evenly Spaced Points on the Utopia Line for a Bi-Objective Problem

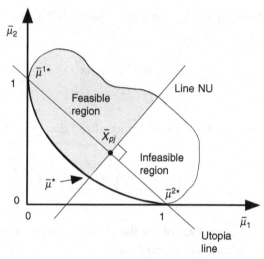

Figure 17.5. Graphical Representation of the Normalized Normal Constraint Method for Bi-Objective Problems

We now present the normalized normal constraint method by defining a seven-step process for its application. To understand the idea of the NC method, consider Figs. 17.4 and 17.5 for the bi-objective case. In Fig. 17.4, we observe the feasible region and the corresponding Pareto frontier. We also note that the two anchor points are obtained by successively minimizing the first and second design metrics (**Problem PUi**). A line joining the two anchor points is drawn, and is called the Utopia line. The Utopia line is divided into $m_1 - 1$ segments, resulting in m_1 points. In Fig. 17.5, one of the generic points intersecting the segments is used to define a normal to the Utopia line. This normal line is used as an additional constraint that progressively reduces the feasible region, and generates successive Pareto solutions, as indicated in Fig. 17.5. If we minimize $\bar{\mu}_2$, the resulting optimum point is $\bar{\mu}^*$. By translating the normal line, a corresponding set of solutions will be generated.

Importantly, we note that the generation of the set of Pareto points is performed in the normalized objective space, resulting in critically beneficial scaling properties. We now proceed to define the seven-step process that formalizes the preceding description.

1. ANCHOR POINTS. Obtain the two anchor points, denoted by μ^{1*} and μ^{2*}, resulting from solving **Problem PU1** and **PU2**, respectively. The line joining these two points is the Utopia line.

2. OBJECTIVES MAPPING/Normalization. To avoid scaling deficiencies, the optimization takes place in the normalized design metric space (design objective space). Let $\bar{\mu}$ be the normalized form of μ. We define the Utopia point, μ^u, as

$$\mu^u = [\mu_1(x^{1*})\ \mu_2(x^{2*})]^T \tag{17.15}$$

and we let l_1 and l_2 be the distances between μ^{2*} and μ^{1*}, and the Utopia point, μ^u, respectively. We have

$$l_1 = \mu_1(x^{2*}) - \mu_1(x^{1*}) \tag{17.16}$$

$$l_2 = \mu_2(x^{1*}) - \mu_2(x^{2*}) \tag{17.17}$$

Using the above definitions, the normalized design metrics can be evaluated as

$$\bar{\mu} = \left\{ \frac{\mu_1(x) - \mu_1(x^{1*})}{l_1}\ \ \frac{\mu_2(x) - \mu_2(x^{2*})}{l_2} \right\} \tag{17.18}$$

Following the normalization of the design metrics, we can generate the Pareto points, as indicated in Figs. 17.4 and 17.5.

3. UTOPIA LINE VECTOR. Define $\bar{N}_1$ as the vector from $\bar{\mu}^{1*}$ to $\bar{\mu}^{2*}$, yielding

$$\bar{N}_1 = \bar{\mu}^{2*} - \bar{\mu}^{1*} \tag{17.19}$$

4. NORMALIZED INCREMENTS. Compute a normalized increment, δ_1, along the direction of $\bar{N}_1$ for a prescribed number of solutions, m_1, as

$$\delta_1 = \frac{1}{m_1 - 1} \tag{17.20}$$

5. GENERATE UTOPIA LINE POINTS. Evaluate a set of evenly distributed points on the Utopia line as (see Fig. 17.4)

$$\bar{X}_{pj} = \alpha_{1j}\bar{\mu}^{1*} + \alpha_{2j}\bar{\mu}^{2*} \tag{17.21}$$

where

$$0 \leq \alpha_{1j} \leq 1 \tag{17.22}$$

$$\sum_{k=1}^{2} \alpha_{kj} = 1 \tag{17.23}$$

Please note that α_{ij} is incremented by δ_1 between 0 and 1 (Fig. 17.4), and we use values of j where $j \in \{1, 2, ..., m_1\}$.

6. PARETO POINTS GENERATION. Using the set of evenly distributed points on the Utopia line, generate a corresponding set of Pareto points by solving a succession of optimization runs of Problem P2. Each optimization run corresponds to a point on the Utopia line. Specifically, for each generated point on the Utopia line, solve for the jth point.

Problem P2 (for j^{th} point) is defined as follows:

$$\min_{x} \bar{\mu}_2 \tag{17.24}$$

subject to

$$g_j(x) \leq 0, \ (1 \leq j \leq r) \tag{17.25}$$

$$h_k(x) = 0, \ (1 \leq k \leq s) \tag{17.26}$$

$$x_{li} \leq x_i \leq x_{ui} \ (1 \leq i \leq n_x) \tag{17.27}$$

$$\bar{N}_1^T (\bar{\mu} - \bar{X}_{pj}) \leq 0 \tag{17.28}$$

$$\bar{\mu} = [\bar{\mu}_1(x) \ \bar{\mu}_2(x)] \tag{17.29}$$

7. PARETO DESIGN METRICS VALUES. Evaluate the non-normalized design metrics that correspond to each Pareto point. This evaluation can be done in two ways. First, since the function $\mu(x)$ is known, a direct evaluation is possible. Alternatively, if the normalized design metrics were saved from Step 6, the non-normalized design metrics can be obtained through an inverse mapping of (12) by using the relation

$$\mu = [\bar{\mu}_1 l_1 + \mu_1(x^{1*}) \ \bar{\mu}_2 l_2 + \mu_2(x^{2*})]^T \tag{17.30}$$

Up to this point, we have not considered the possibility that some of the points generated in some pathological cases might be dominated by other points in the set. This important situation is examined in Sec. 17.5, where a Pareto filter is developed.

Normalized Normal Constraint for n-Objective Case

In this section, the development of the normalized normal constraint method is presented for a general multiobjective case. This development will be terse to avoid repetition with respect to the bi-objective case. The basic steps are similar to those of the bi-objective case.

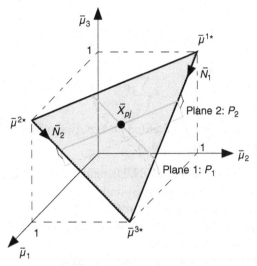

Figure 17.6. Utopia Hyperplane for a Three-Objective Case

1. ANCHOR POINTS. Obtain the anchor points, μ_i. for $i \in \{1, 2, ..., n\}$, which are obtained by solving **Problem PUi**. We define the hyperplane, which is comprised of all the anchor points. This plane is called the Utopia hyperplane (or Utopia plane). Figure 17.6 illustrates the Utopia plane for three design metrics. Recall that the optimum design variables obtained from solving **Problem PUi** are denoted by x^{i*}.

2. OBJECTIVES MAPPING/Normalization. To avoid scaling deficiencies, the optimization is performed in the normalized design metric space. In order to obtain the required mapping parameters, we need to define two points: the Utopia point and the Nadir point, which are respectively evaluated as follows:

$$\mu^u = [\mu_1(x^{1*})\ \mu_2(x^{2*})\ \cdots\ \mu_n(x^{n*})]^T \tag{17.31}$$

$$\mu^N = [\mu_1^N\ \mu_2^N\ \cdots\ \mu_n^N]^T \tag{17.32}$$

where

$$\mu_i^N = \max[\mu_i(x^{1*}), \mu_i(x^{2*}), \cdots, \mu_i(x^{n*})], i \in \{1, 2, \cdots, n\} \tag{17.33}$$

We define the matrix L as

$$L = \left\{ \begin{array}{c} l_1 \\ l_2 \\ \vdots \\ l_n \end{array} \right\} = \mu^N - \mu^u \tag{17.34}$$

which leads to the normalized design metrics as

$$\bar{\mu}_i = \frac{\mu_i - \mu_i(x^{i*})}{l_i}, i = 1, 2, ..., n \tag{17.35}$$

3. UTOPIA PLANE VECTORS. Define the direction, $\bar{N}_k$ from $\bar{\mu}^{k*}$ to $\bar{\mu}^{n*}$ for $k \in \{1, 2, ..., n-1\}$ as

$$\bar{N}_k = \bar{\mu}^{n*} - \bar{\mu}^{k*} \tag{17.36}$$

4. NORMALIZED INCREMENTS. Compute a normalized increment, δ_k, along the direction $\bar{N}_k$ for a prescribed number of solutions, m_k, along the associated $\bar{N}_k$ direction.

$$\delta_k = \frac{1}{m_k - 1}, \qquad (1 \le k \le n-1) \tag{17.37}$$

Care must be taken in choosing the number of points, m_k, for each direction $\bar{N}_k$. To ensure an even distribution of points on the n-dimensional Utopia plane, the following relationship can be used. Given a specified number of points, m_1, along the vector $\bar{N}_1$, m_k is given as

$$m_k = \frac{m_1 \|\bar{N}_k\|}{\|\bar{N}_1\|} \tag{17.38}$$

5. GENERATE HYPERPLANE POINTS. Evaluate a set of evenly distributed points on the Utopia hyperplane as

$$\bar{X}_{pj} = \sum_{k=1}^{n} \alpha_{kj} \bar{\mu}^{k*} \tag{17.39}$$

where

$$0 \le \alpha_{kj} \le 1, \sum_{k=1}^{n} \alpha_{kj} = 1 \tag{17.40}$$

Figure 17.6 describes how generic points are generated in the Utopia plane, where two planes serve as constraints (see Eq. 17.45). Figure 17.7 shows the resulting uniformly distributed points on the Utopia plane for a three-dimensional case in the normalized objective space.

6. PARETO POINTS GENERATION. We generate a set of well-distributed Pareto solutions in the normalized objective space. For each value of $\bar{X}_{pj}$

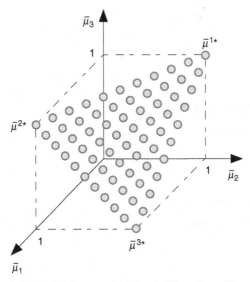

Figure 17.7. Evenly Spaced Points on the Utopia Plane for a Three-Objective Case

generated in Step 5, we obtain the corresponding Pareto solution by solving Problem P3:

$$\min_x \bar{\mu}_n \tag{17.41}$$

subject to

$$g_j(x) \leq 0, \ (1 \leq j \leq r) \tag{17.42}$$

$$h_k(x) = 0, \ (1 \leq k \leq s) \tag{17.43}$$

$$x_{li} \leq x_i \leq x_{ui} \ (1 \leq i \leq n_x) \tag{17.44}$$

$$\bar{N}_k^T (\bar{\mu} - \bar{X}_{pj}) \leq 0, \ (1 \leq k \leq n-1) \tag{17.45}$$

$$\bar{\mu} = \{\bar{\mu}_1(x), ..., \bar{\mu}_n(x)] \tag{17.46}$$

In solving Problem P3 using a gradient-based algorithm, the initial point used contributes to the efficiency of the Pareto frontier generation. In the case of the normalized normal constraint method, a good choice for a starting point is the point $\bar{X}_{pj}$. This automated scheme works well in practice.

7. PARETO DESIGN METRICS VALUES. The design metrics values for the Pareto solutions obtained in Step 6 can be obtained using the unscaling equation

$$\mu_i = \bar{\mu}_i l_i + \mu_i(x^{i*}), i = 1, 2, ..., n \tag{17.47}$$

17.5 Pareto Filter

As indicated earlier, under certain circumstances, the normal constraint method can generate non-Pareto solutions. This unfortunate situation can occur, for example, in the case of a feasible region depicted in Fig. 17.8. In such cases, we propose using a

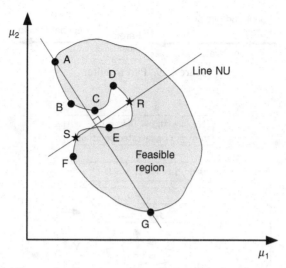

Figure 17.8. Normal Constraint Generates a Non-Pareto Solution Under a Contrived Feasible Region

Pareto filter. A Pareto filter is an algorithm that, given a set of points in objective space, produces a subset of the given points for which none will be dominated by any other. That is, the filter eliminates all dominated points from the given set.

To facilitate the ensuing discussion, it is important to differentiate between local Pareto optimality and global Pareto optimality. We note that a global Pareto optimal point is also a local Pareto optimal point, but the reverse is not typically true. Two important definitions based on the concept of *weak domination* follow.

Definition 1: A design metric vector μ^* is globally Pareto optimal if there does not exist another design metric vector μ such that $\mu_j \leq \mu_j^*$ for all $j \in \{1, 2, ..., n\}$, and $\mu_j < \mu_j^*$ for at least one index of $j, j \in \{1, 2, ..., n\}$ in the feasible design space.

Definition 2: A design metric vector μ^* is locally Pareto optimal if there does not exist another design metric vector μ such that $\mu_j \leq \mu_j^*$ for all $j \in \{1, 2, ..., n\}$, and $\mu_j < \mu_j^*$, for at least one index of $j, j \in \{1, 2, ..., n\}$ in a neighborhood of μ^*.

Consider Fig. 17.8, in which a highly concave feasible region is depicted. Arcs AB and FG represent the (global) Pareto frontier. Regions BC and DE are local Pareto frontiers. Arcs CD and EF are neither globally nor locally Pareto. We now make the important note that generating Pareto points using the normal constraint method with the anchor points A and G will yield non-Pareto solutions. Point S, for example, a non-Pareto solution, would be generated by the normal constraint method when the Line NU is used as the normal constraint. Using a different optimization starting point in a gradient-based scheme, we could obtain the point R. The NBI method would also suffer from this deleterious behavior.

In light of the fact that certain methods yield non-Pareto solutions, it is important to develop a means to avoid retaining dominated points in the set of solutions from which an optimum will be chosen. The Pareto filter does just that. The filter works

Figure 17.9. Pareto Filter

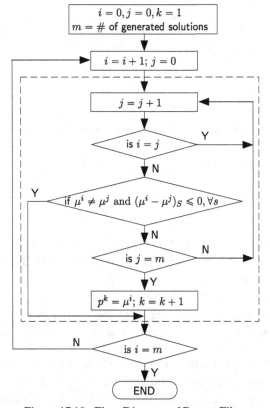

Figure 17.10. Flow Diagram of Pareto Filter

by comparing a point on the Pareto frontier with every other generated point. If a point is not globally Pareto optimal, it is eliminated.

Figure 17.9 provides a functional diagram of the Pareto filter. A four-step process of the Pareto filter algorithm is presented in the following pseudo-code.

1. Initialize
   Initialize the algorithm indices and variables: $i = 0$, $j = 0$, $k = 1$, and $m =$ number of generated solutions; $m = f(m_k)$.
2. Set $i = i + 1$; $j = 0$.
3. (enclosed in the dashed box): Eliminate non-global Pareto points by doing the following: $j = j + 1$
   If $i = j$ go to the beginning of Step 3
   Else continue

If $\mu^i \neq \mu^j$ and $(\mu^i - \mu^j)_s \geq 0, \forall s$
   then μ^i is not a global Pareto point.
   Go to Step 4.
Else if $j = m$
   Then i is a global Pareto point.
   $p^k = \mu^i$
   $k = k + 1$
   Go to Step 4.
Else go to the beginning of Step 3.
4. If $i \neq m$, go to Step 2, else end.

17.6 Examples

In this section, we use the normalized normal constraint method to generate Pareto frontiers in three examples.

The first example deals with scaling issues when one design metric is orders of magnitude larger than the other, and compares the NC and (Weighted Sum) WS methods. The second example demonstrates a case in which non-Pareto points are generated, and compares the behaviors of the NC and NBI methods for the same problem. In the third example, a truss problem is used, in which we deal with a concave Pareto frontier and compare the relative behaviors of the NC and WS methods.

Example: Consider the multiobjective optimization problem below. Generate its Pareto frontier using the normalized normal constraint method.

$$\min_x \left\{ \begin{matrix} \mu_1 \\ \mu_2 \end{matrix} \right\} \tag{17.48}$$

subject to

$$\mu_1 = x_1 \tag{17.49}$$

$$\mu_2 = x_2 \tag{17.50}$$

$$\left(\frac{x_1 - 20}{20} \right)^8 + \left(\frac{x_2 - 1}{1} \right)^8 \leq 1 \tag{17.51}$$

The normalized normal constraint method leads to the following single criterion optimization problem:

$$\min_x \bar{\mu}_2 \tag{17.52}$$

subject to

$$\mu_1 = x_1 \tag{17.53}$$

$$\mu_2 = x_2 \tag{17.54}$$

$$\left(\frac{x_1 - 20}{20} \right)^8 + \left(\frac{x_2 - 1}{1} \right)^8 \leq 1 \tag{17.55}$$

$$\bar{N}_1 (\bar{\mu} - \bar{X}_{pj})^T \leq 0 \tag{17.56}$$

The following MATLAB code generates 21 evenly distributed points on the Pareto frontier (included in the book website www.cambridge.org/Messac).

1. **Main file**

```
clear;clc;
%Define side-constraints
lb=[];ub=[];
%Define initial guess
x0=[20 1];
%Numbers of evenly distributed points
m=21;
%Define options for FMINCON
options=optimset('display','on','algorithm','active-set');

%Initialize points for normalized NC Method
xp=[0,0];
N=[1,1];
%Optimization for separate objectives
x1=fmincon(@NNC_obj1_eg1,x0,[],[],[],[],lb,ub, ...
    @NNC_con_eg1,options,xp,N);
x2=fmincon(@NNC_obj2_eg1,x0,[],[],[],[],lb,ub, ...
    @NNC_con_eg1,options,xp,N);

A1=NNC_obj1_eg1(x1,xp,N);B1=NNC_obj2_eg1(x1,xp,N);
A2=NNC_obj1_eg1(x2,xp,N);B2=NNC_obj2_eg1(x2,xp,N);
N=[A2-A1 B2-B1];
delta=1/(m-1);

%initial index
k=1;
for i=0:m-1
    alpha=i*delta;
    xp=[alpha*A1+(1-alpha)*A2,alpha*B1+(1-alpha)*B2];
    x=fmincon(@NNC_obj2_eg1,x0,[],[],[],[],lb,ub, ...
        @NNC_con_eg1,options,xp,N);
    x0=x;
    A(k)=abs(N(1))*NNC_obj1_eg1(x,xp,N);
    B(k)=abs(N(2))*NNC_obj2_eg1(x,xp,N);
    k=k+1;
end
%%%%%%%%%% Plot Pareto Frontier %%%%%%%%%%%%
figure
plot(A,B,'.')
```

```
axis([-1 20 -.05 1])
xlabel('\mu1')
ylabel('\mu2')
```

2. **First Objective function file**

```
function f=NNC_obj1_eg1(x,xp,N)
f=x(1)/abs(N(1)); %Normalized Form
end
```

3. **Second Objective function file**

```
function f=NNC_obj2_eg1(x,xp,N)
f=x(2)/abs(N(2)); %Normalized Form
end
```

4. **Constraint function file**

```
function [c,ceq]=NNC_con_eg1(x,xp,N)
x1=x(1);
x2=x(2);
c(1)=((x1-20)/20)^8+((x2-1)/1)^8-1;
ceq=[];
if xp(1)~=0
c(2)=[1,-1]*((([x1 x2]-xp)./abs(N))';
end
```

Figure 17.11 shows the Pareto frontier generated by the above MATLAB code.

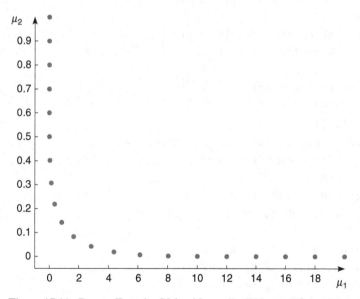

Figure 17.11. Pareto Frontier Using Normalized Normal Constraint

Example: Consider the multiobjective optimization problem below. Use the normalized normal constraint method and the Pareto filter to generate the Pareto frontier.

$$\min_{x} \left\{ \begin{array}{c} \mu_1 \\ \mu_2 \end{array} \right\} \tag{17.57}$$

subject to

$$\mu_1 = x_1 \tag{17.58}$$

$$\mu_2 = x_2 \tag{17.59}$$

$$5e^{-x_1} + 2e^{-0.5(x_1-3)^2} \le x_2 \tag{17.60}$$

The NBI method is used to generate points on the Pareto frontier. A total of 60 evenly spaced points on the Utopia line are used. Figure 17.12(a) depicts the points on the Pareto frontier. In the figure, some points are not globally Pareto optimal.

1. **Main file**

```
clear;clc;
%Define side-constraints
lb=[0 0 -inf];ub=[5 5 inf];
%Define initial guess
x0=[5 0 1];
%Numbers of evenly distributed points
m=61;
%Define options for FMINCON
options=optimset('display','off','algorithm','active-set');

%Initialize points for NC Method
xp=[0,0];
N=[1,1];
%Optimization for separate objectives
x1=fmincon(@NNC_obj1_eg2,x0,[],[],[],[],lb,ub, ...
    @NNC_con_eg2,options,xp,N);
x2=fmincon(@NNC_obj2_eg2,x0,[],[],[],[],lb,ub, ...
    @NNC_con_eg2,options,xp,N);

A1=NNC_obj1_eg2(x1,xp,N);
B1=NNC_obj2_eg2(x1,xp,N);
A2=NNC_obj1_eg2(x2,xp,N);
B2=NNC_obj2_eg2(x2,xp,N);
N=[A2-A1 B2-B1];
delta=1/(m-1);
```

```
    k=1;    %initial index
    for i=0:m-1
        alpha=i*delta;
        xp=[alpha*A1+(1-alpha)*A2,alpha*B1+(1-alpha)*B2];
        x=fmincon(@NNC_obj2_eg2,x0,[],[],[],[],lb,ub, ...
            @NNC_con_eg2,options,xp,N);
        x0=x;
        A(k)=NNC_obj1_eg2(x,xp,N);
        B(k)=NNC_obj2_eg2(x,xp,N);
        k=k+1;
    end
    figure
    plot(A,B,'.');
    xlabel('\mu1');
    ylabel('\mu2');
    title('Normal Boundary Intersection');
```

2. **First Objective function file**

```
function f=NNC_obj1_eg2(x,xp,N)
f=x(1);
end
```

3. **Second Objective function file**

```
function f=NNC_obj2_eg2(x,xp,N)
f=x(2);
end
```

4. **Objective function file for subproblem**

```
function f=NNC_obj3_eg2(x,xp,N)
t=x(3);
f=-t;
end
```

5. **Constraint function file**

```
function [c,ceq]=NNC_con_eg2(x,xp,N)
x1=x(1);
x2=x(2);
t=x(3);
c(1)=5*exp(-x1)+2*exp(-0.5*(x1-3)^2)-x2;
ceq=[];
if xp~=0
ceq=xp+t.*[-1, N(1)/N(2)]-[NNC_obj1_eg2(x,xp,N), ...
NNC_obj2_eg2(x,xp,N)];
end
```

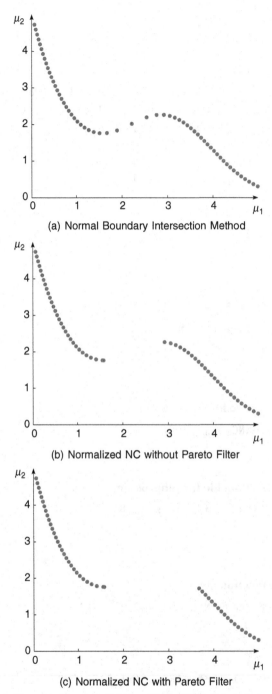

(a) Normal Boundary Intersection Method

(b) Normalized NC without Pareto Filter

(c) Normalized NC with Pareto Filter

Figure 17.12. Pareto Frontier Generated Using Normal Boundary Intersection and Normalized Normal Constraint

This example can also be solved by the normalized normal constraint method, with or without using the Pareto filter. Figures 17.6 and 17.6 are generated using the normalized normal constraint method. Figure 17.6 shows the points generated before the Pareto filter is used. Some points are not globally Pareto optimal. Figure 17.6 shows the points on the Pareto frontier after the Pareto filter is used. The points on Fig. 17.6 are globally Pareto optimal.

1. **Main file**

```
clear;clc;
%Define side-constraints
lb=[0 0];ub=[5 5];
%Define initial guess
x0=[5 0];
%Numbers of evenly distributed points
m=61;
%Define options for FMINCON
options=optimset('display','off','algorithm','active-set');

%Initialize points for NC Method
xp=[0,0];
N=[1,1];
%Optimization for separate objectives
x1=fmincon(@NNC_obj1_eg2,x0,[],[],[],[],lb,ub, ...
    @NNC_con_eg2,options,xp,N);
x2=fmincon(@NNC_obj2_eg2,x0,[],[],[],[],lb,ub, ...
    @NNC_con_eg2,options,xp,N);

A1=NNC_obj1_eg2(x1,xp,N);B1=NNC_obj2_eg2(x1,xp,N);
A2=NNC_obj1_eg2(x2,xp,N);B2=NNC_obj2_eg2(x2,xp,N);
N=[A2-A1 B2-B1];
delta=1/(m-1);

k=1;    %initial index
for i=0:m-1
    alpha=i*delta;
    xp=[alpha*A1+(1-alpha)*A2,alpha*B1+(1-alpha)*B2];
    x=fmincon(@NNC_obj2_eg2,x0,[],[],[],[],lb,ub, ...
        @NNC_con_eg2,options,xp,N);
    x0=x;
    A(k)=abs(N(1))*NNC_obj1_eg2(x,xp,N);
    B(k)=abs(N(2))*NNC_obj2_eg2(x,xp,N);
    k=k+1;
end
```

```
figure
plot(A,B,'.');
xlabel('\mu1');
ylabel('\mu2');
title('Normal Constraint Method without Using Pareto
      Filter');
%%%%%%%%%% Pareto Filtering %%%%%%%%%%%%%%%%
[mA,mB]=pareto_filter(A,B);
%%%%%%%%%% Plot Pareto Frontier %%%%%%%%%%%%
figure
plot(mA,mB,'.');
xlabel('\mu1');
ylabel('\mu2');
title('Normal Constraint Method Using Pareto Filter');
```

2. **First Objective function file**

```
function f=NNC_obj1_eg2(x,xp,N)
f=x(1)/abs(N(1)); %Normalized Form
end
```

3. **Second Objective function file**

```
function f=NNC_obj2_eg2(x,xp,N)
f=x(2)/abs(N(2)); %Normalized Form
end
```

4. **Constraint function file**

```
function [c,ceq]=NNC_con_eg2(x,xp,N)
f1=x(1);f2=x(2);
c(1)=5*exp(-x(1))+2*exp(-0.5*(x(1)-3)^2)-x(2);
ceq=[];
if xp~=0
c(2)=[1,-1]*(([f1 f2]-xp)./abs(N))';
end
```

5. **Pareto Filter**

```
function [mA,mB]=pareto_filter(A,B)
k=1;
m=length(A);
for i=1:m
    for j=i+1:m
        if A(i)<A(j)||B(i)>B(j)
            break;
        end
    end
    if j==m
        p(k)=i;
```

```
            k=k+1;
        end
    end
    mA=A(p);
    mB=B(p);
```

17.7 Summary

This chapter presents two Pareto frontier generation methods: the normal boundary intersection (NBI) method and the normalized normal constraint (NNC) method. These two methods have the ability to generate a well distributed set of Pareto points even in numerically demanding (illconditioned) situations. It is shown that, in non-convex Pareto frontier cases, the NBI method may generate non-Pareto point where the NNC method will avoid these points. All Pareto points generated by NBI will also be generated by NNC. In addition, NBI involves equality constraints while NNC involves inequality constraints, which are generally computationally favorable. This chapter also introduces the notion of a Pareto filter, which performs the function of eliminating all but the global Pareto solutions when given a set of candidate solutions.

17.8 Problems

17.1 Consider the first multiobjective optimization problem given in Sec. 17.6. Reproduce the results using the normalized normal constraint method. Also solve it using the weighted sum method. Show the figures of the Pareto frontier generated by the two methods. Turn in your M-file and results.

17.2 Consider the second multiobjective optimization problem given in Sec. 17.6. Reproduce the results using the normalized normal constraint method and the Pareto filter. Turn in your M-file and results.

17.3 Solve the following three-bar truss optimization problem.
We consider a three-bar truss structure from the following paper written by J Koski in 1985: Defectiveness of weighting methods in multicriterion optimization of structures. Commun. Appl. Numer. Methods 1, 333ˉC337.
The structure and the loading conditions of the problem is provided in Fig. 17.13. For this particular problem, the design metrics are: (1) the volume of the structure, and (2) a linear combination of the displacements at node P, Δ. The design metrics are to be minimized.
The cross sectional areas of the three-bar truss are the design variables, which are allowed to vary between 0.1 and 2 cm^2. The stresses in each bar are limited to 200 MPa. The length L is fixed to 100 cm. The forces, F, which are applied at node P, have the same value of 20 kN. The modulus of elasticity of the material used is 200 GPa. Koski (1985) used a linear combination of the displacement

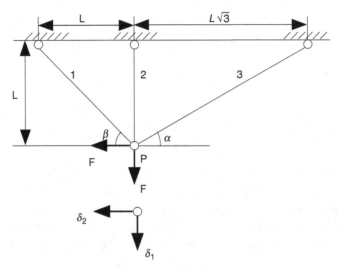

Figure 17.13. Three-Bar Truss Under Static Loading

design metric, Δ, at node P to yield nonconvexity. The coefficients of δ_1 and δ_2 are 0.25 and 0.75, respectively.

Generate the Pareto frontier using the normalized normal constraint method with the Pareto filter.

17.4 Understand how the normalized normal constraint method [3] works, and use that method to duplicate the Pareto frontier shown in Fig. 16.8(a) from Chapter 16.

17.5 Solve Problem 17.4 using a new version of the NNC method recently reported in Ref. [4]. Discuss any possible benefits or drawbacks.

BIBLIOGRAPHY OF CHAPTER 17

[1] I. Das and J. E. Dennis. Normal-boundary intersection: A new method for generating the Pareto surface in nonlinear multicriteria optimization problems. *SIAM J. OPTIM.*, 8(3):631–657, August 1998.

[2] F. Logist and J. Van Impe. Novel insights for multi-objective optimisation in engineering using Normal Boundary Intersection and (Enhanced) Normalised Normal Constraint. *Structural and Multidisciplinary Optimization*, 45(3):417–431, 2012.

[3] A. Messac, A. Ismail-Yahaya, and C. A. Mattson. The normalized normal constraint method for generating the Pareto Frontier. *Structural and Multidisciplinary Optimization*, 25(2):86–98, 2003.

[4] B. J. Hancock and C. A. Mattson. The smart normal constraint method for directly generating a smart Pareto set. *Structural and Multidisciplinary Optimization*, 48(4):763–775, 2013.

18

Physical Programming for Multiobjective Optimization

18.1 Overview

Engineering design problems are *multiobjective* in nature. These problems usually optimize two or more conflicting objectives – simultaneously. An approach to multiobjective problem formulation combines the multiple objectives into a single objective function, also known as the *Aggregate Objective Function* (AOF). This AOF is solved to obtain one Pareto solution. One of several challenges in the area of multiobjective optimization is to judiciously construct an AOF that satisfactorily models the designer's preferences. This chapter provides a concise presentation of the *Physical Programming* method, which defines a framework to effectively incorporate the designer's preferences into the AOF (Ref. [1]).

Several methods to solve multiobjective optimization problems have been discussed in Chapter 6, such as the weighted sum method, compromise programming, and goal programming. These weight-based approaches require the designer to specify numerical weights in defining the AOF. This process can be ambiguous. For example, consider the following: (1) How does the designer specify weights in weight-based approaches? (2) Do the weights reflect the designer's preferences accurately? If the designer chooses to increase the importance of a particular objective, by how much should he/she increase the weight? Is 25% adequate? Or is 200% adequate? (3) Does the AOF denote a true mathematical representation of the designer's preferences?

The above questions begin to explain that the problem of determining "good weights" can be difficult and dubious. Because of this ambiguity, the weight selection process is often a computational bottleneck in large scale multiobjective design optimization problems. The above discussion paves the way for a multiobjective problem formulation framework that alleviates these ambiguities: *Physical Programming* (PP) developed by Messac [2].

Physical Programming systematically develops an AOF that effectively reflects the designer's wishes. This approach eliminates the need for iterative weight setting, which alleviates the above discussed ambiguities. Instead of choosing weights, the designer chooses ranges of desirability for each objective. The PP method formulates the AOF from these ranges of desirability, while yielding interesting and useful properties for the AOF.

The **PhysPro** software embodies the Physical Programming method, which is described below and fully presented in Refs. [2, 3]. For information regarding **PhysPro**, please visit www.physpro.com (Ref. [4]).

Next, Linear Physical Programming (LPP) is studied in detail, followed by Nonlinear Physical Programming (NPP).

18.2 Linear Physical Programming (LPP)

18.2.1 Classification of Preferences: Soft and Hard

Using Physical Programming, the designer can express preferences for each design objective with more flexibility, as opposed to specifying *maximize, minimize, greater than, less than*, or *equal to*, which are the only choices available in conventional optimization approaches. Using the PP approach, a designer can express preferences with respect to each design objective using four different classes.

Figure 18.1 illustrates the four classes available in LPP. A generic design objective, μ_i, is represented on the horizontal axis, and the function to be minimized for that objective, z_i, henceforth called the *preference function* or the *class function*, is represented on the vertical axis. Each class consists of two subclasses, *hard* and *soft*, referring to the sharpness of the preference. These subclasses are also illustrated in Fig. 18.1, and are characterized as follows:

1. Soft Classes:
   a) Class 1S: Smaller-is-better (minimization)
   b) Class 2S: Larger-is-better (maximization)
   c) Class 3S: Value-is-better
   d) Class 4S: Range-is-better

2. Hard Classes:
   a) Class 1H: Must be smaller
   b) Class 2H: Must be larger
   c) Class 3H: Must be equal
   d) Class 4H: Must be in range

Physical Programming offers a flexible lexicon to express *ranges of desirability* for both hard and soft classes. The lexicon consists of six ranges of desirability for classes 1S and 2S, ten ranges for the class 3S, and eleven ranges for the class 4S.

18.2.2 Ranges of Desirability for Various Classes

Following are the definitions of the differing ranges of desirability under LPP, with which a designer can express his/her preferences. To explain, consider the case of class 1S shown in Fig. 18.1. The ranges of desirability are defined as follows in order of decreasing preference.

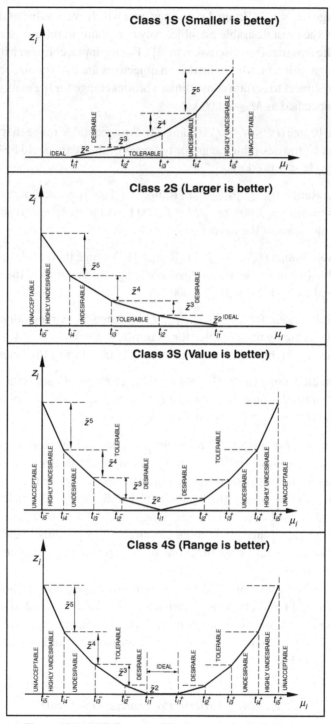

Figure 18.1. LPP Ranges of Preferences for Soft Classes

1. Ideal Range ($\mu_i \leq t_{i1}^+$) (Range 1): A range over which every value of the criterion is *ideal*, or the most desirable possible. Any two points in this range are of equal value to the designer (see discussion in [3]). For example, consider a hypothetical beam design problem, where the design objectives are to minimize the mass and deflection subject to certain constraints. The ideal range for the mass of the beam could be specified as $M \leq 2,000$ kg.

2. Desirable Range ($t_{i1}^+ \leq \mu_i \leq t_{i2}^+$) (Range 2): An acceptable range that is desirable (for example, the desirable range for the mass of the beam could be specified as $2,000$ kg $\leq M \leq 3,000$ kg).

3. Tolerable Range ($t_{i2}^+ \leq \mu_i \leq t_{i3}^+$) (Range 3): This is an acceptable, tolerable range (for example, $3,000$ kg $\leq M \leq 3,500$ kg could be specified as a tolerable range for the mass of the beam).

4. Undesirable Range ($t_{i3}^+ \leq \mu_i \leq t_{i4}^+$) (Range 4): A range that, while acceptable, is undesirable (for example, the undesirable range for the mass of the beam could be specified as $3,500$ kg $\leq M \leq 4,000$ kg).

5. Highly Undesirable Range ($t_{i4}^+ \leq \mu_i \leq t_{i5}^+$) (Range 5): A range that, while still acceptable, is highly undesirable (for example, $4,000$ kg $\leq M \leq 4,500$ kg could be specified as the highly undesirable range for the mass of the beam).

6. Unacceptable Range ($\mu_i \geq t_{i5}^+$) (Range 6): The range of values that the design objective must not take (the range $M \geq 4,500$ kg could be specified as the unacceptable range for the mass of the beam).

The range-defining parameters t_{i1}^+ through t_{i5}^+, defined above for soft classes, are physically meaningful constants that are specified by the designer to quantify the preferences associated with the ith design objective. For example, the set of t_{ij}^+ values specified above for the mass of the beam are $[2,000, 3,000, 3,500, 4,000, 4,500]$.

In the case of hard classes, only two ranges are defined, *acceptable* and *unacceptable*. All soft class functions become constituent components of the AOF to be minimized, and all the hard class functions simply appear as constraints in the LPP problem formulation.

The preference functions map the design objectives into non-dimensional, strictly positive real numbers. This mapping transforms disparate design objectives with different physical meanings onto a dimensionless scale through a unimodal convex function. The preference functions are piecewise linear and convex in the LPP method, as seen in Fig. 18.1.

18.2.3 Inter-Criteria Preferences: OVO Rule

Specifying *intra-criterion* preferences (preferences within one objective) using Physical Programming [3] has been explained. In order to completely formulate the multiobjective optimization problem, the designer also needs to specify the inter-criteria preferences (preferences among several objectives). The PP method operates within an *inter-criteria* heuristic rule, called the *One vs. Others* rule (OVO). The inter-criteria preference for each soft criterion, μ_i, is defined as follows. Consider the following two options:

1. Full improvement of μ_i across a given range of preference, and
2. Full reduction of all the other criteria across the next better range of preference.

The PP method then formulates the AOF so that Option 1 is preferred over Option 2. The OVO rule has a built-in preemptive nature whereby the worst criterion tends to be minimized first.

For example, consider a multiobjective problem with ten objectives. According to the OVO rule, it is preferable for a single objective to improve over a full tolerable range, than it is for the remaining nine objectives to improve over the full desirable range. The next subsection explains how the OVO rule is implemented in the LPP method.

18.2.4 LPP Class Function Definition

The class function maps the design objectives into non-dimensional, strictly positive real numbers that reflect the designer's preferences. To accomplish the above, the class function, z_i, is required to possess the following properties.

1. A lower value of the class function is preferred over a higher value thereof (see Fig. 18.1). Irrespective of the class of a criterion (1S, 2S, 3S, or 4S), the ideal value of the criterion always corresponds to the minimum value of the class function, which is zero.
2. A class function is strictly positive ($z_i \geq 0$).
3. A class function is continuous, piecewise linear, and convex.
4. The value of the class function at a given range limit, $z_i(t_{is}^+)$, is always fixed (see Fig. 18.1). From criterion to criterion, only the location of the limits (t_{is}^+) changes, but not the corresponding z_i values. Because of this property, as one travels across all the criteria, and observes a given range type, the change in the class function value, $\tilde{z}_i$, will always be of the same magnitude (see Fig. 18.1). This property of the class function results in a normalizing effect, eliminating numerical conditioning problems that arise because of improper scaling between design objectives of disparate magnitudes.
5. The magnitude of the class function's vertical excursion across any range must satisfy the OVO rule. (This property is represented in Eq. 18.3). Observe in Fig. 18.1 that the value of $\tilde{z}^2$ (desirable) is less than that of $\tilde{z}^5$ (highly undesirable). This is in keeping with the OVO rule.

Based on the above properties, the mathematical relations used in the LPP algorithm are now presented. From Property (4.), the following relation holds

$$z^s = z_i(t_{is}^+) = z_i(t_{is}^-); \quad \forall i; \quad (2 \leq s \leq 5); \quad z^1 = 0 \tag{18.1}$$

where s and i denote a generic range intersection and the soft criterion number, respectively.

The change in z_i across the sth range is given by

$$\tilde{z}^s = z^s - z^{s-1}; \quad (2 \leq s \leq 5); \quad z^1 = 0 \tag{18.2}$$

The OVO rule is enforced by the equation

$$\tilde{z}^s > (n_{sc} - 1)\tilde{z}^{s-1}; \quad (3 \leq s \leq 5); \quad (n_{sc} > 1) \tag{18.3}$$

or,

$$\tilde{z}^s = \beta(n_{sc} - 1)\tilde{z}^{s-1}; \quad (3 \leq s \leq 5); \quad n_{sc} > 1; \quad \beta > 1 \tag{18.4}$$

where n_{sc} denotes the number of soft criteria, and β is used as a convexity parameter. To use Eq. 18.4, the value of $\tilde{z}^2$ needs to be specified. Assume $\tilde{z}^2$ to be equal to a small positive number (say 0.1) in practice. Eq. 18.4 by itself does not guarantee convexity of the class function. The convexity also depends on the targets chosen by the decision maker.

The relations that specifically enforce convexity of the class function are defined by the following quantities:

$$\tilde{t}_{is}^+ = t_{is}^+ - t_{i(s-1)}^+; \quad (2 \leq s \leq 5) \tag{18.5}$$

$$\tilde{t}_{is}^- = t_{is}^- - t_{i(s-1)}^-; \quad (2 \leq s \leq 5) \tag{18.6}$$

Note that the above equations define the length of the sth range of the ith criterion. Using the above definition, the magnitude of the slope of the class function is given by

$$w_{is}^+ = \frac{\tilde{z}^s}{\tilde{t}_{is}^+}; \quad (2 \leq s \leq 5) \tag{18.7}$$

$$w_{is}^- = \left| \frac{\tilde{z}^s}{\tilde{t}_{is}^-} \right|; \quad (2 \leq s \leq 5) \tag{18.8}$$

Note that these slopes change from range to range, and from criterion to criterion. The convexity requirement can be enforced by using the relation

$$\tilde{w}_{min} = \min_{i,s}(\tilde{w}_{is}^+, \tilde{w}_{is}^-) > 0 \quad (2 \leq s \leq 5) \tag{18.9}$$

where

$$\tilde{w}_{is}^+ = w_{is}^+ - w_{i(s-1)}^+; \quad (2 \leq s \leq 5) \tag{18.10}$$

$$\tilde{w}_{is}^- = w_{is}^- - w_{i(s-1)}^-; \quad (2 \leq s \leq 5) \tag{18.11}$$

$$w_{i1}^- = w_{i1}^+ = 0; \quad (2 \leq s \leq 5) \tag{18.12}$$

The quantities $\tilde{w}_{is}^+$ and $\tilde{w}_{is}^-$ defined above are the slope increments of the class function between the different ranges of desirability. Equation 18.9 implies that if all the incremental weights are positive, the class function (which is piecewise linear) will be convex. The LPP weight algorithm can be used to define the class function using the equations given in this subsection.

18.2.5 LPP Weight Algorithm

The LPP weight algorithm is given below.

1. Initialize: $\beta = 1.1$; $w_{i1}^- = w_{i1}^+ = 0$; $\tilde{z}^2 =$ small positive number, say 0.1. $i = 0$; $s = 1$; $n_{sc} =$ number of soft criteria.
2. Set $i = i + 1$.
3. Set $s = s + 1$.
4. Evaluate in the same order: $\tilde{z}^s$, $\tilde{t}_{is}^+$, $\tilde{t}_{is}^-$, w_{is}^+, w_{is}^-, $\tilde{w}_{is}^+$, $\tilde{w}_{is}^-$, and $\tilde{w}_{min}$.
5. If $\tilde{w}_{min}$ is less than some chosen small positive value (say 0.01), increase β. Set $i = 0, s = 1$, and go to Step 2..
6. If $s \neq 5$, go to Step 3..
7. If $i = n_{sc}$, terminate. Otherwise, go to Step 2..

A MATLAB code that uses this algorithm to compute weights, given the preference values for each criterion, is given in the book website (www.cambridge.org/Messac). Once the weights are obtained from the above algorithm, the piecewise linear class function can be defined for each criterion.

The formulation of the LPP problem involves the presence of numerous weights because of the piecewise linear nature of the class function. *However, the designer need not choose these weights.* All the required weights are automatically evaluated by the LPP weight algorithm. The LPP AOF is defined using *deviational variables*, denoted by d_{is}^- and d_{is}^+. A deviational variable is defined as the deviation of the ith design criterion from its sth range intersection. The class function for soft classes can then be defined in terms of the deviational variables as

$$z_i = \sum_{s=2}^{5} \{\tilde{w}_{is}^- d_{is}^- + \tilde{w}_{is}^+ d_{is}^+\} \tag{18.13}$$

18.2.6 LPP Problem Formulation

The LPP application procedure consists of four distinct steps.

1. Specify the class type for each design objective (1S-4H).
2. Provide the ranges of desirability (t_{is}^+, or t_{is}^-, or both) for each class (see Fig. 18.1). The designer specifies five limits for classes 1S or 2S, nine limits for the class 3S, and ten limits for the class 4S. For hard classes, the designer specifies one limit for classes 1H, 2H, and 3H, and two limits for 4H (see Fig. 18.1).

3. Use the LPP weight algorithm to obtain the incremental weights, $\tilde{w}_{is}^+$ and $\tilde{w}_{is}^-$. *Note that the designer does not need to explicitly define the class function z_i.*
4. Solve the following linear programming problem.

$$\min_{d_{is}^-,d_{is}^+,x} J = \sum_{i=1}^{n_{sc}} \left[\sum_{s=2}^{5} \{\tilde{w}_{is}^- d_{is}^- + \tilde{w}_{is}^+ d_{is}^+\} \right] \tag{18.14}$$

subject to

$$\mu_i - d_{is}^+ \le t_{i(s-1)}^+; \quad d_{is}^+ \ge 0; \quad \mu_i \le t_{i5}^+ \qquad (1S, 3S, 4S) \tag{18.15}$$

$$\mu_i + d_{is}^- \ge t_{i(s-1)}^-; \quad d_{is}^- \ge 0; \quad \mu_i \ge t_{i5}^- \qquad (2S, 3S, 4S) \tag{18.16}$$

$$\mu_j \le t_{j,max} \qquad (1H) \tag{18.17}$$

$$\mu_j \ge t_{j,min} \qquad (2H) \tag{18.18}$$

$$\mu_j = t_{j,val} \qquad (3H) \tag{18.19}$$

$$t_{j,min} \le \mu_j \le t_{j,max} \qquad (4H) \tag{18.20}$$

$$x_{min} \le x \le x_{max} \tag{18.21}$$

where $i = \{1, 2, ..., n_{sc}\}$, $s = \{2, ..., 5\}$, $j = \{1, 2, ..., n_{hc}\}$, n_{hc} is the number of hard classes, x is the design variable vector, and $\mu_i = \mu_i(x)$.

A recent application of LPP is provided in Ref. [5]. The Nonlinear Physical Programming (NPP) method is described next.

18.3 Nonlinear Physical Programming (NPP)

The NPP method can be more advantageous than the LPP method for solving certain optimization problems. The piecewise linear nature of the class function in LPP may lead to computational difficulties because of the discontinuities in the class function derivatives at the intersection of the range limits. The NPP method alleviates this difficulty by defining a class function that is smooth across all range limits. However, the NPP method can be computationally expensive, since it is formulated as a nonlinear optimization problem.

This section provides a brief discussion of the NPP method. Interested readers can refer to [2] for a more detailed description of the NPP method. First, the similarities and differences between LPP and NPP will be identified.

18.3.1 LPP vs. NPP

The LPP and NPP methods share certain similarities, as listed below.

1. The class and the subclass definitions are the same in LPP and NPP.
2. The PP lexicon and the classification of preferences are the same for NPP and LPP, with one exception; the analog of *ideal* (LPP) is *highly desirable* (NPP).

Figure 18.2 provides the classification of the design objectives, and the ranges of differing preferences for soft classes in NPP.

3. The OVO rule is defined in the same manner in LPP and NPP.

The difference between NPP and LPP can be observed by comparing the class function plot in Fig. 18.2 with that in Fig. 18.1. In the case of LPP, the class functions are piecewise linear. For NPP, they are nonlinear and smooth.

The class functions in NPP are defined using a special class of *splines*. A detailed discussion regarding the mathematical development of these splines can be found in [2]. A summary of the mathematical background for NPP is presented next.

18.3.2 NPP Class Function Definition

A suitable class function in NPP must possess the following properties.

1. All soft class functions must:
   a) be strictly positive,
   b) have continuous first derivatives, and
   c) have strictly positive second derivatives (implying convexity of the class function).

2. All the above defined properties must hold for any practical choice of range limits.

The NPP class function (Fig. 18.2) in the highly desirable range is defined by a decaying exponential function; while in all the other ranges, the class functions are defined by spline segments [2]. A detailed description of the class function properties and definition is provided in [2].

18.3.3 NPP Problem Model

The following steps are used to generate the NPP problem.

1. Specify the class type for each design objective (1S - 4H).
2. Provide the ranges of desirability for each design objective (see Fig. 18.2).
3. Solve the constrained nonlinear minimization problem that is given by

$$\min_{x} \ J = \log_{10} \left[\frac{1}{n_{sc}} \sum_{i=1}^{n_{sc}} z_i[\mu_i(x)] \right] \text{ (soft classes)} \tag{18.22}$$

subject to

$$\mu_i(x) \leq t_{i5}^+ \qquad\qquad \text{(1S)} \qquad\qquad (18.23)$$

$$\mu_i(x) \geq t_{i5}^- \qquad\qquad \text{(2S)} \qquad\qquad (18.24)$$

$$t_{i5}^- \leq \mu_i(x) \leq t_{i5}^+ \qquad\qquad \text{(3S, 4S)} \qquad\qquad (18.25)$$

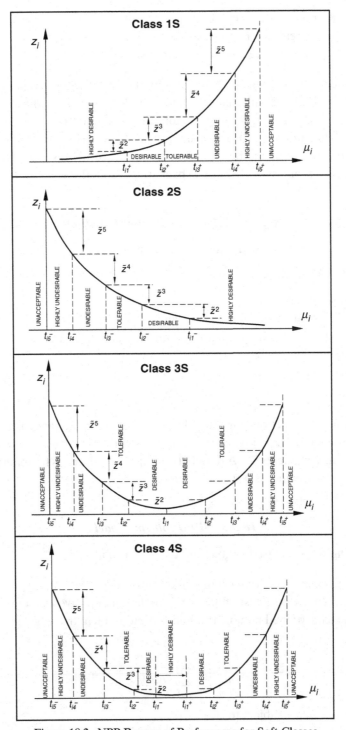

Figure 18.2. NPP Ranges of Preferences for Soft Classes

$$\mu_i(x) \leq t_{i,max} \qquad \text{(1H)} \qquad (18.26)$$

$$\mu_i(x) \geq t_{i,min} \qquad \text{(2H)} \qquad (18.27)$$

$$\mu_i(x) = t_{i,val} \qquad \text{(3H)} \qquad (18.28)$$

$$t_{i,min} \leq \mu_i(x) \leq t_{i,max} \qquad \text{(4H)} \qquad (18.29)$$

$$x_{j,min} \leq x_j \leq x_{j,max} \qquad \qquad (18.30)$$

where $t_{i,min}$, $t_{i,max}$, and $t_{i,val}$ represent the specified preferences values for the ith hard objective; and $x_{j,min}$ and $x_{j,max}$ are the minimum and the maximum values for x_j, respectively. The ranges of desirability, t_{i5}^{+} and t_{i5}^{-}, are provided by the designer, and n_{sc} is the number of soft objectives. The hard classes are treated as constraints, while the soft classes are part of the objective function. Plans are for a limited edition of the NPP software to be provided in the book website (`www.cambridge.org/Messac`).

18.4 Comparison of LPP with Goal Programming

The flexibility offered by the LPP method is now compared to that offered by goal programming, previously discussed in Chapter 6. As shown in Fig 18.3, the GP method offers limited flexibility, with the option of choosing two weights and a target value for each objective. The LPP method, in contrast, allows the designer to choose up to ten physically meaningful target values or ranges of desirability for each design objective. While the designer is required to choose the weights in the GP method, the LPP method completely eliminates the often ambiguous task of choosing weights.

In Fig. 18.4, three-dimensional visualizations of the AOF for the GP and the LPP methods are presented. The XY plane of each figure provides the contour plots of the AOF for each method. In typical GP form, there are two-sided goals/criteria, yielding an intersection of four planes. Also, note that the contour plots of the GP AOF are quadrilaterals.

The AOF surface for LPP is obtained by the intersection of 81 planes (for the 4-S criterion), which reflects a more realistic preference. Observe the multi-faceted contours of the AOF for the LPP method.

The effectiveness of LPP comes from the well defined class function, which tailors itself to the complex nature of the designer's choices. A numerical example to illustrate the LPP approach is presented.

18.5 Numerical Example

This example solves the optimization problem using the LPP method, and compares the results to those obtained by the GP method. A company manufactures two products, A and B. The ideal production levels per month for A and B are 25 units and 10 units, respectively. The profit per unit sold for A and B are $12k and $10k, respectively. Under these conditions, the total monthly profit is $400k. The company needs to make a profit of at least $580k to stay in business. The designer has certain

Table 18.1. *Preference Ranges for μ_1 and μ_2*

Preference level	μ_1	μ_2
Ideal	< 25	< 10
Desirable	25 - 31	10 - 18
Tolerable	31 - 36	18 - 26
Undesirable	36 - 44	26 - 33
Highly Undesirable	44 - 50	33 - 40
Unacceptable	> 50	> 40

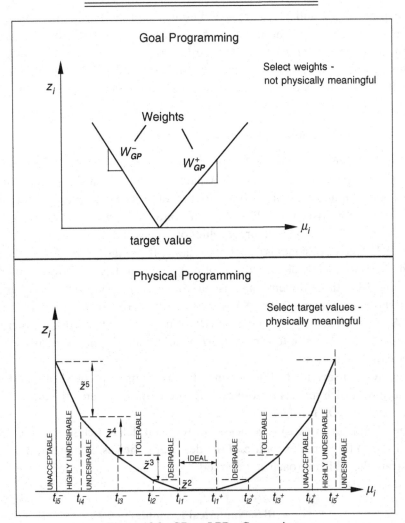

Figure 18.3. GP vs. LPP—Comparison

target preferences for the production levels for A and B, given in Table 18.1. Define μ_1 and μ_2 as the two design criteria, which denote the production levels of products A and B, respectively. The profit constraint function is given as

$$12\mu_1 + 10\mu_2 \geq 580 \qquad (18.31)$$

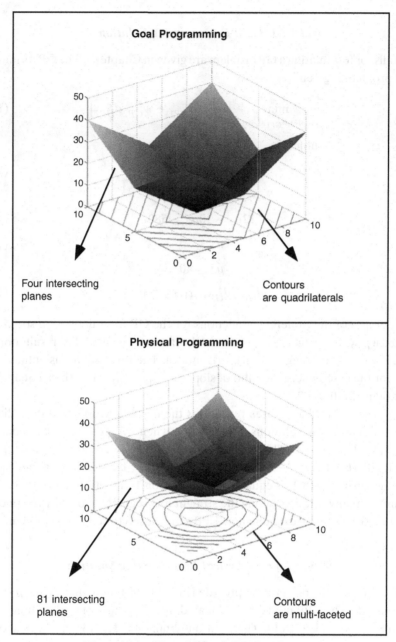

Figure 18.4. GP vs. LPP—AOF Visualization

18.5.1 Goal Programming Solution

The details for formulating a GP problem are given in Chapter 6. The GP formulation for this problem is given by

$$\min_{\mu_1,\mu_2,d_{GP1}^+,d_{GP2}^+} \left[w_{GP1}^+ d_{GP1}^+ + w_{GP2}^+ d_{GP2}^+ \right] \tag{18.32}$$

subject to

$$\mu_1 - d_{GP1}^+ \leq 25 \tag{18.33}$$

$$\mu_2 - d_{GP2}^+ \leq 10 \tag{18.34}$$

$$12\mu_1 + 10\mu_2 \geq 580 \tag{18.35}$$

$$\mu_1 \leq 50 \tag{18.36}$$

$$\mu_2 \leq 40 \tag{18.37}$$

$$d_{GP1}^+, d_{GP2}^+, \mu_1, \mu_2 \geq 0 \tag{18.38}$$

The slopes of the preference functions for the GP formulation are specified by w_{GP1}^+ and w_{GP2}^+. The target for μ_1 is 25, and the target for μ_2 is 10. The results obtained using GP are shown in Fig. 18.5 (a), (b), and (c). The three solutions obtained with GP are for the cases where the ratio of slopes w_{GP1}^+/w_{GP2}^+ is less than, equal to, and greater than $12/10 = 1.2$.

In Fig. 18.5, the shaded area represents the feasible region, and the solid dots represent the optimum solutions. The solution when $w_{GP1}^+/w_{GP2}^+ < 1.2$ is the point P $=(40, 10)$ in Fig. 18.5 (a). The solution when $w_{GP1}^+/w_{GP2}^+ > 1.2$ is the point Q $= (25, 28)$ in Fig. 18.5 (c). In Fig. 18.5 (b), when $w_{GP1}^+/w_{GP2}^+ = 1.2$, the slope of the objective function given in Eq. 18.32 is equal to that of the constraint given in Eq. 18.35. There are infinitely many solutions along the straight line segment shown by the thick line in Fig. 18.5 (b). We now examine how LPP can be used to solve this problem.

18.5.2 Linear Physical Programming Solution

From the values of the preferences provided in Table 18.1, note that μ_1 and μ_2 belong to the class 1S. The LPP model is formulated using the linear programming model, given in Sec. 18.2.6, Eq. 18.14. The solution obtained is R $= (31, 20.8)$, as shown in Fig. 18.5 (d).

Compare the solutions P and Q obtained by the GP method, and the solution R obtained by the LPP method. The solution obtained with GP is highly sensitive to the weights chosen for each objective. For the point P $= (40, 10)$ (see Fig. 18.5 (a)), μ_1 lies in the undesirable range, while μ_2 lies in the desirable range. For the point Q $= (25, 28)$ (see Fig. 18.5 (c)), μ_1 lies in the desirable range, while μ_2 lies in the undesirable range. The solutions P and Q lie in the undesirable ranges because the GP problem formulation does not fully represent the designer's preferences given in Table 18.1. The LPP method, on the other hand, utilizes all the information

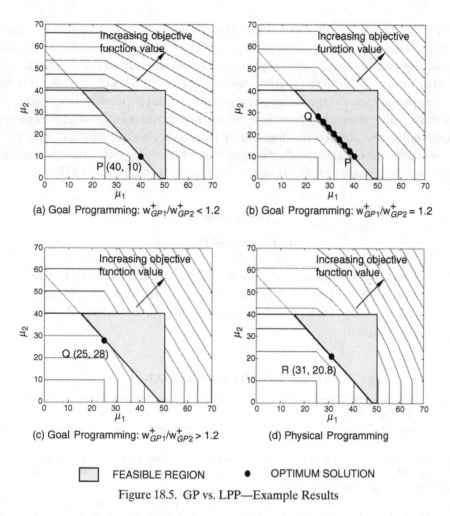

(a) Goal Programming: $w^+_{GP1}/w^+_{GP2} < 1.2$

(b) Goal Programming: $w^+_{GP1}/w^+_{GP2} = 1.2$

(c) Goal Programming: $w^+_{GP1}/w^+_{GP2} > 1.2$

(d) Physical Programming

☐ FEASIBLE REGION   ● OPTIMUM SOLUTION

Figure 18.5. GP vs. LPP—Example Results

provided by the designer in Table 18.1 to formulate the problem. With the LPP method, observe that the optimum point $R = (31, 20.8)$ (Fig. 18.5 (d)) lies on the desirable/tolerable boundary for μ_1, and within the tolerable range for μ_2.

In addition, it is interesting to contrast the contours of the LPP AOF against those of the GP AOF in Fig. 18.5. The shape and the number of sides of these contours are significantly different. These observations can be better understood from Fig. 18.4, where a three-dimensional representation is provided.

18.6 Summary

Most numerical optimization algorithms are developed for application to single objective problems. To pose the multiobjective problem in a single objective framework, the designer needs to effectively *aggregate* the criteria into a single AOF. In doing so, he/she has to model the intra-criterion and inter-criteria preferences into the AOF. This chapter presented the Physical Programming framework for

AOF formulation. The PP method provides a framework to unambiguously incorporate the designer's preferences into the AOF. The PP method precludes the need for the designer to specify physically meaningless weights. The PP algorithm generates the weights of the class function based on the designer's preferences, allowing the designer to focus on specifying physically meaningful preference values for each objective. This renders the PP method unique, and provides an effective framework for multiobjective decision-making. The PP method has been applied to a wide variety of applications, such as product design, multiobjective robust design, production planning, and aircraft structure optimization (see Refs. [6, 7, 8, 9, 10, 11]).

18.7 Problems

Graduate Level Problems

18.1   Read the paper: Messac, A., "From Dubious Construction of Objective Functions to the Application of Physical Programming" (Ref. [12]). Prepare a two page summary of this paper in your own words, emphasizing the key messages and approach.

BIBLIOGRAPHY OF CHAPTER 18

[1] K. E. Lewis, W. Chen, and L. C. Schmidt. *Decision Making in Engineering Design.* ASME Press, 2006.
[2] A. Messac. Physical Programming: Effective optimization for computational design. *AIAA Journal*, 34(1):149–158, 1996.
[3] A. Messac, S. Gupta, and B. Akbulut. Linear Physical Programming: A new approach to multiple objective optimization. *Transactions on Operational Research*, 8:39–59, 1996.
[4] PhysPro. www.physpro.com.
[5] K. K. Pochampally and S. M. Gupta. Use of linear physical programming and Bayesian updating for design issues in reverse logistics. *International Journal of Production Research*, 50(5):1349–1359, 2012.
[6] B. H. Wilson, C. Erin, and A. Messac. Optimal design of a vibration isolation mount using Physical Programming. *Journal of Dynamic Systems, Measurement, and Control*, 121(2):171–178, 1999.
[7] A. Messac and A. Ismail-Yahaya. Multiobjective robust design using Physical Programming. *Structural and Multidisciplinary Optimization*, 23(5):357–371, 2002.
[8] A. Maria, C. A. Mattson, A. Ismail-Yahaya, and A. Messac. Linear Physical Programming for production planning optimization. *Engineering Optimization*, 35(1):19–37, 2003.
[9] A. Messac, W. M. Batayneh, and A. Ismail-Yahaya. Production planning optimization with Physical Programming. *Engineering Optimization*, 34(4):323–340, 2002.
[10] M. P. Martinez, A. Messac, and M. Rais-Rohani. Manufacturability-based optimization of aircraft structures using Physical Programming. *AIAA Journal*, 39(3):517–525, 2001.
[11] A. Messac and P. D. Hattis. Physical programming design optimization for high speed civil transport. *Journal of Aircraft*, 33(2):446–449, 1996.
[12] A. Messac. From the dubious construction of objective functions to the application of Physical Programming. *AIAA Journal*, 38(1):155–163, January 2000.

19

Evolutionary Algorithms

19.1 Overview

This book has presented various algorithms and applications where the optimizer was primarily gradient-based (*i.e.*, the search direction is governed by gradient and/or Hessian information). This chapter introduces an entirely different class of optimization algorithms called the *evolutionary algorithms* (EA). Evolutionary algorithms imitate natural selection processes to develop powerful computational algorithms to select optimal solutions. *Genetic algorithms* (GA), *simulated annealing* (SA), *ant colony optimization* (ACO), *particle swarm optimization* (PSO), and *tabu search* (TS) are some of the popular techniques that fall under the umbrella of evolutionary algorithms.

The motivation for using biologically-inspired computational approaches stems from two key observation. *First*, the mathematical optimization algorithms in solving complex problems in engineering, computing, and other fields suffer strong limitations. The common challenges in these areas revolve around the lack of mathematical models that define the physical phenomena, discontinuous functions, and high non-linearity. *Second*, many complex problems encountered in engineering already exist in nature in some relevant form. Optimization is inherent in nature, such as in the process of adaptation performed by biological organisms in order to survive. Engineers and scientists continue to explore the various efficient problem-solving techniques employed by nature to optimize natural systems.

The relative advantages and limitations of EAs vs. traditional optimization methods are as follows:

1. Traditional algorithms typically generate a single candidate optimum at each iteration that progresses toward the optimal solution. Evolutionary algorithms generate a population of points at each iteration. The best point in the population approaches an optimal solution.

2. Traditional algorithms calculate the candidate optimal point at the next iteration by a deterministic computation. EAs usually select the next population by a combination of operations that use random number generators.

3. Traditional algorithms require gradient and/or Hessian information to proceed, while EAs usually require only function values. As a result, EAs can solve a variety of optimization problems in which the objective function is not smooth and potentially discontinuous.

4. To their disadvantage, EAs often require more function evaluations than do gradient-based methods, particularly for single-objective optimization. The computation time associated with EAs is longer than that of the gradient-based methods.

5. Evolutionary algorithms are stochastic methods that typically involve random choices. Therefore, different runs of the same EA code may yield different optimal solutions.

6. Evolutionary algorithms do not have proofs of convergence to an optimal solution; unlike gradient based methods, where at least a local optimum is guaranteed upon convergence. It has been observed in practice that evolutionary algorithms, if employed with careful settings, have the potential to yield globally optimal solutions. Because of the inherent randomness in the search process of EAs, a much larger solution space is generally explored when compared to traditional methods, which are limited in their search scope.

This chapter focusses on genetic algorithms (Sec. 19.2) because they are the most popular evolutionary algorithms used in the design optimization community. Using a simplified example, the basic concept of a genetic algorithm is explained. Multiobjective optimization with GAs is discussed in Sec. 19.3 using an example. A brief overview of other evolutionary algorithms is provided with pertinent references in Sec. 19.4. A summary of the chapter is provided in Sec. 19.5.

19.2 Genetic Algorithms

This section explains the basics of how a GA works. Practical software implementations are outside the scope of this introductory chapter. MATLAB provides the "Genetic Algorithms and Direct Search Solvers" that provides software implementations of optimization using GAs. Genetic algorithms have been used to solve a wide range of problems involving continuous and discrete variables. For example, the use of GAs is popular for the optimization of laminate composite structures [1], multiobjective optimization (presented later), and structural and design problems. A simplified version of a GA is demonstrated next using an example.

19.2.1 Basics of Genetic Algorithms

A genetic algorithm repeatedly modifies a set or *population* of solutions or *individuals* in the course of its entire run. At each step, the genetic algorithm selects individuals from the current population to be *parents* based on certain criteria (discussed shortly). The parents are then used to produce the next generation of individuals,

	Initial Population				Reproduction				
	Initial random population	decimal equivalent	function value	Initial random population (parent)	Reproduction option (random choice)		New population (child)	decimal equivalent	function value
1	1 0 1 0 1	21	441	1 0 1 0 1	*crossover with 2 – bits 1, 2, 3*		1 1 0 0 1	25	625
2	1 1 0 0 1	25	625	1 1 0 0 1	*crossover with 1 – bits 1, 2, 3*		1 0 1 0 1	21	441
3	0 1 0 0 1	9	81	0 1 0 0 1	*elite*		0 1 0 0 1	9	81
4	1 1 1 0 1	29	841	1 1 1 0 1	*crossover with 5 – bit 2*		1 0 1 0 1	21	441
5	1 0 1 1 1	23	529	1 0 1 1 1	*crossover with 4 – bit 2*		1 1 1 1 1	31	961
6	1 0 0 0 0	16	256	1 0 0 0 0	*Mutation (bits 1, 3)*		0 0 1 0 0	4	16

Figure 19.1. Basics of Genetic Algorithms

called *children*. Over successive generations, the population "evolves" toward an optimal solution, or a set of Pareto optimal solutions (in the case of a multiobjective problem).

This section describes how GAs work by minimizing the function $f(x) = x^2$, $0 < x < 40$. The objective function is also known as the *fitness function* in the GA literature. A similar example that further helps us to understand how GAs work is available in Ref. [2].

In its simplest form, a GA implementation involves the following tasks.

1. **Encoding:** Encoding is a method that represent individuals in evolutionary algorithms. Typically, individuals are coded as a fixed length string (*e.g.*, a binary number with 0's or 1's). This string is also known as a *chromosome*. Other variable string length encodings are also possible [3, 2]. Coding is a representation of a number as a string of 0's and 1's.

   Example: We could use a string length of 5 to code a number in binary (*e.g.*, 10001). This string can be de-coded into a base 10 decimal number as

   $$1 \times 2^4 + 0 \times 2^3 + 0 \times 2^2 + 0 \times 2^1 + 1 \times 2^0 = 16 + 1 = 17 \qquad (19.1)$$

2. **Initial population:** The algorithm begins by generating a random initial population. Important initialization choices, such as the number of individuals in each population and the number of bits in the encoding, must be made by the user. These choices govern the performance of the GA. A detailed discussion can be found in [2].

   Example: Assume six individuals in the population for the present example. Each individual (represented by a row in Fig. 19.1) is randomly generated by a series of five fair coin flips, heads=1 and tails=0. Note that five coin flips are needed because we chose to encode each individual using a 5-bit binary string. Say the initial population generated is 10101, 11001, 01001, 11101, 10111, and 10000 (Column 2 in Fig. 19.1). In the decimal system, the initial population is the following set of numbers {21, 25, 9, 29, 23, 16} (Column 3 in

Fig. 19.1). Now evaluate the fitness function value ($f(x) = x^2$) at each of the individuals in the initial population. The fitness function value set is {441, 625, 81, 841, 529, 256} (Column 4 in Fig. 19.1).

Now proceed to the next step.

3. **Reproduction:** A new generation, called *child*, in the genetic algorithm is created by reproduction from the previous generation, called the *parent*. The notion of "survival of the fittest" is usually used in genetic algorithms. There are three main mechanisms used to create a new generation. Different implementations of GAs use different combinations of the three ideas below.

   a) *Elitism*: In this approach, the individuals with the best fitness values in the current generation are guaranteed to survive in the next generation.

      Example: In the example considered earlier, of the five individuals in the parent population, {10101, 11001, 01001, 11101, 10111}, the third individual, 01001, had the lowest function value for the current minimization problem. This individual is considered *elite*, and will become part of the next generation.

   b) *Crossover:* In this technique, some bits of the encoded string of one parent individual are exchanged with the corresponding bits of another parent individual. A series of random choices are made for this mechanism, which are explained with the following example.

      Example: Assume that the elite individual, 01001, is part of the next generation. First, choose which individual is to be crossed over with which individual (*e.g.*, Individual 1 with Individual 2, Individual 1 with Individual 3, or Individual 1 with Individual 4). This choice is usually made randomly. Assume that Individual 1, 10101, is crossed over with Individual 2, 11001; and Individual 4, 11101, is crossed over with Individual 5, 10111.
      Next, decide how many bits in the individuals will be exchanged. For example, will it be 1, 2, 3, or 4 bits? This choice is also made randomly. Assume that 3 bits in the string will be exchanged for Individuals 1 and 2, and one bit will be exchanged for Individuals 4 and 5 (shown by the grey shaded regions in Rows 1, 2, 4, and 5 in Fig. 19.1).
      The final choice to be made (again implemented using random numbers) is the positions of the bits. For example, exchange the first three bits or the last three bits. For this example, the first three bits will be exchanged for Individuals 1 and 2, and Bit 2 for Individuals 4 and 5 (shown by the grey shaded regions in Rows 1, 2, 4, and 5 in Fig. 19.1).

   c) *Mutation:* Unlike the crossover operation (which requires two parents), mutation children are generated from a single parent by randomly reversing some bits from 0 to 1, or vice versa. In most GA implementations, a probability value for a mutation to occur is assumed.

Example: Make the random choice that Individual 6 goes through a mutation on Bits 1 and 3. As shown in Fig. 19.1, the bits are reversed for these two positions, leading to a new child (noted by the grey shaded regions in the last row in Fig. 19.1).

Column 7 in Fig. 19.1 presents the child population generated by the above three mechanisms. The decimal equivalents are reported in Column 8.

4. The function values of the new population that is generated are computed, as shown in Fig. 19.1. Using a combination of the above reproduction options, the algorithm proceeds further until a desired stopping criterion is achieved. Examples of stopping criteria include the number of generations, time limit, and function tolerance.

Example: Observe the function values of the child population presented in the last column of Fig. 19.1. The best individual in the child generation shows a decrease in the function value when compared to the parent generation.

Options available in the MATLAB Genetic Algorithm and Direct Search Solvers are specified within the Global Optimization Toolbox. The GA Solver will be used to demonstrate how a software implementation of GA works.

19.2.2 Options in MATLAB

The MATLAB GA Solver, within the Global Optimization Toolbox, has software implementations of several GA capabilities. The ga command provides nonlinear constrained optimization capabilities, and the gamultiobj command allows for multiobjective optimization (we will study this later). In this subsection, we will solve an example problem using the traditional gradient based solvers in MATLAB (fmincon) and the genetic algorithm routine.

We note that the genetic algorithm by itself allows only for unconstrained optimization. Nonlinear constraints are usually incorporated using penalty schemes, which were discussed in Chapter 13. Consider the following constrained optimization problem.

$$\min_{x} \ \{f = x_1^2 + 10x_2^2 - 3x_1x_2\} \tag{19.2}$$

subject to

$$2x_1 + x_2 \geq 4 \tag{19.3}$$

$$x_1 + x_2 \geq -5 \tag{19.4}$$

$$-5 \leq x_1, x_2 \leq 5 \tag{19.5}$$

Let us now discuss how this problem can be setup in MATLAB. There are two options possible: (1) we can invoke the ga command from the graphical user interface of the GA Solver (we will explore this option in the next section), or (2) we can call the `ga` command from an M-file. We will need to generate three M-files: a main file, an objective function file, and a constraint function file. Note that this file structure is similar to the one we used with `fmincon`. The `main.m` file contains the initializations, bounds, options, and the `ga` command. The `confun.m` file contains the nonlinear inequality and equality constraints. The `objfun.m` file contains the objective or the fitness function definition. The files are shown below.

1. **Main file**

```
clc
clear all
close all
global fcount     % to count function evaluations
lb=[-5 -5];       %Lower bound
ub=[5 5];         %Upper bound
A = [];           %LHS matrix for linear inequalities
b = [];           %RHS vector for linear inequalities
Aeq = [];         %LHS matrix for linear equalities
beq = [];         %RHS vector for linear equalities
%If you do not specify options, defaults are used.
fcount = 0;       %Initialize function count
nvars = 2;        % Number of design variables

[x,fval] = ga(@objfun,nvars,A,b,Aeq,beq,lb,ub,@confun);
% More arguments are available, check \Matlab\ help

display(fcount)
display(x)
```

2. **Objective function file**

```
function f= objfun(x)
global fcount
fcount = fcount+1;
f= x(1)^2+10*x(2)^2-3*x(1)*x(2);
```

3. **Constraint function file**

```
function [c,ceq]= confun(x)
c(1) = 4-2*x(1)-x(2);
c(2) = -5 - x(1) - x(2);
ceq = [];
```

Some important observations that can be made from this example are:

1. Each run of the genetic algorithm may yield a slightly different result. This is to be expected, since the genetic algorithm is stochastic in nature and involves random operators. Multiple runs are required to build confidence in the solution. The

solution for this problem is $x_1 = 1.825, x_2 = 0.3593$, and is generated correctly by the GA. Interestingly, solving this optimization problem using `fmincon` yields the same results. (Think of why that might be?)

2. GAs require a large number of function evaluations. For this example, `ga` requires approximately 4,000 function evaluations, while `fmincon` requires only 16 evaluations. Since this is a relatively simple problem, the issue of function evaluations may seem trivial. In computationally expensive models, however, the associated burden can become significant.

3. There are several settings within the `ga` command in MATLAB that can improve the computational performance of the `ga`.

Another popular application of GAs is to obtain Pareto optimal sets for multiobjective problems, which we discuss next.

19.3 Multiobjective Optimization Using Genetic Algorithms

Several approaches for solving multiobjective optimization problems were discussed in Chapter 6. Most of the previously studied methods involved the weight-based aggregation of the objectives into a single function. One of the most significant drawbacks of the weight-based methods is the need to specify appropriate weights, which is often a significant challenge.

The motivation for using evolutionary algorithms to solve multiobjective problems is two-fold: **(1)** EAs work with a population of candidate solutions, and use the concept of non-domination; thereby allowing for a series of Pareto solutions to be found in one converged run. This in in contrast to applying traditional techniques where the Pareto solutions are found sequentially, one run at a time. **(2)** EAs are significantly less sensitive to the shape of the Pareto frontier (convex or concave) or to discontinuous Pareto fronts.

There are many implementations of multiobjective genetic algorithms (MOGA) available in the literature [4, 5, 6, 7, 8]. The first practical implementation of a MOGA was called vector evaluated genetic algorithm (VEGA) by Schaffer [9]. A drawback of the VEGA approach is its bias toward some Pareto solutions. A so-called non-dominated sorting procedure [5, 10] was later implemented by several researchers to overcome the drawbacks of VEGA. A ranking procedure is adopted to rank individuals in a population. An individual, a, is said to dominate another individual, b, if a is strictly better than b in at least one objective, and a is no worse than b in all objectives. A distance measure is used to compare individuals with equal rank.

MATLAB provides a multiobjective optimization algorithm based on GAs called `gamultiobj`. An example is provided to illustrate how the `gamultiobj` works.

19.3.1 Example

Recall the following optimization problem from the exercises of Chapter 6 (Problem 6.2). The Pareto frontier for this problem is non-convex, and cannot be generated

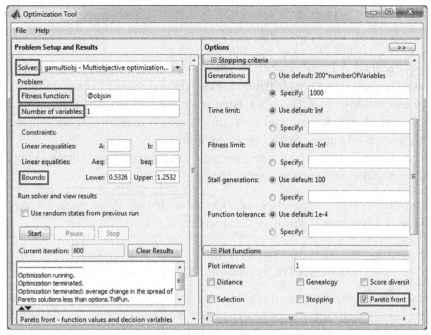

Figure 19.2. The MATLAB Genetic Algorithm and Direct Search Toolbox

using a weighted-sum method. This multiobjective problem can be solved using the MATLAB Genetic Algorithm Solver.

$$\min_{x} \ \{f_1 = \sin x, f_2 = 1 - \sin^7 x\} \tag{19.6}$$

subject to

$$0.5326 \leq x \leq 1.2532 \tag{19.7}$$

The `gamultiobj` accepts only linear equality constraints, linear inequality constraints, and bounds on the design variables. We now explore the graphical user interface (GUI) of the MATLAB GA Solver, within the Global Optimization Toolbox. Figure 19.2 provides a screen shot of the optimization tool that has various options. This screen can be opened by typing `optimtool('gamultiobj')` in the Command Window. Alternatively, you can type `optimtool` in the Command Window, followed by choosing the `gamultiobj` solver option from the top left dropdown menu. Before you can use this tool, you need to create a file that contains the objective function `objsin.m` that defines the two objectives, as shown below.

```
function f = objsin(x)
f(1) = sin(x);
f(2) = 1 - sin(x)^7;
```

In the GUI for the GA Solver, the user can provide the function handle for the fitness function, as shown in Fig. 19.1. Please keep in mind that the `objsin.m` file

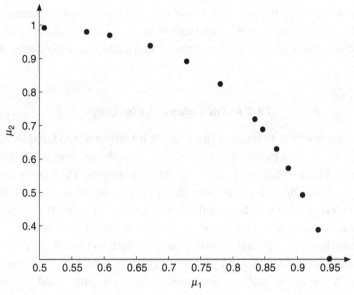

Figure 19.3. Pareto Front for Multiobjective Genetic Algorithm Example

must be saved in the current directory in MATLAB. In Fig. 19.2, note the selections highlighted by boxes. There are several other options in the window. Based on these options, the solutions may change. For the latest help on the additional features, refer to MATLAB help.

Figure 19.3 depicts the Pareto frontier for this problem as generated by the MATLAB Multiobjective Genetic Algorithm Solver. When the Parallel Computing Toolbox is available in the GUI for the GA Solver, it is possible to set the option as "in parallel" for evaluating fitness and constraint functions. By using the built-in parallel computing capabilities or defining a custom parallel computing implementation of the optimization problem, it is possible to significantly decrease solution time.

19.4 Other Evolutionary Algorithms

This section summarizes other major metaheuristic and/or non-gradient-based algorithms. Similar to Genetic Algorithms, some of these algorithms are also inspired by natural phenomena such as the social behavior of animals (*e.g.*, particle swarm optimization (Sec. 19.4.4), ant colony optimization (Sec. 19.4.1), and predator-prey optimization [11, 12]). These algorithms are often jointly classified as *evolutionary algorithms* in the literature; however, all of them do not necessarily mimic evolutionary phenomena (*e.g.*, swarm-based algorithms). Thus, a more technically appropriate name for this class of algorithms may be "*nature-inspired optimization algorithms*." There also exist metaheuristic algorithms that are inspired by human behavior and human-developed processes (*e.g.*, tabu search (Sec. 19.4.3) and simulated annealing (Sec. 19.4.2)). Overall, these classes of algorithms are generally intended to solve highly nonlinear optimization problems, which involve non-convex, discontinuous, or multi-modal criteria functions. A brief introduction to four of these major metaheuristic algorithms is provided below.

19.4.1 Ant Colony Optimization

The ant colony optimization algorithm (ACO), introduced by Marco Dorigo in 1992 in his Ph.D. thesis, is a probabilistic technique for solving computational problems that can be reduced to finding good paths through graphs. They are inspired by the behavior of ants in finding paths from the colony to food. In their path to search for food, ants deposit a substance called phermone that helps them smell or identify the path for later use. In a group of ants, each ant goes in search of food in a random direction. The ant that finds the shortest path to food returns to the colony in the shortest time. Deciding which path to take can be based on the amount of phermone. A larger phermone concentration along a path usually implies that a higher number of ants used the path, inferring that the path is likely shorter. Longer paths are progressively abandoned. The subsequent ant searches then use the phermone along the shortest path to direct their search. More details and examples are provided in [13].

19.4.2 Simulated Annealing

This method mimics the metallurgical process of annealing: heating a material and slowly lowering the temperature to decrease defects, thus minimizing the system energy. At each iteration of the simulated annealing algorithm, a new point is randomly generated. The distance of the new point from the current point, or the extent of the search, is based on a probability distribution with a scale proportional to the temperature. The algorithm accepts all the new points that lower the objective function; but also, with a certain probability, points that raise the objective function. By accepting points that raise the objective function, the algorithm minimizes the likelihood of being trapped in local minima, and is capable of greater global exploration. An annealing schedule is selected to systematically decrease the temperature as the algorithm proceeds. As the temperature decreases, the algorithm reduces the extent of its search to converge to a minimum. More details and examples are provided in [14].

19.4.3 Tabu Search

Tabu Search is a mathematical optimization method belonging to the class of local search techniques. Tabu Search enhances the performance of a local search method by using memory structures: once a potential solution has been determined, it is marked as "taboo" so that the algorithm does not visit that possibility repeatedly. More details and examples are provided in [15].

19.4.4 Particle Swarm Optimization (PSO)

The Particle Swarm Optimization (PSO) algorithm imitates the dynamics of swarm behavior observed in nature (*e.g.*, a flock of geese or a swarm of bees). In PSO, an initial set of randomly generated individuals is used, with each as a candidate solution. These individuals are also known as particles; hence the name particle swarm. Over a sequence of iterations, the population of particles searches for (or ideally converges to) the global optimum, where the motion of each particle in the design space is guided by a velocity update equation inspired by perceived swarm behavior. In this strategy, the locations of best fitness are remembered or recorded by each individual. An individual's best solution or success is called the particle best, and this information is shared with the neighbors. A swarm is typically modeled by particles in a multidimensional space, where each particle has a position and a velocity at each iteration. More details and examples are provided in [16, 14, 17]. With the above basic overview of certain key evolutionary algorithms, we conclude this chapter.

19.5 Summary

In this chapter, we presented the basics of popular evolutionary algorithms. The important differences between traditional optimization algorithms and evolutionary algorithms are discussed. We primarily focus on genetic algorithms because of their popularity in the design optimization community. We also illustrate the use of the MATLAB genetic algorithm tools for single objective and multiobjective optimizaton using examples. A brief review of other popular non-gradient-based algorithms is also presented.

19.6 Problems

19.1 Using the MATLAB Genetic Algorithms Solver, reproduce the results shown in Fig. 19.3.

19.2 Reproduce the results presented in Sec. 19.2.2 using both the `ga` and the `fmincon` commands. Perform the comparison for the number of function evaluations.

19.3 Solve the following problem using the multiobjective optimization tool in the GA Solver in MATLAB.

$$\min_x \ \{f_1, f_2\} \tag{19.8}$$

where

$$f_1 = \sum_{i=1}^{n-1} -10 \exp\{0.2\sqrt{x_i^2 + x_{i+1}^2}\} \tag{19.9}$$

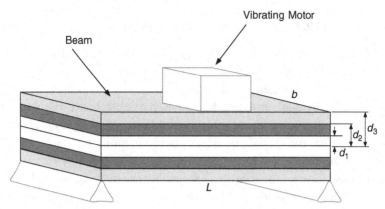

Figure 19.4. Sandwich Beam Designed with Vibrating Motor

$$f_2 = \sum_{i=1}^{n} \{|x|^{0.8} + 5(\sin(x_i))^3\} \tag{19.10}$$

subject to

$$-5 \le x_1, x_2, x_3 \le 5 \tag{19.11}$$

Try solving the above problem using the weighted sum method (Use fmincon). Can you obtain a good representation of the Pareto frontier? Explain why.

19.4 Recall that we solved the following problem earlier. Let us now employ GAs to solve this problem.

A vibratory disturbance (at v Hz) is imparted from the motor onto the beam. The beam is of length, L, and width, b. The variables, d_1 and d_2, respectively, locate the contact of materials one and two, and two and three. The variable, d_3, locates the top of the beam. The mass density, Young's modulus, and cost per unit volume for materials one, two, and three, are respectively denoted by the triplets (ρ_1, E_1, c_1), (ρ_2, E_2, c_2), and (ρ_3, E_3, c_3).

The overall objective is to design the preceding sandwich beam in such a way as to passively minimize the vibration of the beam that results from the disturbance ($v = 10$Hz). Minimizing the vibration will require maximizing the fundamental frequency, Fr, of the beam. The optimal solution should be such that the fundamental frequency is maximized economically (*i.e.*, at minimum cost, c). The following aggregate objective function is provided to you.

$$f = -50Fr^2 + 100C^2 \tag{19.12}$$

In the design of the plant, the quantities of interest are as follows:

$$Fr = (\pi/2L^2)(EI/\mu)^{\frac{1}{2}} \tag{19.13}$$

$$EI = (2b/3)[E_1 d_1^3 + E_2 (d_2^3 - d_1^3) + E_3 (d_3^3 - d_2^3)] \tag{19.14}$$

$$\mu = 2b[\rho_1 d_1 + \rho_2(d_2 - d_1) + \rho_3(d_3 - d_2)] \qquad (19.15)$$

$$C = 2bL[c_1 d_1 + c_2(d_2 - d_1) + c_3(d_3 - d_2)] \qquad (19.16)$$

$$M = \mu L \qquad (19.17)$$

$$w_2 = d_2 - d_1 \qquad (19.18)$$

$$w_3 = d_3 - d_2 \qquad (19.19)$$

$$x = [d_1 \quad d_2 \quad d_3 \quad b \quad L] \qquad (19.20)$$

The constraints are given as follows.

$$0.01 \le d_1 \le 0.299 \qquad (19.21)$$

$$0.3 \le d_3 \le 0.6 \qquad (19.22)$$

$$0.681 \le b \qquad (19.23)$$

$$3.999 \le L \qquad (19.24)$$

$$Fr \ge 155.6 \qquad (19.25)$$

$$C \le 1054 \qquad (19.26)$$

$$\mu L \le 1845 \qquad (19.27)$$

$$d_2 - d_1 \le 0.01 \qquad (19.28)$$

$$d_3 - d_2 \le 0.36 \qquad (19.29)$$

The other constants are $\rho_1 = 100 \,\mathrm{kg/m^3}$, $\rho_2 = 2,770 \,\mathrm{kg/m^3}$, $\rho_3 = 7,780 \,\mathrm{kg/m^3}$, $E_1 = 1.6 \times 10^9$ Pa, $E_2 = 70 \times 10^9$ Pa, $E_3 = 200 \times 10^9$ Pa, $c_1 = 500$ $/m^3, $c_2 = 1,500$ $/m^3, and $c_3 = 800$ $/m^3.

(1) Solve the above optimization problem using fmincon.
(2) Solve the optimization problem using the genetic algorithm solver in MATLAB. Run the program 20 times. Do you obtain the same answer for all the runs? Explain why or why not?
(3) You may have noticed that for some runs, ga terminates with an error that the nonlinear constraint file is not returning a real value. Determine why that happens by closely examining your code. How would you fix this issue? Subsequently, re-run the ga routine the number of times needed to ensure the issue has been fixed.
(4) As we studied in Chapter 7, we need to be careful about numerical scaling issues for the situations you encountered in the previous question. Examine whether the current problem has such scaling issues. To make our program more robust, fix the scaling issue as discussed in Chapter 7.
(5) How many function evaluations does the genetic algorithm need? How many function evaluations does fmincon need?

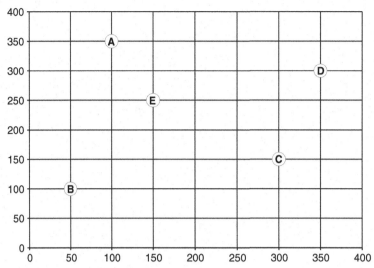

Figure 19.5. Traveling Salesman Problem: Locations of Cities

19.5 Consider the following optimization problem. Note that the objective function is discontinuous.

$$\min_{x} \quad f \tag{19.30}$$

where

$$f(x) = \begin{cases} -\exp\{-\{\frac{x}{20}\}^2\}; & x \le 20 \\ -\exp(-1) + (x - 20)(x - 22); & x > 20 \end{cases}$$

subject to

$$-10 \le x \le 25 \tag{19.31}$$

(1) Plot this function to study the local/global optima for this problem.
(2) Use `fmincon` to minimize the function. Do you obtain the global optimum? Try different starting points.
(3) Now try the `ga` command. Do you obtain the global minima? Note that you need to change the "initial range" option to help the `ga` command.
(4) Compare the number of function evaluations for each optimization as required by the genetic algorithm and `fmincon`.

19.6 A salesman has to travel through several cities (to sell his/her product) in such a way that the total traveling expense is minimized. In this case, the traveling expense is directly proportional to the distance traveled (Traveling Salesman Problem). Solve the Traveling Salesman Problem (TSP) using the MATLAB GA function, for a particular case where the number of cities is equal to five, as shown in Fig. 19.5. The locations of the cities are also given in the figure, and the salesman can start from any of the cities.

1. Formulate the optimization problem.
2. Determine the optimum route for the traveling salesman (using MATLAB GA).
3. If the salesman has to return to the starting city at the end of his journey, will the optimum route change? (Justify the answer through optimization)

BIBLIOGRAPHY OF CHAPTER 19

[1] Z. Gürdal, R. T. Haftka, and P. Hajela. *Design and Optimization of Laminated Composite Materials*. John Wiley and Sons, 1999.

[2] D. E. Goldberg. *Genetic Algorithms in Search, Optimization, and Machine Learning*. Addison-Wesley Publishing Company, Inc, 1989.

[3] N. N. Schraudolph and R. K. Belew. Dynamic parameter encoding for genetic algorithms. *Machine Learning*, 9(1):9–21, 1992.

[4] E. Zitzler and L. Thiele. *Multiobjective Optimization Using Evolutionary Algorithms – A Comparative Case Study*. Lecture Notes in Computer Science. Springer Berlin Heidelberg, 1998.

[5] N. Srinivas and K. Deb. Multi-objective function optimization using non-dominated sorting genetic algorithms. *Evolutionary Computation*, 2(3):221–248, 1994.

[6] C. M. Fonseca and P. J. Fleming. Genetic algorithms for multiobjective optimization: Formulation, discussion, and generalization. In *Proceedings of the 5th International Conference on Genetic Algorithms*, pages 416–423, Urbana-Champaign, IL, USA, June 1993.

[7] J. Horn, N. Nafploitis, and D. E. Goldberg. A niched Pareto genetic algorithm for multi-objective optimization. In *Proceedings of the first IEEE Conference on Evolutionary Computation*, pages 82–87, Orlando, FL, USA, June 1994.

[8] C. A. Coello Coello, G. B. Lamont, and D. A. Van Veldhuizen. *Evolutionary Algorithms for Solving Multi-objective Problems*. Springer, 2nd edition, 2007.

[9] J. D. Schaffer. Multiple objective optimization with vector evaluated genetic algorithms. In *Proceedings of the First International Conference on Genetic Algorithms*, pages 93–100, Pittsburg, PA, USA, July 1985.

[10] K. Deb, S. Agarwal, A. Pratap, and T. Meyarivan. A fast elitist non-dominated sorting genetic algorithm for multi-objective optimization: NSGA-II. *IEEE Transactions on Evolutionary Computation*, 6(2):182–197, 2002.

[11] S. Chowdhury, G. S. Dulikravich, and R. J. Moral. Modified predator-prey algorithm for constrained and unconstrained multi-objective optimisation. *International Journal of Mathematical Modelling and Numerical Optimisation*, 1(1):1–38, 2009.

[12] G. Venter and J. Sobieszczanski-Sobieski. Particle swarm optimization. *AIAA Journal*, 41(8):1583–1589, 2003.

[13] L. N. De Castro and F. J. Von Zuben. *Recent Developments in Biologically Inspired Computing*. Idea Group Publishing, 2005.

[14] M. Tsuzuki and T. DeCastro Martins. *Simulated Annealing: Strategies, Potential Uses and Advantages*. Nova Science Publishers, 2014.

[15] F. Glover and M. Laguna. *Tabu Search*. Kluwer Academic Publishers, 1997.

[16] M. Clerc. *Particle Swarm Optimization*. ISTE Ltd, 2006.

[17] S. Chowdhury, W. Tong, A. Messac, and J. Zhang. A mixed-discrete particle swarm optimization algorithm with explicit diversity-preservation. *Structural and Multidisciplinary Optimization*, 47(3):367–388, 2013.

Author Index

Subject Index

Printed in the United States
By Bookmasters